RESIDUE REVIEWS

VOLUME 79

Triphenyltin compounds
and their degradation products

By

Rudolf Bock

Translation by Inge and Heinz Trebitz

RESIDUE REVIEWS

Residues of Pesticides and Other
Contaminants in the Total Environment

Editor
FRANCIS A. GUNTHER

Assistant Editor
JANE DAVIES GUNTHER

Riverside, California

VOLUME 79

Springer Science+Business Media, LLC

1981

Coordinating Board of Editors

© 1981 by Springer Science+Business Media New York
Softcover reprint of the hardcover 1st edition 1981
Library of Congress Catalog Card Number 62-18595.
The use of general descriptive names, trade names, trademarks, etc. in this publication, even if the former are not especially identified, is not to be taken as a sign that such names, as understood by the Trade Marks and Merchandise Marks Act, may accordingly be used freely by anyone.

New York: 175 Fifth Avenue, New York, N.Y. 10010
Heidelberg: 6900 Heidelberg 1, Postfach 105 280, West Germany

ISBN 978-1-4612-5879-7   ISBN 978-1-4612-5877-3 (eBook)
DOI 10.1007/978-1-4612-5877-3

Foreword

Worldwide concern in scientific, industrial, and governmental communities over traces of toxic chemicals in foodstuffs and in both abiotic and biotic environments has justified the present triumvirate of specialized publications in this field: comprehensive reviews, rapidly published progress reports, and archival documentations. These three publications are integrated and scheduled to provide in international communication the coherency essential for nonduplicative and current progress in a field as dynamic and complex as environmental contamination and toxicology. Until now there has been no journal or other publication series reserved exclusively for the diversified literature on "toxic" chemicals in our foods, our feeds, our geographical surroundings, our domestic animals, our wildlife, and ourselves. Around the world immense efforts and many talents have been mobilized to technical and other evaluations of natures, locales, magnitudes, fates, and toxicology of the persisting residues of these chemicals loosed upon the world. Among the sequelae of this broad new emphasis has been an inescapable need for an articulated set of authoritative publications where one could expect to find the latest important world literature produced by this emerging area of science together with documentation of pertinent ancillary legislation.

The research director and the legislative or administrative advisor do not have the time even to scan the large number of technical publications that might contain articles important to current responsibility; these individuals need the background provided by detailed reviews plus an assured awareness of newly developing information, all with minimum time for literature searching. Similarly, the scientist assigned or attracted to a new problem has the requirements of gleaning all literature pertinent to his task, publishing quickly new developments or important new experimental details to inform others of findings that might alter their own efforts, and eventually publishing all his supporting data and conclusions for archival purposes.

The end result of this concern over these chores and responsibilities and with uniform, encompassing, and timely publication outlets in the field of environmental contamination and toxicology is the Springer-Verlag (Heidelberg and New York) triumvirate:

Residue Reviews (vol. 1 in 1962) for basically detailed review articles concerned with any aspects of residues of pesticides and other chemical contaminants in the total environment, including toxicological considerations and consequences.

Bulletin of Environmental Contamination and Toxicology (vol. 1 in
    1966) for rapid publication of short reports of significant advances
    and discoveries in the fields of air, soil, water, and food contami-
    nation and pollution as well as methodology and other disciplines
    concerned with the introduction, presence, and effects of toxicants
    in the total environment.

Archives of Environmental Contamination and Toxicology (vol. 1 in
    1973) for important complete articles emphasizing and describing
    original experimental or theoretical research work pertaining to the
    scientific aspects of chemical contaminants in the environment.

Manuscripts for *Residue Reviews* and the *Archives* are in identical
formats and are subject to review, by workers in the field, for adequacy
and value; manuscripts for the *Bulletin* are not reviewed and are published
by photo-offset to provide the latest results without delay. The individual
editors of these three publications comprise the Joint Coordinating Board
of Editors with referral within the Board of manuscripts submitted to one
publication but deemed by major emphasis or length more suitable for
one of the others.

                                                Coordinating Board of Editors

Preface

That residues of pesticide and other contaminants in the total environment are of concern to everyone everywhere is attested by the reception accorded previous volumes of "Residue Reviews" and by the gratifying enthusiasm, sincerity, and efforts shown by all the individuals from whom manuscripts have been solicited. Despite much propaganda to the contrary, there can never be any serious question that pest-control chemicals and food-additive chemicals are essential to adequate food production, manufacture, marketing, and storage, yet without continuing surveillance and intelligent control some of those that persist in our foodstuffs could at times conceivably endanger the public health. Ensuring safety-in-use of these many chemicals is a dynamic challenge, for established ones are continually being displaced by newly developed ones more acceptable to food technologists, pharmacologists, toxicologists, and changing pest-control requirements in progressive food-producing economies.

These matters are of genuine concern to increasing numbers of governmental agencies and legislative bodies around the world, for some of these chemicals have resulted in a few mishaps from improper use. Adequate safety-in-use evaluations of any of these chemicals persisting into our foodstuffs are not simple matters, and they incorporate the considered judgments of many individuals highly trained in a variety of complex biological, chemical, food technological, medical, pharmacological, and toxicological disciplines.

It is hoped that "Residue Reviews" will continue to serve as an integrating factor both in focusing attention upon those many residue matters requiring further attention and in collating for variously trained readers present knowledge in specific important areas of residue and related endeavors involved with other chemical contaminants in the total environment. The contents of this and previous volumes of "Residue Reviews" illustrate these objectives. Since manuscripts are published in the order in which they are received in final form, it may seem that some important aspects of residue analytical chemistry, biochemistry, human and animal medicine, legislation, pharmacology, physiology, regulation, and toxicology are being neglected; to the contrary, these apparent omissions are recognized, and some pertinent manuscripts are in preparation. However, the field is so large and the interests in it are so varied that the editors and the Advisory Board earnestly solicit suggestions of topics and authors to help make this international book-series even more useful and informative.

"Residue Reviews" attempts to provide concise, critical reviews of timely advances, philosophy, and significant areas of accomplished or needed endeavor in the total field of residues of these and other foreign chemicals in any segment of the environment. These reviews are either general or specific, but properly they may lie in the domains of analytical chemistry and its methodology, biochemistry, human and animal medicine, legislation, pharmacology, physiology, regulation, and toxicology; certain affairs in the realm of food technology concerned specifically with pesticide and other food-additive problems are also appropriate subject matter. The justification for the preparation of any review for this book-series is that it deals with some aspect of the many real problems arising from the presence of any "foreign" chemicals in our surroundings. Thus, manuscripts may encompass those matters, in any country, which are involved in allowing pesticide and other plant-protecting chemicals to be used safely in producing, storing, and shipping crops. Added plant or animal pest-control chemicals or their metabolites that may persist into meat and other edible animal products (milk and milk products, eggs, etc.) are also residues and are within this scope. The so-called food additives (substances deliberately added to foods for flavor, odor, appearance, etc., as well as those inadvertently added during manufacture, packaging, distribution, storage, etc.) are also considered suitable review material. In addition, contaminant chemicals added in any manner to air, water, soil or plant or animal life are within this purview and these objectives.

Manuscripts are normally contributed by invitation but suggested topics are welcome. Preliminary communication with the editors is necessary before volunteered reviews are submitted in manuscript form.

Department of Entomology   F.A.G.
University of California   J.D.G.
Riverside, California
January 26, 1981

RESIDUE REVIEWS

VOLUME 79

Triphenyltin compounds
and their degradation products

By

Rudolf Bock* **

Contents

* Chemin de Béranges 141, CH-1814 La Tour de Peilz, Switzerland.

** Translation into English by Inge Trebitz and Heinz Trebitz, American Hoechst Corporation, Route 202/206 North, Somerville, N. J. 08876.

I. Introduction/History

Organotin compounds have been widely used as pesticides during the last 20 years. Because of the rapidly growing importance of this group, many studies have been undertaken to establish synthetical routes and to test the physical, chemical, and most of all, the biochemical behavior of several of these compounds, especially of triphenyltin derivatives.

For more than 100 years, sporadic reports about the physiological effects of organotin compounds have been published in the chemical literature: Irritating effects of ethyltin iodide (FRANKLAND 1853) and of triethyltin chloride (BUCKTON 1858 and 1859) on eyes and mucous membranes of human beings; toxic effects of ethyltin compounds in dogs and frogs (JOLYET and CAHOURS 1869); toxic effects of triethyltin acetate in dogs, rabbits, and frogs (WHITE 1881 and 1886); highly increased toxicity of triphenyltin bromide compared with tetraphenyltin in mice (KRAUSE 1929, COLLIER 1929); repelling moths with tetraphenyltin (HARTMANN *et al.* 1925); biocidal effects of triethyltin and triphenyltin compounds (Tisdale 1943); and, finally, insecticidal and ovicidal qualities of trialkyltin chlorides (Kraak 1947).

Only through systematic research, however, by Van Der Kerk and co-workers since about 1950, who followed the much earlier recommendation of Hedges to do comprehensive work in the field of organotin compounds, the fungicidal effects of certain tinorganic derivatives became more important and areas for their practical application were opened up (Van Der Kerk 1952, 1960, 1961, 1970, and 1975; Van Der Kerk and Luijten 1954, 1956 a and c, and 1969; Van Der Kerk *et al.* 1958, 1962, and 1969).

In the beginning, application of these compounds as pesticides in agriculture appeared to be impractical because of their considerable phytotoxicity. This difficulty was first resolved with "Brestan"®, a Hoechst AG product using triphenyltin acetate as active ingredient. In this product, effectiveness against several fungi was combined with sufficiently low phytotoxicity (Haertel and Baumann 1956; Haertel 1958 a and b, and 1962; compare also Kaars Sijpesteijn 1959).

Organic compounds of tin show, contrary to organometallic compounds of other elements with fungicidal qualities (e.g., arsenic, cadmium, copper, mercury, zinc), a very important difference: the final product of their decomposition is inorganic tin (IV), which is harmless to warm-blooded animals even in large doses, whereas the ions and inorganic compounds of the above-mentioned elements are more or less toxic. Therefore, the agricultural application of these tinorganic compounds cannot result in an accumulation of a harmful metal in the soil.

The following summary comprises triphenyltin compounds of the type $(C_6H_5)_3SnX$ where X can be any inorganic or organic anion. The most important representatives of this group are triphenyltin chloride $(C_6H_5)_3SnCl$, the acetate $(C_6H_5)_3SnOCOCH_3$, and the hydroxide $(C_6H_5)_3SnOH$. Additional potential contaminants or decomposition products appear to be: tetraphenyltin $(C_6H_5)_4Sn$, as well as the di- and monophenyltin compounds $(C_6H_5)_2SnCl_2$ and $(C_6H_5)SnCl_3$, and, furthermore, the oxygen-containing compounds $(C_6H_5)_3SnOSn(C_6H_5)_3$, $(C_6H_5)_2SnO$, and $(C_6H_5)SnOOH$. In addition, there are the sulfide $[(C_6H_5)_3Sn]_2S$ and some complex compounds of triphenyltin chloride which have also been proposed as pesticides.

Excluded from this overview are phenyltin compounds that contain, apart from phenyl, other alkyl or aryl groups directly attached to tin ("asymmetric" or "mixed" organotin compounds), further phenyltin compounds with Sn-Sn-links [e.g., $(C_6H_5)_3Sn-Sn(C_6H_5)_3$], phenyltin compounds with Sn-H-links, and phenyltin compounds of the general formula $(C_6H_5)_2Sn.$[1]

---

[1] Recent experiments have shown that these compounds contain tetravalent tin; they are polymers.

The term "inorganic" will be used here for all derivatives of tin that do not contain Sn-C-links, even though the tin might be linked to organic anions such as tartrate, citrate, or ethylenediamine tetraacetate.

The International Standardization Organization (ISO) has proposed the name "Fentin" for the triphenyltin group.

A number of experimental and commercial products are listed with their producers in Table I.

II. Synthesis of phenyltin compounds

a) Synthesis of tetraphenyltin

1. Grignard-synthesis.—As far as presently known, tetraphenyltin is the most important starting material for the manufacture of other phenyltin compounds. On a technical scale, it is synthesized exclusively through Grignard-reaction, using a modification of an old procedure by PFEIFFER and SCHNURMANN (1904) with chlorobenzene and tin tetrachloride as starting materials (VAN DER KERK et al. 1969, BOKRANZ and PLUM 1971, JOHNSON 1962):

$$SnCl_4 + 4\,(C_6H_5)MgCl \rightarrow (C_6H_5)_4Sn + 4\,MgCl_2 \qquad (1)$$

Since diethyl ether, the commonly used solvent in Grignard-synthesis, is impractical for large batches it was proposed to carry out the reaction in a mixture of a hydrocarbon (e.g., toluene) and diethyl ether at a ratio of 9 to 1 (RAMSDEN and DAVIDSON 1950, RAMSDEN 1953). Subsequently, the ether was substituted by tetrahydrofuran or similar aliphatic oxygen

Table I. *Trade names and producers of triphenyltin compounds.*

Compound	Name (% a.i.)	Producer
Acetate	Brestan (20%)	Hoechst AG
	Brestan conc. (60%)	Hoechst AG
	Breston Super (60% + 20% of Maneb)	Hoechst AG
	Batasan	—
	Bedilan	—
	Suzu	
Hydroxide	Du-Ter (47.5%)	Philips-Duphar
	Exithane 50	Bayer AG
	Farmatin	—
	Suzu H	—
	Tinicide (50%)	Luxembourg Chem. Co., Tel Aviv
	Fennite	Fisons Pest Control
	Tubotin	May and Baker
	Vitospor	Vitax
Chloride	Brestanol (40%)	Hoechst AG
	Tinmate	Nichino

compounds; this allows the compound to work at higher temperatures, which leads to a faster and more uniform reaction (RAMSDEN 1956 and 1957, RAMSDEN et al. 1957). A further improvement is the multi-step reaction starting with diphenyltin chloride (LEWIS 1955).

In other variations oxygen-free solvents, for example decalin (OKHYLOBYSTIN et al. 1961, ZAKHARKIN et al. 1962) or toluene (ECKSTEIN and EJMOCKI 1964) are used. There the magnesium has to be activated with iodine and the reaction is started with bromobenzene. Other possibilities are the reaction of a Mg/Sn-alloy with bromobenzene (POLKINHORNE and TAPLEY 1953) or a synthesis where chlorobenzene serves as both the reagent and the solvent (BORBÉLY-KUSZMANN and NAGY 1962).

The Grignard-reactions have to proceed smoothly and without interruptions; important are an effective initiator, purity and size of the Mg filings, and purity of the solvents. Further, the system should be self-stabilizing, which can be achieved if the arylchloride has a lower boiling point than the other components present; if the reaction becomes too vigorous, part of the arylchloride will evaporate; consequently, the reaction rate in the liquid phase is retarded (CRAMER 1959, TICHY 1961).

The advantages of the Grignard-synthesis lie in high yields, high purity of the products, and low corrosion to the equipment. Disadvantages are the necessity to use expensive starting materials (Mg metal, $SnCl_4$) and the formation of the useless $MgCl_2$. Further, high-purity solvents are required which are difficult to recycle.

2. **Wurtz-synthesis.**—The synthesis of tetraphenyltin according to WURTZ can be described by the following equation (HARRIS 1944, VAN DER KERK and LUIJTEN 1956 c):

$$SnCl_4 + 4\ C_6H_5Cl + 8\ Na \rightarrow (C_6H_5)_4Sn + 8\ NaCl \qquad (2)$$

However, under these highly reductive conditions several side reactions may occur:

$$2\ C_6H_5Cl + 2\ Na \rightarrow C_6H_5\text{-}C_6H_5 + 2\ NaCl \qquad (3)$$

$$(C_6H_5)_2SnCl_2 + 2\ Na \rightarrow (C_6H_5)_2Sn + 2\ NaCl \qquad (4)$$

$$2\ (C_6H_5)_3SnCl + 2\ Na \rightarrow (C_6H_5)_3Sn\text{-}Sn(C_6H_5)_3 + 2\ NaCl \qquad (5)$$

$$SnCl_4 + 2\ Na \rightarrow SnCl_2 + 2\ NaCl \qquad (6)$$

$$SnCl_2 + 2\ Na \rightarrow Sn + 2\ NaCl \qquad (7)$$

The most inconvenient reduction of $SnCl_4$ to $SnCl_2$ and Sn metal can be avoided by starting with $(C_6H_5)_2SnCl_2$ instead of tin tetrachloride. This compound, together with $(C_6H_5)Na$ in a benzene solution at approximately 5° to 6°C, reacts to form tetraphenyltin without considerable losses. Part of the tetraphenyltin is removed from the process, the remainder is reconverted with $SnCl_4$ to $(C_6H_5)_2SnCl_2$ (see equation 10 below—METAL and THERMIT 1949, REINDL and BOIDOL 1957). The reaction can be conducted in a similar manner, using $(C_6H_5)_3SnCl$ as a starting material (OAKES 1964).

3. **Synthesis using aluminum-organic compounds.**—The synthesis using Al-organic compounds follows the equation:

$$4 \ (C_6H_5)_3Al_2Cl_3 + 3 \ SnCl_4 \rightarrow 3 \ (C_6H_5)_4Sn + 8 \ AlCl_3 \qquad (8)$$

For high yields in this reaction, the by-product $AlCl_3$ has to be removed by complex formation (for example with NaCl or diethyl ether). If not, an equilibrium is reached which disfavors the formation of tetraphenyltin (WITTENBERG 1960 and 1962).

For other synthetic routes which, however, never gained practical importance, compare the summaries of JONES and GILMAN (1954), INGHAM et al. (1960), DUB (1961), and GMELIN (1975). An electrochemical synthesis has been described by KOBETZ and PINKERTON (1960).

Tetraphenyltin can be purified by recrystallization from chloroform (POLIS 1889), pyridine (PFEIFFER and SCHNURMANN 1904), benzene (PETERSON et al. 1965), or xylene (RAMSDEN 1956).

b) Synthesis of phenyltin halogenides

1. **Grignard-synthesis.**—The reaction of phenylmagnesium-chloride with $SnCl_4$ proceeds in steps, and the yields of the different phenyltin chlorides can be affected by varying the reaction conditions (Table II). However, in each case the main product contains a considerable amount of the other phenyltin chlorides (RAMSDEN 1956). In the production of phenyltin chlorides the C_6H_5MgCl should be added to the solution of $SnCl_4$; if in reverse the tin tetrachloride is added to the solution of phenylmagnesium-chloride, the formation of $(C_6H_5)_4Sn$ is favored (RAMSDEN 1957).

Table II. *Sequential reaction of $SnCl_4$ with C_6H_5MgCl* (according to RAMSDEN 1956).

Moles C_6H_5MgCl/ mole $SnCl_4$	Principal product
1	$(C_6H_5)SnCl_3$
2	$(C_6H_5)_2SnCl_2$
3	$(C_6H_5)_3SnCl$
4	$(C_6H_5)_4Sn$

2. **Kocheshkov-synthesis.**—A technically important access to phenyltin chlorides is the reaction of tetraphenyltin with $SnCl_4$, which basically has been described by LADENBURG (1871), but was applied to tetraphenyltin only by KOCHESHKOV[2] (1929) (KOCHESHKOV et al. 1934). Together, the

---

[2] In older literature: KOZESCHKOW.

two starting materials are heated for several hours to temperatures of approximately 160° to 220°C, which yields mixtures of the 3 phenyltin chlorides. However, depending on the ratio of starting materials used, one compound will prevail (equations 9 to 11):

$$3 (C_6H_5)_4Sn + SnCl_4 \rightarrow 4 (C_6H_5)_3SnCl \qquad (9)$$

$$(C_6H_5)_4Sn + SnCl_4 \rightarrow 2 (C_6H_5)_2SnCl_2 \qquad (10)$$

$$(C_6H_5)_4Sn + 3 SnCl_4 \rightarrow 4 (C_6H_5)SnCl_3 \qquad (11)$$

Apart from that, different phenyltin chlorides can react with each other, but their reaction rate is lower (OAKES 1964):

$$(C_6H_5)_4Sn + (C_6H_5)SnCl_3 \rightarrow (C_6H_5)_3SnCl + (C_6H_5)_2SnCl_2 \quad (12)$$

$$(C_6H_5)_4Sn + (C_6H_5)_2SnCl_2 \rightarrow 2 (C_6H_5)_3SnCl \qquad (13)$$

The conditions chosen for each reaction are slightly different from each other (Table III); yields for the most important reaction (9) have been found to be approximately 80% $(C_6H_5)_3SnCl$ with 2 to 5% $(C_6H_5)_2SnCl_2$ and about 10 to 15% unchanged $(C_6H_5)_4Sn$ (REINDL and GELBERT 1959).

The reaction can also be conducted in xylene solution at temperatures of approximately 140° to 150°C or in ethylene chloride under pressure at 120°C (REINDL and GELBERT 1959). The conversion of tetraphenyltin with $SnCl_4$ to $(C_6H_5)_2SnCl_2$ at ambient temperatures under the influence of UV-radiation seems to be of no technical importance (RAZUVAEV 1950).

The conversion as described by KOCHESHKOV is a variation of a general reaction which may be described as an exchange or distribution of functional groups (CHALLENGER and PRITCHARD 1924, CALINGAERT and BEATTY 1939, CALINGAERT et al. 1939). If univalent ligands are exchanged at a central atom of valence n, the system can be described with $n - 1$ equilibrium constants, which amount in the KOCHESHKOV-reaction to 3 (MOEDRITZER 1966).

If a reaction is allowed to reach equilibrium a statistical distribution of the reaction products may result. In other cases, the formation of a certain compound may be favored or much suppressed.

Equilibriums of exchange reactions, among others, have been studied for some alkyltin compounds; however, to date the KOCHESHKOV-reaction has not been investigated.

3. Reaction of tetraphenyltin with halogens.—The reaction of $(C_6H_5)_4Sn$ with chlorine can be described by the following equations with $SnCl_4$ being the final product:

$$(C_6H_5)_4Sn + Cl_2 \rightarrow (C_6H_5)_3SnCl + C_6H_5Cl \qquad (14)$$

$$(C_6H_5)_3SnCl + Cl_2 \rightarrow (C_6H_5)_2SnCl_2 + C_6H_5Cl \qquad (15)$$

$$(C_6H_5)_2SnCl_2 + Cl_2 \rightarrow (C_6H_5)SnCl_3 + C_6H_5Cl \qquad (16)$$

Table III. *Reaction conditions for the synthesis of phenyltin chlorides*
(according to KOCHESHKOV 1929).

Reaction	Conditions
3 $(C_6H_5)_4Sn$ + $SnCl_4$ → 4 $(C_6H_5)_3SnCl$	During 30 min, heat to 205°C, hold for 3 hr at 205° to 210°C, then for 3 hr at 150° to 160°C (KOCHESHKOV *et al.* 1934 and 1936).
	Heat slowly to 220°C, hold at 220°C for 1 hr (GILMAN & ROSENBERG 1952).
	During 2 hr heat to 220°C, hold for 20 hr at 220° to 240°C (D'ANS & ZIMMER, 1952).
	During 3 hr heat to 205° to 215°C, hold for 3 hr at 180° to 190°C (LUIJTEN & VAN DER KERK 1955, VAN DER KERK & LUIJTEN 1956 b).
	Heat to 210°C and hold for 4 hr (KRIEGSMANN & GEISSLER 1963).
	During 1 hr heat to 210°C, hold at that temp. for 2 hr (FREITAG 1972).
$(C_6H_5)_4Sn$ + $SnCl_4$ → 2 $(C_6H_5)_2SnCl_2$	220°C, 1.5 hr (KOCHESHKOV 1929).
	Reflux for 3 hr in xylene-solution (CHALLENGER & ROTHSTEIN 1934, REINDL & GELBERT 1959).
	Heat to 200°C and hold 1 to 2 hr (JOHNSON & CHURCH 1948).
	Heat to 220°C and hold for 8 hr (ZIMMER & SPARMANN 1954).
	During 1 hr heat to 180°C, hold for 2 hr (GILMAN & GIST 1957).
	Heat to 240°C and hold for 1 hr (POLKINHORNE & TAPLEY 1955).
$(C_6H_5)_4Sn$ + 3 $SnCl_4$ → 4 $(C_6H_5)SnCl_3$	Heat to 210° to 220°C and hold for 1.5 to 2 hr (KOCHESHKOV 1929).
	1 hr to 215°C, then 2 hr to 220° to 230°C (D'ANS & ZIMMER 1952).
	Heat slowly to 210°C, hold at 210° to 220°C 1.5 hr (LUIJTEN & VAN DER KERK 1955).
	Agitate for 3 hr at 150°C (GILMAN & GIST 1957).
$(C_6H_5)_3SnCl$ + $(C_6H_5)SnCl_3$ → 2 $(C_6H_5)_2SnCl_2$	Under vacuum heat to 175° to 225°C (JOHNSON & CHURCH 1948).

This reaction series results in mixtures of various phenyltin chlorides rather than yielding a single product.

Bromine in $CHCl_3$ or other solutions will replace simultaneously two phenyl groups in tetraphenyltin; only by working in pyridine can triphenyltin bromide be isolated as the first reaction product (KRAUSE 1918). The reaction of tetraphenyltin with iodine in a $CHCl_3$-solution at 35° to 40°C results in triphenyltin iodide in a yield of 80% (LE QUAN MINH 1968).

4. **Reaction of tetraphenyltin with HCl.**—Tetraphenyltin can be converted to triphenyltin chloride by feeding gaseous HCl into a suspension of $(C_6H_5)_4Sn$ in $CHCl_3$ (BAEHR 1940):

$$(C_6H_5)_4Sn + HCl \rightarrow (C_6H_5)_3SnCl + C_6H_6 \qquad (17)$$

Similarly to those described previously, this reaction does not stop at the triphenyltin compound, but continues with further formation of benzene. However, by adding HCl at the correct rate, $(C_6H_5)_3SnCl$ can be obtained in good yields.

Neither the conversion with halides nor the one with HCl nor various other procedures (compare INGHAM et al. 1960, DUB 1961) have gained any importance.

Mixtures of phenyltin chlorides, obtained in the KOCHESHKOV-synthesis or by other methods, can be separated by vacuum distillation (D'ANS and ZIMMER 1952, JOHNSON and CHURCH 1948, RAMSDEN 1956 and 1957). At a pressure of 0.5 Torr the boiling points of tri-, di-, and monophenyltin chloride differ from each other by approximately 45° to 50°C (FREITAG 1972).

Other recovery methods use the difference in solubility. Unreacted tetraphenyltin, for example, can be removed by dissolving the reaction mixture in diethyl ether (KOCHESHKOV et al. 1936) or methanol (SANTO 1961, METAL and THERMIT 1961). Most of the $(C_6H_5)_4Sn$ remains undissolved in both solvents.

If the reaction according to KOCHESHKOV is carried out in xylene solution most of the unconverted tetraphenyltin remains with the residue and can be removed by filtration (REINDL and GELBERT 1959).

Tri- and diphenyltin chloride can be purified by recrystallization from the following solvents:

$(C_6H_5)_3SnCl$:	ethanol	(KOCHESHKOV et al. 1936)
	diethyl ether	(ARONHEIM 1878)
	petroleum ether	(BAEHR 1940, GILMAN and MELVIN 1949)
	xylene	(RAMSDEN 1956)
$(C_6H_5)_2SnCl_2$:	petroleum ether	(ARONHEIM 1878, GILMAN and GIST 1957)
	xylene	(RAMSDEN 1956)

Other methods of purification use the relative insolubility of triphenyltin fluoride (KRAUSE and BECKER 1920, INGHAM et al. 1960, NEUMANN 1967) and that of the ammonia adduct of the triphenyltin chloride, $(C_6H_5)_3SnCl \cdot 2\ NH_3$ which can be precipitated from the solution in diethyl ether with gaseous NH_3 (INGHAM et al. 1960).

c) Synthesis of phenyltin hydroxides and oxides

Phenyltin hydroxides and oxides are formed from solutions of the chlorides in organic solvents with aqueous alkali:

$$(C_6H_5)_3SnCl + MOH \rightarrow (C_6H_5)_3SnOH + MCl$$
$$- H_2O \rightarrow (C_6H_5)_3SnOSn(C_6H_5)_3 \qquad (18)$$

$$(C_6H_5)_2SnCl_2 + 2\,MOH \rightarrow [(C_6H_5)_2Sn(OH)_2] + 2\,MCl$$
$$- H_2O \rightarrow (C_6H_5)_2SnO \qquad\qquad (19)$$

$$(C_6H_5)SnCl_3 + 3\,MOH \rightarrow [(C_6H_5)Sn(OH)_3] + 3\,MCl$$
$$- H_2O \rightarrow (C_6H_5)SnOOH \qquad\qquad (20)$$

Triphenyltin hydroxide can be obtained from the solution of the chloride in diethylether with aqueous NaOH (KOCHESHKOV et al. 1936), from the ethanol solution with aqueous ammonia (FRIEBE and KELKER 1963), or from the acetone solution with aqueous KOH (KUSHLEFSKY et al. 1963). The hydroxide precipitates or may be obtained by evaporating the organic solvent. The material can be recrystallized from aqueous ethanol (SCHMITZ-DUMONT 1941); diluting the ethanol with 5 to 10% water has been recommended (FRIEBE and KELKER 1963, KUSHLEFSKY et al. 1963); 90% methanol may also be used (FREITAG 1972).

In the equilibrium between triphenyltin hydroxide and bis-triphenyltin oxide (hexaphenyldistannoxane) formation of the oxide is rather favored. Already at approximately 60°C, water is lost from the hydroxide; this also happens if this compound is left standing over silica gel or just by dissolving the hydroxide in organic solvents such as benzene, chloroform, acetonitrile, and others which are free of water (SCHMITZ-DUMONT 1941, FRIEBE and KELKER 1963). The oxide forms when $(C_6H_5)_3SnCl$ is treated with aqueous alkali, if the chloride is dissolved in an organic solvent, which takes up less than 0.45 g of $H_2O/100$ ml (VAN RIJ 1960).

In another synthesis, the hydroxide is heated with toluene until all the water has been removed (KUSHLEFSKY et al. 1963). For production on a technical scale, triphenyltin chloride in a benzene solution may be reacted with slaked lime; the water dissolved in the organic phase can be removed by distillation (DOERFELT and GELBERT 1959). However, in order to avoid disproportionation, the oxide should not be heated over 100°C (compare section III g).

Hexaphenyldistannoxane can be recrystallized from ethanol or acetonitrile (SCHMITZ-DUMONT 1941); it can also be purified by precipitation of its carbonate from chloroform or benzene solution with CO_2. The precipitate is then heated to 60° to 80°C, decomposing the carbonate into bis-triphenyltin oxide and CO_2 (LUIJTEN 1960, FRIEBE and KELKER 1963).

Diphenyltin oxide is produced by treating solutions of diphenyltin dichloride in an organic solvent with aqueous NaOH, KOH, or ammonia; the oxide precipitates (rather than the hydroxide which is unstable) (compare summary by INGHAM et al. 1960).

Similarly, $(C_6H_5)SnOOH$ can be obtained from reacting phenyltin trichloride with alkali or ammonia. Here it is important to avoid a large excess of the precipitating agent which would redissolve the phenylstannic acid.

d) Synthesis of other phenyltin compounds

Triphenyltin acetate can be obtained from the hydroxide by heating it with glacial acetic acid to 130° (LUIJTEN and VAN DER KERK 1955, VAN DER KERK and LUIJTEN 1956 a) or from hexaphenyldistannoxane with acetic anhydride (KAARS SIJPESTEIJN et al. 1962). Another paper describes the reaction of triphenyltin chloride with sodium acetate (HEROK and GOETTE 1963).

A common method for introducing acid groups into triphenyltin compounds is the reaction of triphenyltin bromide or iodide with silver salts (most likely one can also start from triphenyltin chloride). In this manner, the acetate (KAARS SIJPESTEIJN et al. 1962), chloroacetate, benzoate, or oxalate (SRIVASTAVA and TANDON 1963 and 1964) can be produced.

Triphenyltin compounds of phosphoric and thiophosphoric acids can be synthesized by reacting triphenyltin halogenides, dissolved in an organic solvent, with the sodium, potassium, or ammonium salt of the respective acid. Further, the free acids may be dissolved in benzene and reacted with $(C_6H_5)_3SnOH$; the water formed during the reaction is distilled off (KUBO 1965).

Bis-triphenyltin sulfide is formed by feeding gaseous H_2S into a solution of the chloride in ethanol (INGHAM et al. 1960).

Formation of water-soluble triphenyltin compounds from the chloride or acetate can be effected by heating their solution in methanol with triethanolamine (INABA and WATANABE 1963); the components react as follows:

$$(C_6H_5)_3SnCl + N\begin{matrix} CH_2CH_2OH \\ -CH_2CH_2OH \\ CH_2CH_2OH \end{matrix} \rightarrow (C_6H_5)_3Sn-{}^+N\begin{matrix} CH_2CH_2OH \\ -CH_2CH_2OH \\ CH_2CH_2OH \end{matrix} + Cl^- \quad (21)$$

In another synthesis (PARKIN and POLLER 1976, POLLER 1976), sucrose molecules are linked via phthalic acid with triphenyltin hydroxide:

$$(22)$$

e) Purification of triphenyltin compounds

In addition to recrystallization some other methods have been described for producing pure triphenyltin compounds on a bench scale:

(a) Dissolving the reaction mixture (chlorides) in hot methanol, filtering off the tetraphenyltin, cooling, and precipitating of $(C_6H_5)_3SnCl$ by addition of water (D'ANS and ZIMMER 1952).

(b) Dissolving the reaction mixture (chlorides) in diethyl ether, filtering off the tetraphenyltin and treating the solution with aqueous $0.1N$ NaOH-solution. Di- and monophenyltin hydroxides will precipitate; the triphenyltin hydroxide remaining in the ether can be converted to chloride by using HCl (BOCK et al. 1962).

(c) Dissolving the reaction mixture (chlorides) in benzene, extracting the di- and monophenyltin compounds with caustic aqueous tartrate solution, and finally extracting the triphenyltin compounds with aqueous $0.2N$ H_2SO_2 (tetraphenyltin remains in the organic phase). From the aqueous solution, the triphenyltin hydroxide is precipitated with NaOH or the acetate with acetic acid (BOCK et al. 1962).

f) Synthesis of radioactively-labeled phenyltin compounds

Phenyltin compounds can be labeled with a radioactive tin isotope, with ^{14}C, or with tritium.

In labeling with radioactive tin, the isotope ^{113}Sn is the most suitable one; it is obtained through an (n, γ)-reaction from ^{112}Sn, of which however, only 1% is contained in natural tin. To produce tin compounds with high specific activity it is recommended, therefore, to start with ^{112}Sn-enriched preparations.

During the activation of natural as well as enriched tin, radioactive antimony-isotopes are formed as by-products which have to be removed since they will react together with the tin in the synthesis of labeled organotin compounds. The activated tin is chlorinated and the $SnCl_4$ is repeatedly distilled under vacuum, while some $SbCl_3$ is added (HEROK and GOETTE 1963).

As described above, labeled triphenyltin chloride is synthesized via the tetraphenyltin compound $(C_6H_5)_4$ ^{113}Sn, which is then reacted with $SnCl_4$ (HEROK and GOETTE 1963). From the $(C_6H_5)_3$ $^{113}SnCl$ thus obtained, the radioactively-labeled lower phenyltin homologues can be produced by reaction with $(C_6H_5)_2SnCl_2$ or $(C_6H_5)SnCl_3$, respectively [heat for 3 hr to about 180°C (FREITAG 1972)].

^{113}Sn is converted to ^{113m}In by electron capture, whereby a γ-radiation of 255 KeV appears; the ^{113m}In is converted to the stable ^{113}In, yielding a γ-radiation of 393 KeV:

$$^{113}Sn \xrightarrow{\;115\text{ days}\;} {}^{113m}In \xrightarrow{\;100\text{ min}\;} {}^{113}In \qquad (23)$$

The half-life periods show that the radioactive equilibrium is reached in about 19 hr.

Triphenyltin acetate can be labeled with ^{14}C within the phenyl- or the acetate group. The phenyl labeling is achieved by using ^{14}C-bromobenzene with the Grignard-synthesis; for labeling the acetate-group, $(C_6H_5)_3SnCl$ is reacted with ^{14}C-sodium acetate (BARNES *et al.* 1973.).

Tritium-labeled tetraphenyltin is produced by keeping $(C_6H_5)_4Sn$ for 4 days in an atmosphere of tritium (CARSON *et al.* 1962).

III. Chemical and physical behavior of phenyltin compounds

a) Acid-base-reactions

1. Triphenyltin compounds.—$(C_6H_5)_3SnX$ compounds, where X may be any anion, are, as mentioned already, converted rapidly and quantitatively into the hydroxide by alkali:

$$(C_6H_5)_3SnX + MOH \rightarrow (C_6H_5)_3SnOH + MX \qquad (24)$$

The reaction can also be reversed and proceeds with ease:

$$(C_6H_5)_3SnOH + HX \rightarrow (C_6H_5)_3SnX + H_2O \qquad (25)$$

The hexaphenyldistannoxane formed from the hydroxide by loss of water, also reacts easily with acids:

$$(C_6H_5)_3SnOSn(C_6H_5)_3 + 2\ HX \rightarrow 2\ (C_6H_5)_3SnX + H_2O \qquad (26)$$

This last reaction can, for example, be achieved by shaking a benzene solution of the oxides with aqueous acid.

2. Diphenyltin compounds.—$(C_6H_5)_2SnX_2$ compounds react similarly. However, the hydroxide quickly loses water forming the oxide. $(C_6H_5)_2$-SnO is difficult to dissolve in aqueous acids; therefore, a quantitative reconversion to the starting material, for example $(C_6H_5)_2SnCl_2$, cannot be achieved with certainty (FREITAG 1972). A suspension of diphenyltin oxide in diethyl ether has been converted to the dichloride by treating it with gaseous HCl (CHAMBERS and SCHERER 1926).

3. Monophenyltin compounds.—$(C_6H_5)SnX_3$ compounds are also converted with alkali or ammonia to the hydroxide which immediately loses water to form phenylstannonic acid $(C_6H_5)SnOOH$. Right after precipitation this compound dissolves in acid and in strongly alkaline solutions (to $C_6H_5SnOONa$ with alkali); from the latter, it can be precipitated again with CO_2. Phenylstannonic acid precipitates age quickly, and after a short while of drying the material in the open it is difficult to dissolve in acids (FREITAG 1972). Phenylstannonic acid is insoluble in common organic solvents.

b) Hydrolysis and decomposition with benzene formation

1. Triphenyltin compounds.—Triphenyltin chloride in the presence of water quickly hydrolyzes to the hydroxide (PRINCE 1958 and 1959). The acetate in a water slurry at 20°C hydrolyzes in a few hr almost completely (BAUMANN 1958); the kinetics of the hydrolysis were studied using material with a ^{14}C-labeled acetate group. The presence of acetic acid in the solution slows down the hydrolysis (BARNES *et al.* 1973, Fig. 1). The acetate hydrolyzes in dilute hydrochloric acid solution (pH 3), but not the chloride; both compounds are converted to the hydroxide at pH 8.3 (BROWN 1973).

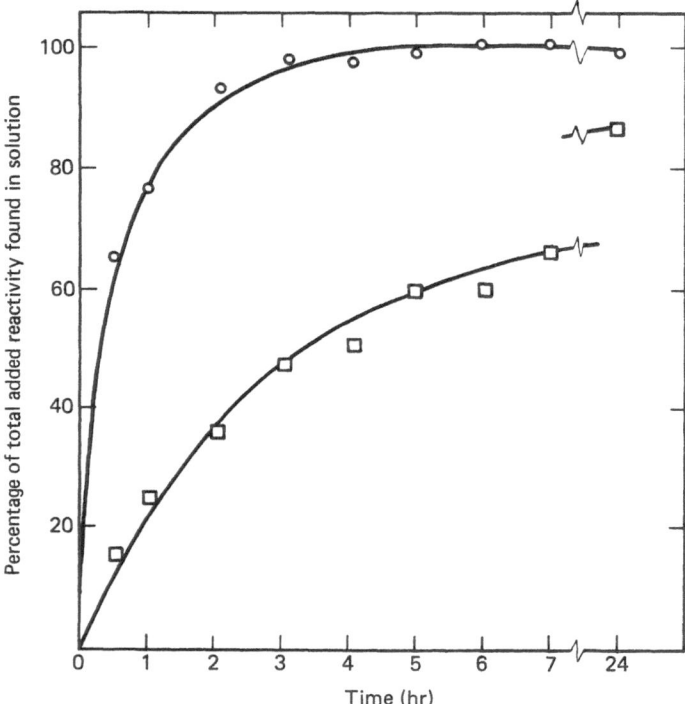

Fig. 1.   Hydrolysis of triphenyltin 1,2[^{14}C]-acetate (from BARNES *et al.* 1973):
O–O = water and □–□ = acetic acid (1%).

In aqueous solutions (or slurries) and at room temperature triphenyl compounds decompose at a slow rate, losing the phenyl groups as benzene. This may be concluded from the long-lasting biological activity of the acetate in spring water (several wk—DESCHIENS and FLOCH 1962 and 1963, STRUFE 1968). If boiled, the $(C_6H_5)_3Sn^+$-content is diminished within 2 hr (depending on the pH of the liquid) to about 40 to 70% of the original concentration (SCHMITT-STRECKER 1976; Fig. 2).

By contrast, heating of triphenyltin compounds with acid or alkali leads to quick decomposition. Solutions of the sulfate in diluted sulfuric acid (pH 1) can be decomposed completely by boiling them for about 15 min (BOCK *et al.* 1962); strong acids or bases cause decomposition

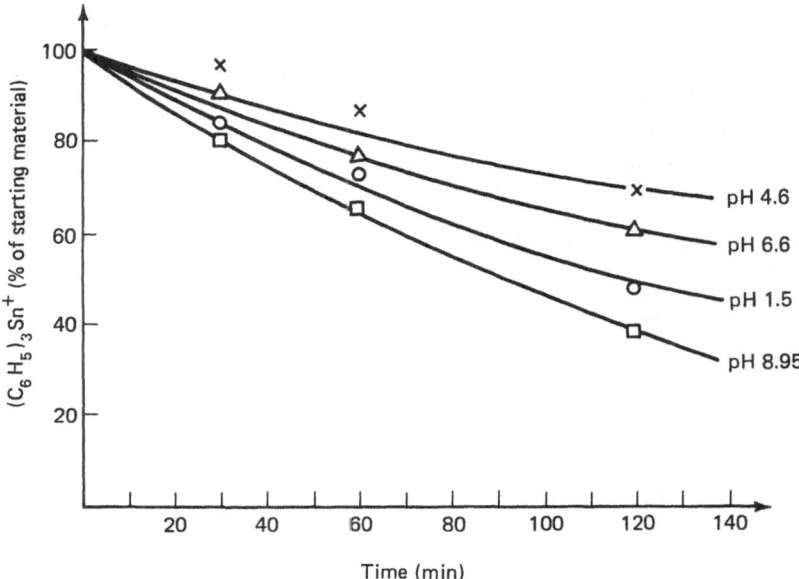

Fig. 2.   Decomposition of triphenyltin acetate by boiling of water slurries (SCHMITT-
STRECKER 1976).

already at normal or slightly raised temperatures (BUERGER 1961 a, BOCK
et al. 1962).

2. Diphenyltin compounds.—Diphenyltin dichloride initially hydrolyzes to $(C_6H_5)_2Sn(OH)Cl$ (ARONHEIM 1878), then the oxide is
formed via dimeric intermediates (ALLESTON et al. 1963). The rate of
hydrolysis has not been studied.

Diphenyltin compounds, too, split off phenyl groups as benzene
under the influence of water, acids, or bases. The reaction proceeds at
least as fast as with the triphenyltin compounds, probably even faster.

Diphenyltin chloride can be decomposed at 150° to 190°C with chelating agents such as acetylacetone, salicylaldehyde, and 8-hydroxy-quino-
line. The resulting tin compounds are of the type $SnCl_2(Ch)_2$ (Ch =
chelating agent) (MARTIN et al. 1966 and 1967).

3. Monophenyltin compounds.—Those giving off benzene decompose
faster than the respective diphenyltin compounds (FREITAG 1972).

c) Complex formation

Following the behavior of $SnCl_4$, organotin halogenides can also form
complexes with many organic ligands (CAHOURS 1862). Initially, only
1:2 complexes with hexacoordinate tin were synthesized, for example

$(C_6H_5)_2SnCl_2$ · 2 pyridine or $(C_6H_5)_3SnCl$ · 2 pyridine hydrochloride (PFEIFFER 1910, PFEIFFER et al. 1911, BEATTIE 1963). Later on, also 1:1 compounds with pentacoordinate tin were obtained (BEATTIE et al. 1962, HULME 1963, GROSJEAN et al. 1963, POLLER 1965). Besides aliphatic amines and aminehydrochlorides, mostly molecules with N-O, S-O, or P-O groups were used as electron donors.

Table IV. *Complexes of phenyltin halogenides with pentacoordinate tin (examples).*

Tin compound	Ligand	Complex
$(C_6H_5)_3SnCl$	Dimethylsulfoxide	$(C_6H_5)_3SnCl$ · $OS(CH_3)_2$
$(C_6H_5)_3SnCl$	Quinoline-N-oxide	$(C_6H_5)_3SnCl$ · ONC_9H_7
$(C_6H_5)_3SnCl$	Tetramethylammonium chloride	$[(C_6H_5)_3SnCl_2]$ · $[(CH_3)_4N]$
$(C_6H_5)_3SnCl$	Triphenylphosphine oxide	$(C_6H_5)_3SnCl$ · $OP(C_6H_5)_3$
$(C_6H_5)_3SnCl$	Decyltriphenylphosphonium bromide	$[(C_6H_5)_3SnClBr][P(C_6H_5)_3C_{10}H_{21}]$
$(C_6H_5)_2SnCl_2$	Tetraphenylarsonium chloride	$[(C_6H_5)_2SnCl_3][(C_6H_5)_4As]$
$(C_6H_5)SnCl_3$	Tetraphenylarsonium chloride	$[(C_6H_5)SnCl_4][(C_6H_5)_4As]$

These complexes are usually formed in exothermic reactions at high yields; they can be recrystallized and even be distilled without decomposition (BOEHRINGER SOHN 1965, LANGER and BLUT 1966, JERCHEL 1969). Some of these dissociate in suitable solvents such as nitrobenzene or acetonitrile, as can be shown by conductometric and potentiometric methods (SCHROEDER et al. 1966, TAGLIAVINI and ZANELLA 1968, ZANELLA and TAGLIAVINI 1968).

d) Oxidation and reduction

Strong oxidation agents such as $KMnO_4$, $HClO_4$, KIO_3, Cl_2, Br_2, I_2, and others will decompose triphenyltin compounds, forming inorganic tin (IV), sometimes through transient compounds. Di- and monophenyltin compounds react in the same way. In the presence of reducing agents, several products are formed. The reaction with sodium in inert solvents yields hexaphenylditin $(C_6H_5)_3Sn-Sn(C_6H_5)_3$, which, under continued treatment with sodium can be converted to triphenyltin sodium $(C_6H_5)_3$-SnNa. This, together with NH_4Br, will form the hydride $(C_6H_5)_3SnH$, which can also be obtained from triphenyltin chloride and $LiAlH_4$. Diphenyltin compounds react in a similar manner.

The reduction products may be interesting from the point of synthetic and theoretic chemistry; however, for this study they are not important. For more information refer to summaries by VAN DER KERK et al. (1958) and INGHAM et al. (1960).

If heated with strong reducing agents like H_3PO_2 and others in aqueous solutions, triphenyltin compounds are completely decomposed (BOCK et al. 1962).

e) Electrochemical behavior; radical formation

1. Electrolytical dissociation.—In organic solvents like pyridine, acetonitrile, nitrobenzene, dimethylformamide, and others, triphenyltin chloride shows only very low electric conductivity possibly resulting from HCl which is split off by residual moisture (THOMAS and ROCHOW 1957 a and b). On the other hand, almost complete dissociation has been observed in a 44% aqueous ethanol-solution (JANSSEN and LUIJTEN 1962).

2. Electrolysis and radical formation.—During the electrolysis of triphenyltinchloride, in methanol solution, $(C_6H_5)_3Sn\text{-}Sn(C_6H_5)_3$ is formed at the cathode (RICCOBONI 1937).

Studies about the behavior of triphenyltin chloride at the dropping mercury electrode have led to further knowledge about the electrolysis (DESSY et al. 1966, VANACHAYANGKUL and MORRIS 1968, BOOTH et al. 1970). According to BOOTH and FLEET (1970), the triphenyltin cation reacts at the cathode in H_2O/ethanol-solution (1:1, with 0.002% Triton X-100 as maximum-suppressor) in a reversible one-electron reaction. First, the radical $(C_6H_5)_3Sn^{\cdot}$ is formed, which is adsorbed at the surface of the mercury. This then reacts either to electrochemically neutral hexaphenylditin, or it is further reduced to the triphenyltin anion. $(C_6H_5)_3Sn^-$ can react with hydrogen-ions, forming the hydride which can be reoxydized at the anode to the triphenyl cation (Fig. 3).

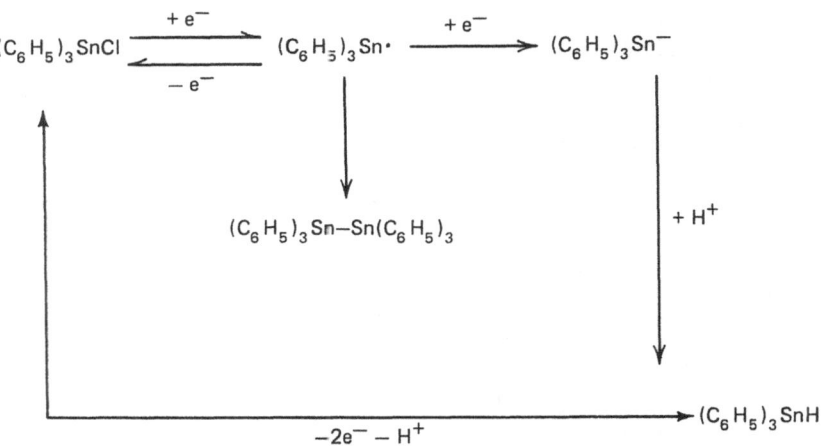

Fig. 3. Electrolytic reduction of triphenyltin chloride at the dropping mercury electrode (BOOTH and FLEET 1970).

In the absence of mercury, the triphenyltin radical seems to be quite unstable; it could not be found in a hexane solution of hexaphenylditin (HAGUE and PRINCE 1966). However, a pale yellow radical, probably $(C_6H_5)_3Sn^{\cdot}$, was observed when $(C_6H_5)_3SnH$ had been exposed to ultraviolet light at temperatures below 160 K (Schmidt *et al.* 1965). Hexa-phenylditin, exposed to UV-light, produced the same radical (Willpatte-Steinert and Nasielski 1970). During reduction of C_6H_5Br at a Sn-cathode, the formation of $(C_6H_5)_4Sn$ was observed (Ulery 1972).

Adding organotin compounds to the alcoholic electrolyte of an alternating current polarographic cell results in a lower capacitance (Jehring and Mehner 1967).

f) Solubilities

Quantitative and semi-quantitative data about the solubility of several phenyltin compounds are shown in Tables V to VII and in Figure 4.

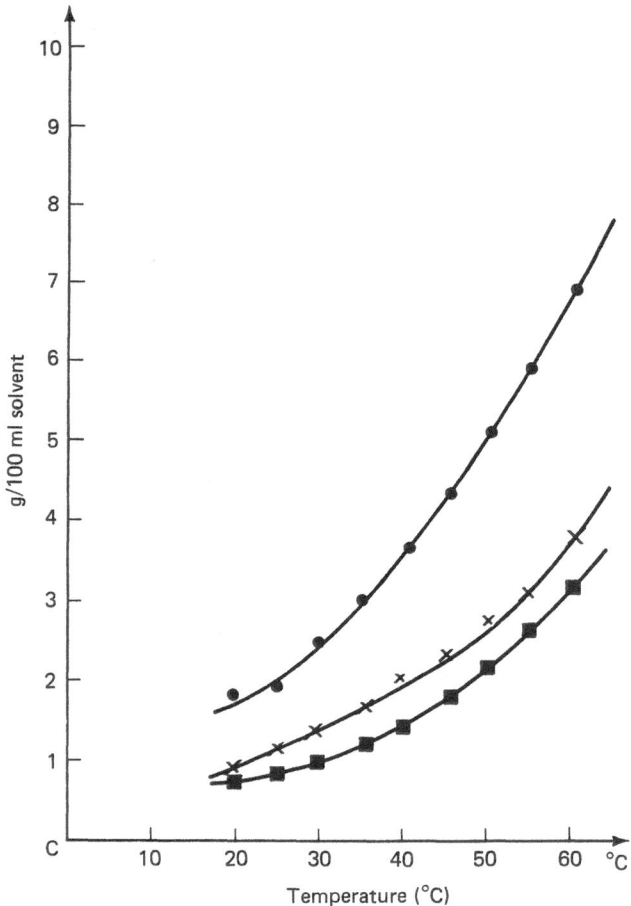

Fig. 4. Temperature dependence of the solubility of tetraphenyltin (after Newkirk 1972): ● = chloroform, × = benzene, and ■ = dioxane.

Table V. *Solubility of tetraphenyltin.*

Solvent	Temp. (°C)	Solubility	References
Seawater	*ca.* 20	<<11 ppm	VIND & HOCHMANN (1962 and 1963)
Diethyl ether	20	1.02 g/L	STROHMEIER & MILTENBERGER (1958)
Dioxane	20	5.29 g/L	STROHMEIER & MILTENBERGER (1958)
Pyridine	20	6.8 g/L	STROHMEIER & MILTENBERGER (1958)
CCl₄	20	5.24 g/L	STROHMEIER & MILTENBERGER (1958)
Heptane	20	0.2 g/L	STROHMEIER & MILTENBERGER (1958)
p-Xylene	20	3.0 g/L	STROHMEIER & MILTENBERGER (1958)
Benzene	20	9.18 g/L	STROHMEIER & MILTENBERGER (1958)
Benzene	20	1.00 g/100 g benzene	NEFEDOV & VARSHAV (1954)
Benzene	30	1.52 g/100 g benzene	NEFEDOV & VARSHAV (1954)
Benzene	40	2.31 g/100 g benzene	NEFEDOV & VARSHAV (1954)
Benzene	50	3.49 g/100 g benzene	NEFEDOV & VARSHAV (1954)
Benzene	60	5.08 g/100 g benzene	NEFEDOV & VARSHAV (1954)
Benzene	70	7.85 g/100 g benzene	NEFEDOV & VARSHAV (1954)

Judging by the author's own qualitative experiences, the figures of the *Codex Committee* (1970) for the solubilities of triphenyltin chloride, -hydroxide and -acetate in water may be too high. Table VIII shows several qualitative observations.

g) *Thermal behavior*

1. **Melting points and vapor pressures.**—Melting points and vapor pressures of several phenyltin compounds have been compiled from the literature in Tables IX and X.

2. **Disproportionation reactions.**—When heated, both triphenyltin hydroxide and hexaphenyldistannoxane yield tetraphenyltin and diphenyltin oxide:

$$2 (C_6H_5)_3SnOH \rightarrow (C_6H_5)_4Sn + (C_6H_5)_2SnO + H_2O \qquad (27)$$

$$(C_6H_5)_3SnOSn(C_6H_5)_3 \rightarrow (C_6H_5)_4Sn + (C_6H_5)_2SnO \qquad (28)$$

Conceivably, reaction (27) does not take place directly, but via the hexaphenyldistannoxane. This disproportionation has been reported to occur already when benzene solutions of the hydroxide are heated to boiling (CHAMBERS and SCHERER 1926), when hexaphenyldistannoxane is heated to >80°C (VAN RIJ 1960) or to 100°C (DOERFELT and GELBERT 1959). In the absence of solvents, both reactions are almost complete at temperatures of 135° to 140°C (SCHMITZ-DUMONT 1941, LUIJTEN and VAN DER KERK 1968).

Table VI. *Solubility of triphenyltin compounds.*

Compound	Solvent	Temp. (°C)	Solubility	References
Triphenyltin chloride	H_2O	20	40 ppm	*Codex Committee* (1970)
	H_2O	30	78 ppm	Kubo (1965)
	seawater	*ca.* 20	<1 ppm	Vind & Hochmann (1962 and 1963)
	CH_3OH	20	*ca.* 7%	Freitag (1972)
	C_2H_5OH	22	2.34%	Kubo (1965)
	acetone	22	>10%	Kubo (1965)
	$CHCl_3$	20	*ca.* 7%	Freitag (1972)
	n-hexane	22	1.36%	Kubo (1965)
	cyclohexane	20	*ca.* 4%	Freitag (1972)
Triphenyltin hydroxide	H_2O	20	8 ppm	*Codex Committee* (1970)
	C_2H_5OH	20	*ca.* 100 g/L	Pieters (1961)
	C_2H_5OH	20	1.0%	*Codex Committee* (1970)
	diethyl ether	20	28 g/L	Pieters (1961)
	diethyl ether	20	2.8%	*Codex Committee* (1970)
	CH_2Cl_2	20	17.1%	*Codex Committee* (1970)
	1,2-dichloro-ethane	20	7.4%	*Codex Committee* (1970)
	benzene	20	41 g/L	Van Rij (1960)
	benzene	20	4.1%	*Codex Committee* (1970)
Hexaphenyl-distannoxane	seawater	*ca.* 20	<1 ppm	Vind & Hochmann (1962)
	benzene	20	630 g/L	Van Rij (1960)
Triphenyltin acetate	H_2O	20	<3.3 ppm	Barnes *et al.* (1973)
	H_2O	20	28 ppm	*Codex Committee* (1970)
	seawater	*ca.* 20	<1 ppm	Vind & Hochmann (1962 and 1963)
	CH_3OH	20	7%	*Codex Committee* (1970)
	C_2H_5OH	*ca.* 20	*ca.* 0.2%	Strufe (1968)
	diethyl ether	20	3.7%	*Codex Committee* (1970)
	CCl_4	20	7.2%	*Codex Committee* (1970)
	benzene	20	5%	*Codex Committee* (1970)

Table VII. *Solubilities of diphenyltin compounds.*

Compound	Solvent	Temp. (°C)	Solubility	References
Diphenyltin dichloride	H_2O	20	*ca.* 50 ppm	Freitag (1972)
	CH_3OH	20	*ca.* 15%	Freitag (1972)
	$CHCl_3$	20	*ca.* 15%	Freitag (1972)
	cyclohexane	20	*ca.* 9%	Freitag (1972)
Diphenyltin oxide	seawater	*ca.* 20	6 ppm	Vind & Hochmann (1962)

Table VIII. *Qualitative observations on the solubilities of phenyltin compounds (solubilities as listed in Tables V, VI, and VII excluded).*

Compound	Easily soluble in	Slightly soluble in	Hardly soluble in
$(C_6H_5)_4Sn$	$CHCl_3$, CS_2, CH_3COOH	ethanol	H_2O, CH_3OH, petroleum ether
$(C_6H_5)_3SnCl$	diethyl ether, benzene	petroleum ether	
$(C_6H_5)_3SnOH$	CH_3OH (90%), C_2H_5OH (90%)	—	$CHCl_3$, cyclohexane
$(C_6H_5)_3SnOSn(C_6H_5)_3$	ethanol, acetonitrile	—	H_2O
$(C_6H_5)_2SnCl_2$	C_2H_5OH, acetone, diethyl ether	petroleum ether	—
$(C_6H_5)_2SnO$		immediately after precipitation: C_2H_5OH, diethyl ether	H_2O
$(C_6H_5)SnCl_3$	H_2O, CH_3OH, C_2H_5OH, acetone, $CHCl_3$, cyclohexane, acids and alkali	petroleum ether	—
$(C_6H_5)SnOOH$	immediately after precipitation: acids, alkali		aged: acids and alkali, all commonly used organic solvents

Table IX. *Melting points of various phenyltin compounds.*[a]

Compound	Melting point ($^\circ$C)
$(C_6H_5)_4Sn$	228–230 (224–226)
$(C_6H_5)_3SnCl$	108 (105–107)
$(C_6H_5)_3SnOH$	120–122 (dec.) (118–120)
$(C_6H_5)_3SnOSn(C_6H_3)_3$	124
$(C_6H_5)_3SnOCOCH_3$	124.5 (120–123)
$(C_6H_5)_3SnSSn(C_6H_5)_3$	143–144
$(C_6H_5)_2SnCl_2$	44 (42–43)
$(C_6H_5)_2SnO$	>360 (345)[b]
$(C_6H_5)SnCl_3$	−32
$(C_6H_5)SnOOH$	>350

[a] The highest melting points found in the literature are listed first; other melting points that are mentioned several times are listed in parentheses.

[b] Varying depending on method of synthesis.

Table X. *Vapor pressures of various phenyltin compounds.*

Compound	Vapor pressures and reference
$(C_6H_5)_4Sn$	$6.32 \cdot 10^{-9}$ Torr at 25°C (CARSON *et al.* 1962)
	760 Torr at >420°C (POLIS 1889)
$(C_6H_5)_3SnCl$	$5 \cdot 10^{-4}$ Torr at 150°–200°C (WITTENBERG 1962)
	0.5 Torr at 185°C (FREITAG 1972)
	13.5 Torr at 240°C (KRAUSE 1918)
$(C_6H_5)_3SnOCOCH_3$	760 Torr at 230°C (VAN DER KERK *et al.* 1958)
$(C_6H_5)_2SnCl_2$	$5 \cdot 10^{-4}$ Torr at 100°–120°C (WITTENBERG 1962)
	0.5 Torr at 135°C (FREITAG 1972)
	2 Torr at 163°–165°C (JITSU *et al.* 1969)
	5 Torr at 180°–185°C (ZIMMER & SPARMANN 1954)
	760 Torr at 333°–337°C (dec.) (ARONHEIM 1878)
$(C_6H_5)SnCl_3$	0.23 Torr at 68°–69°C (WHITESIDES *et al.* 1970)
	0.5 Torr at 90°–92°C (FREITAG 1972)
	1.4 Torr at 96°–97°C (GILMAN & GIST 1957)
	2 Torr at 102°C (WARDELL 1967)
	4.5 Torr at 120°–125°C (POLLER 1962)
	7 Torr at 109°–113°C (JITSU *et al.* 1969)
	10 Torr at 105°–115°C (WITTENBERG 1960)
	10 Torr at 109°–110°C (WITTENBERG 1962)
	10 Torr at 136°C (GRIFFITHS & DERWISH 1959)
	12 Torr at 131°C (LUIJTEN & VAN DER KERK 1955)
	13–15 Torr at 128°–130°C (D'ANS & ZIMMER 1952)
	15 Torr at 128°C (ZIMMER & SPARMANN 1954)
	25 Torr at 142°–143°C (KOCHESHKOV 1929)
	755 Torr at 245°–310°C (dec.) (KOCHESHKOV 1929)

Phenyltin chlorides react in a similar way (ARONHEIM 1878):

$$2\,(C_6H_5)_3SnCl \rightarrow (C_6H_5)_4Sn + (C_6H_5)_2SnCl_2 \qquad (29)$$

$$3\,(C_6H_5)_2SnCl_2 \rightarrow 2\,(C_6H_5)_3SnCl + SnCl_4 \qquad (30)$$

Reaction (29) takes place at temperatures of about 200°C (LUIJTEN and VAN DER KERK 1968). Several other disproportionation reactions have been observed during the synthesis of phenyltin compounds via Sn/Na-alloys (NAD and KOCHESHKOV 1938).

With monophenyltin compounds, disproportionation of the potassium salt of phenylstanonic acid has been reported (INGHAM *et al.* 1960):

$$2\,(C_6H_5)SnOOK \rightarrow (C_6H_5)_2SnO + K_2SnO_3 \qquad (31)$$

3. Thermal decomposition.—Tetraphenyltin is thermally fairly stable. Decomposition temperatures are listed as 348°C (BLAKE *et al.* 1961) and 352°C (TAMBORSKI *et al.* 1965).

Depending on the anion, triphenyltin compounds vary in their stability. Triphenyltin chloride remains unchanged during 10 hr at 95°C (JITSU *et al.* 1967); it will not decompose unless heated beyond 270° to 280° C (D'ANS and ZIMMER 1952).

Data about the thermal decomposition of triphenyltin acetate are somewhat inconsistent. At normal temperatures and dry storage, the compound lasts indefinitely (HAERTEL 1962). Other authors, however, have observed minor decomposition after a few hr at 45°C, which increases at 100°C (JITSU *et al.* 1967). The unfavorable results may be attributed to atmospheric moisture resulting in hydrolysis and loss of acetic acid. Other reports show that during 30 hr at 90°C the acetate does not decompose; within 3 hr at 150°C decomposition amounted to approximately 15% (*Codex Committee* 1970, Fig. 5 and Table XII).

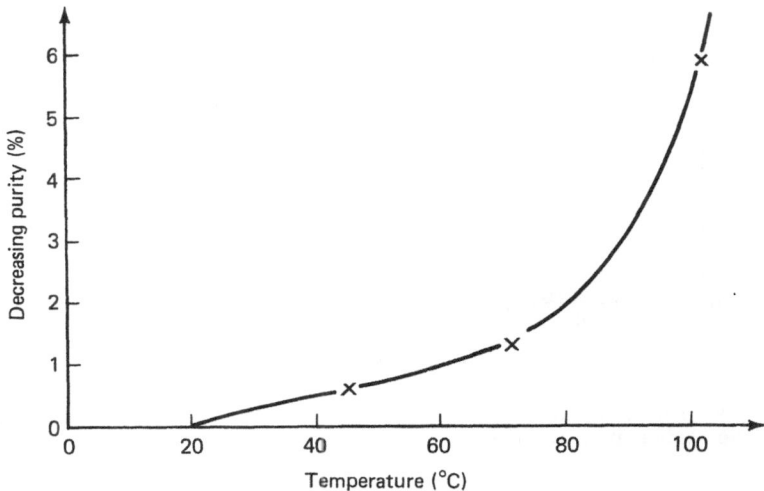

Fig. 5.   Decreasing purity of triphenyltin acetate within 5 hr, depending on temperature (after JITSU *et al.* 1967).

As already mentioned, triphenyltin hydroxide loses water at fairly low temperatures, although it is rather stable below 45°C (JITSU *et al.* 1967; Table XI). The transformation to hexaphenyldistannoxane proceeds rather quickly from 60° to 70°C on (*Codex Committee* 1970; Fig. 6). Hexaphenyldistannoxane will not decompose unless heated to

Table XI. *Purity change of triphenyltin hydroxide, depending on time* (after JITSU *et al.* 1967).

| | Assay of triphenyltin hydroxide (%) | | |
Analytical method	Before experiment	After 10 wk at 28°C	After 5 wk at 28°C & 5 wk at 45°C
Hydroxide determination	100.2	99.8	97.9
Thin-layer chromatography	98.7	98.7	97.3

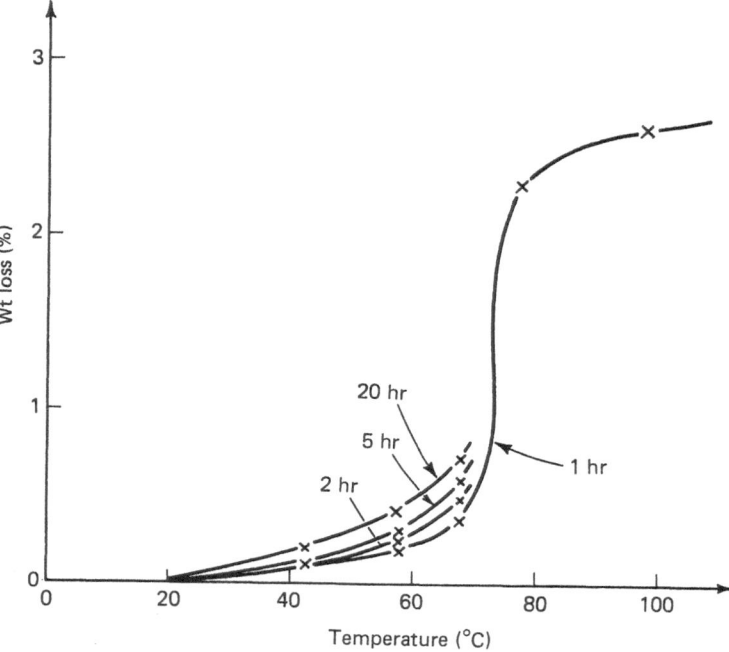

Fig. 6. Wt loss of triphenyltin hydroxide after different heating periods, depending
on temperature (after JITSU et al. 1967).

approximately 250°C (LUIJTEN and VAN DER KERK 1968; compare data
under **III** g **2**, disproportionation).

The results of storage tests of some triphenyltin compounds in sealed
bottles at 45°C are listed in Table XII (JITSU et al. 1967).

h) Spectra

1. Ultraviolet absorption.—Tetraphenyltin and the phenyltin chlorides
show one strong and one weak absorption band in the ultraviolet range

Table XII. *Decreasing purity of various triphenyltin derivatives at*
45°C, depending on time (after JITSU et al. 1967).

Duration of test (in wk)	Active ingredient (%)			
	Acetate	Hydroxide	Chloride	Oxide
0	98.3	98.7	98.1	99.3
1	97.8	98.1	98.0	98.4
2	97.2	97.5	98.0	97.5
3	97.0	97.3	97.9	97.4

(K- and B-band at about 210 and 260 nm, respectively). The longer-wave B-band is attributed to the electrons moving from the ground state A_{1g} to a stable excitation state B_{2u} of the phenyl group; the shorter-wave K-band conforms with the excitation from the original state to state B_{1u} in phenyl groups (MILAZZO 1941, GRIFFITHS and DERWISH 1959, MARROT et al. 1965). A very weak band of phenyltin chlorides at about 290 nm is probably created by excitation of the Sn-Cl-bond.

The UV-spectra of organotin compounds with differing numbers of phenyl groups are very similar to each other; triphenyltin acetate may be used as an example (Fig. 7).

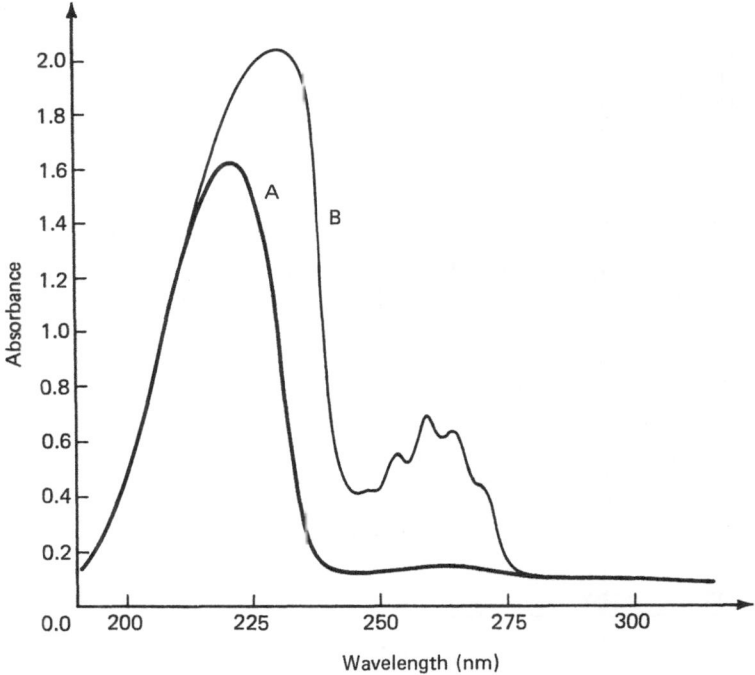

Fig. 7. UV-absorption of triphenyltin acetate in cyclohexane solution (1 cm-cuvette): A = 24.6 μg/ml and B = 246 μg/ml.

2. Infrared absorption.—The IR-absorption bands occurring in compounds of the type $(C_6H_5)_nSnX_{4-n}$ can be attributed to three different functions in the molecule: oscillation of atoms within the phenyl groups, oscillation of the Sn-C- and Sn-X-groups in the long-wave IR, and, finally, oscillation within X, if X consists of several atoms. If not, for example where X is Cl, Br, or others, then only the bands originating in the phenyl groups will occur in the NaCl frequency range of the spectrum.

The IR-spectra of tetraphenyltin and of the phenyltin chlorides are nearly identical in this area; this means that there is no significant interaction between the phenyl groups nor any influence from the halogen, X (see Fig. 8).

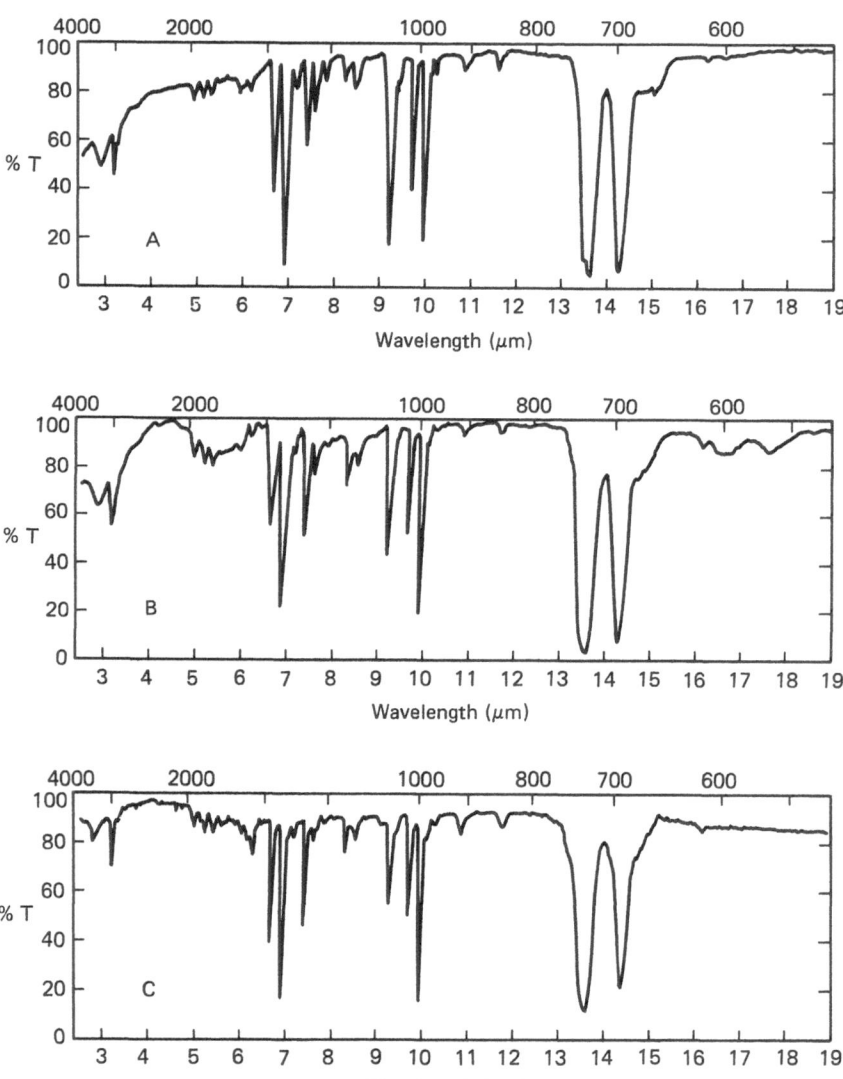

Fig. 8 (a–c). IR-spectra of phenyltin chlorides: A = triphenyltin chloride, B = diphenyltin dichloride, and C = monophenyltin trichloride.

Table XIII. *Far-infrared absorption bands of tetraphenyltin and phenyltin chlorides*[a] (after POLLER 1966).

Assignment	Absorption bands (cm^{-1}) of			
	$(C_6H_5)_4Sn$	$(C_6H_5)_3SnCl$	$(C_6H_5)_2SnCl_2$	$(C_6H_5)SnCl_3$
6 b (B_2)	614	614	614	608
16 b (B_1)	455	451–440	443	439
	443		444	
γ-as(Sn-X)	—	—	364	385–364
γ-s(Sn-X)	—	—	356	
			350	
γ-(Sn-X)	—	346–338	—	—
			279	
γ-as(Sn-Ph)	270	274–265	276	—
			274	
γ-s(Sn-Ph)	263	239	230	
			226	—
			247[b]	
γ-(Sn-Ph)	—	—	245[b]	250
			242[b]	248

[a] Nujol mull.
[b] In solution.

Correlation of different phenyl frequencies follows the rules established for mono-substituted benzene (GRIFFITHS and DERWISH 1960, POLLER 1962, HARRAH et al. 1962, KRIEGSMANN and GEISSLER 1963, MAIRE et al. 1965, DURIG et al. 1966, GMELIN 1975). Oscillations in which the Sn-atom with its greater mass is involved give very low frequencies. In this range, the various phenyltin compounds show marked differences (BROWN et al. 1965, POLLER 1966, DURIG et al. 1966, SRIVASTAVA 1967, SMITH 1968).

Table XIII shows long-wave bands of several phenyltin compounds and their respective assignments. Table XIV lists the Sn-O-oscillations; the OH-oscillation of $(C_6H_5)_3SnOH$ at 898 cm^{-1} is of analytical importance (see chapter VIII, Analytical methods).

3. ¹H-NMR-spectra.—The ¹H-NMR-spectra of both tetraphenyltin and the phenyltin chlorides, consist of two adjacent signals with the intensity ratio 2:3, which belong to the o- and (m+p)-protons of the phenyl groups. In phenyltin trichloride, the signals overlap. The position of the bands depends to some extent on concentration (MAIRE and

Table XIV. *Sn-O-frequencies* (after POLLER 1962).

Compound	γ (cm^{-1})
$(C_6H_5)_3SnOH$	912; 898
$(C_6H_5)_3SnOSn(C_6H_5)_3$	777
$(C_6H_5)_2SnO$	571; 553

28 RUDOLF BOCK

HEMMERT 1963, MAIRE 1967, ANGELETTI and MAIRE 1969, PRESTON *et al.*
1972).

At higher sensitivity, satellites have been found which are caused by
spin-spin-coupling of the *o*-protons with the ^{117}Sn- and ^{119}Sn-isotopes of
tin (VERDONCK and VAN DER KELEN 1965).

i) Photolysis through visible light and ultraviolet radiation

Phenyltin compounds are decomposed slowly by visible light, much
faster by UV-radiation.

Figure 9a shows the decomposition of triphenyltin acetate in aqueous
solution by radiation of a UV-arc lamp (Hanau S 81), depending on
time (STRUFE 1968). After only 1 hr, approximately 90% of the organotin
compound is decomposed. The decomposition takes place even faster in
12% hydrochloric acid where, after only 30 min of radiation, all had
been converted to the inorganic form (WOGGON and JEHLE 1973).

A series of experiments with solid triphenyltin acetate, applied to

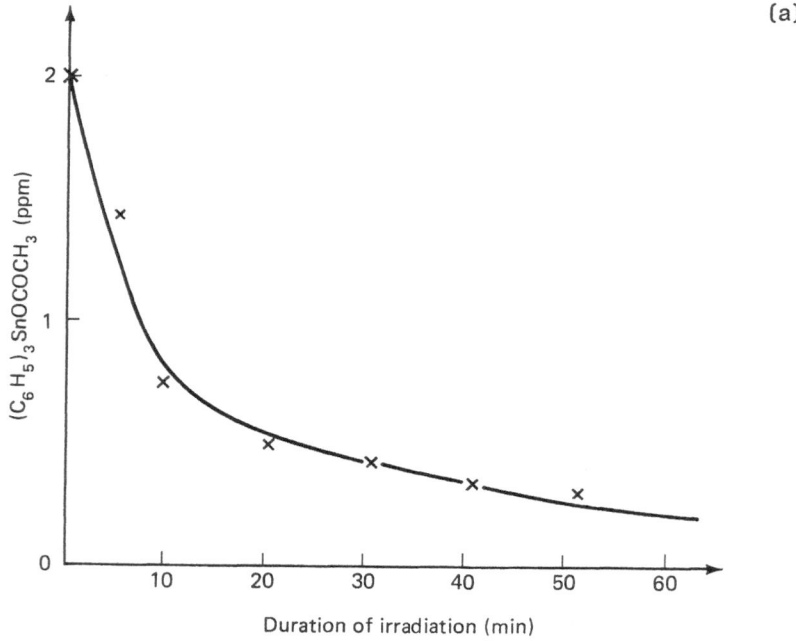

Fig. 9. (a) Decomposition of triphenyltin acetate (in aqueous solution) by UV-
radiation, starting concentration 2 ppm (after STRUFE 1968); (b) decrease
in concentration of ^{113}Sn (calculated as triphenyltin chloride) in the brains
of rats depending on time (after HEATH 1966): A = 185 mg/kg oral in
Ol. arachidis, and B = 5 mg/kg intraperitoneally.

(b)

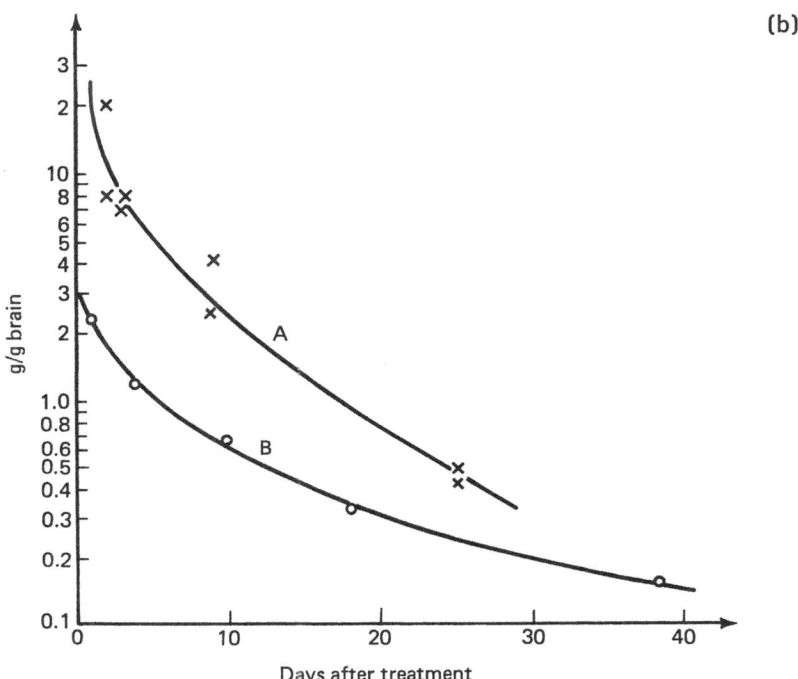

Days after treatment

glass plates as a 5 to 10 μm film, is shown in Table XV. The plates were placed at a distance of 10 cm from a mercury arc lamp. During experiments (4) and (5), the light was filtered, cutting off wavelengths under 350 nm (CHAPMAN and PRICE 1972).

The decomposition to inorganic tin took place via di- and monophenyltin compounds. Apparently the reaction was slower than in the experiments with aqueous solutions described above. This corresponds with results of WOGGON and JEHLE (1973).

In other experiments, triphenyltin compounds were photolyzed on chromatographic paper (CENCI and CREMONINI 1969). The substances were applied as spots of 5 mm diameter, and the strips were either exposed horizontally to full sunlight or placed at a distance of 1 cm from a 30-watt mercury arc lamp. In sunlight, the decomposition was complete after 8 hr; with UV-radiation, it proceeded much faster (Table XVI).

Irradiation of triphenyltin chloride with short-wave UV-light (254 nm, lamp: Sylvania F8 T5/132 B, 8 watt) on a thin-layer chromatographic plate caused complete decomposition within 192 hr; after this time, 30% of the tin was inorganic. With radiation of higher wavelength (350 nm; lamp: Sylvania G8 T5, 8 watt), the process was much slower; after 192

Table XV. *Decomposition of triphenyltin acetate by UV-radiation* (after CHAPMAN and PRICE 1972).

No.	Thickness of layer (μm)	λ (nm)	Duration of irradiation (hr)	$(C_6H_5)_3SnX$ (%)	$(C_6H_5)_2SnX_2$ (%)	$(C_6H_5)SnX_3$ (%)	Inorganic Sn (%)
1	5	254 (12 W/cm²)	5	33	27	24	18
2	5	254	25	18	26	22	36
3	5	254	60	11	31	14	44
4	10	350 (4 W/cm²)	85	63	25	7	6
5	10	350	100	39	20	19	19

Table XVI. *Decomposition of triphenyltin compounds by sunlight and UV-radiation* (after CENCI and CREMONINI 1969).

	Decomposition complete after	
Radiation	$(C_6H_5)_3SnOCOCH_3$	$(C_6H_5)_3SnOH$
Sunlight	8 hr	8 hr
UV-radiation, 366 nm	3 hr	3 hr
UV-radiation, 254 nm	15 min	15 min

hr, most of the triphenyltin chloride was unchanged, and only about 6% had been converted to inorganic tin (MASSAUX 1971).

After irradiation of triphenyltin acetate spotted on a silica gel thin-layer chromatographic plate (Hg-arc lamp 125 watt, main intensity at 365 nm, distance 20 cm), most of the organotin compound was decomposed within 1 hr; the intermediate of this reaction, leading to inorganic tin, was found to be $(C_6H_5)_2Sn^{2+}$ (BARNES *et al.* 1973). Further qualitative experiments showed that the decomposition of triphenyltin compounds by UV-radiation proceeds via the diphenyl- and monophenyltin compounds to inorganic tin (AKAGI and SAKAGAMI 1971).

By UV-radiation, tetraphenyltin on thin-layer chromatographic plates can be decomposed to compounds with fewer than 4 phenyl groups (BRAUN and HEINES 1968), but no experiments were conducted to find out how fast this reaction takes place.

IV. Pharmacology and toxicology of phenyltin compounds; behavior in warm-blooded animals

a) General

Triphenyltin compounds can enter the human body directly through vegetarian food or indirectly through meat products. Therefore, extensive research has been done on their toxicity, excretion, and accumulation and on their metabolism. Interesting in this respect is also the fate of tetraphenyltin introduced as a potential impurity and that of the decomposition products of triphenyltin compounds. Technical triphenyltin acetate contains, besides approximately 90% active ingredient, approximately 7% tetraphenyltin, approximately 2% diphenyltin compounds, and approximately 1% volatiles (KLIMMER 1968).

b) Acute toxicities

1. **Tetraphenyltin.**—The toxicity of tetraphenyltin was first studied by COLLIER (1929) and KRAUSE (1929). By injecting a suspension of the compound in gum Arabic solution into the nape of the neck of mice, a "tolerable dose" of 3.3 mg/20 g of body wt was found. The LD_{50} oral for white mice is 3,000 mg/kg (KOLLA and ZALESOV 1964); for rats, accord-

ing to SCHOLZ (1960 b) more than 10,000 mg/kg. A hen weighing 735 g showed no harmful effects after being fed 5 g (C_6H_5)$_4$Sn (GUTHRIE and HARWOOD 1941).

2. **Diphenyltin compounds.**—As far as presently known, diphenyltin compounds are less toxic than the corresponding triphenyltin compounds. For example, a single oral administration of diphenyltin dichloride to rats at a dose of 160 mg/kg body wt resulted in only temporary symptoms without detectable damage of the tissues; a dose of 256 mg/kg caused some damage in the stomach, but all animals survived. Only after administering 410 mg/kg, several rats died at irregular times after ingestion, possibly because of damage in the gastrointestinal tract (STONER 1966). The LD_{50} oral for diphenyltin diacetate is, according to SCHOLZ (1960 b), 800 mg/kg body wt of rats.

Intraperitonal injection resulted in death in all cases at only 15.3 mg/kg; the dissection showed strong exudates in the abdominal cavity (STONER 1966).

When treated with oral doses of 250 to 300 mg of diphenyltin dichloride/kg, 5.8% of the chickens died (GRABER and GRAS 1966); the LD_{50} is approximately 850 mg/kg (GRABER and GRAS 1965).

The LD_{50} oral of diphenyltin oxide for rats was found to be about 200 mg/kg (CASTEL et al. 1960). However, experiments by SCHOLZ (1960 b) showed an oral LD_{50} of 800 mg/kg for diphenyltin oxide in rats. The mortality rate of chickens after oral doses of 85 to 220 mg/kg was 3% (GRABER and GRAS 1966).

3. **Inorganic tin and benzene.**—All experience so far shows that inorganic tin is not harmful to human beings even in rather high doses. Extensive experiments with various animals demonstrated that Sn(IV) is hardly absorbed by the body. Tests with human beings who drank Sn-containing fruit juices showed no ailments up to 730 ppm of Sn in the juice; only at 1,370 ppm (= 1.7 to 2.7 mg/kg body wt), could toxic effects (nausea, diarrhea) be noticed (BENOY et al. 1971).

Rats were not harmed by feeding them 0.1% Sn(=22 to 33 mg/kg/ day); at 0.3%, a slower growth rate, slight anemia, and liver damage were noticed (DE GROOT et al. 1973).

The FAO/WHO recommends the following maximum tin contents: 250 ppm for canned fruit, vegetables, meat, and fish; 150 to 250 ppm for fruit juices, and 100 ppm for baby food (WOIDICH and PFANNHAUSER 1973). In England the limit for food is 250 ppm Sn (BARNES and STONER 1959).

Benzene, a breakdown product of phenyltin compounds (see below), is toxic for human beings, but at levels which are much higher than what can be found in the human organism due to residual amounts of phenyltin compounds in food. The recommended workplace concentration for benzene in air is 10 ppm.

4. **Triphenyltin compounds.**—COLLIER (1929) and KRAUSE (1929) were first to establish the toxicity of a triphenyltin compound, namely

Table XVII. *Acute toxicity of triphenyltin compounds on mice.*

Sex	Route	Compound	LD$_{50}$ (mg/kg)^a	References
M	oral	acetate	93.3 (52.5–166.1)	Ueda & Iijima (1961)
M	oral	acetate	81.3 (66.6–99.3)	Stoner (1966)
M	oral	acetate	81 (54.7–199.9)	Ueda et al. (1961)
M	oral	chloride	80 (44.5–144)	Ueda et al. (1961)
—	oral	chloride	90	Schicke et al. (1968)
M	oral	hydroxide	245 (208–289)	Philips-Duphar (1964 a)^b
F	oral	hydroxide	209 (158–276)	Philips-Duphar (1964 a)^b
—	oral	hydroxide	245	Hopf et al. (1967)
—	oral	hydroxide	510	Luijten (1960)
—	oral	hydroxide	511; 619	Pieters (1961)
—	oral	hydroxide	80	Schicke et al. (1968)
—	oral	oxide	1,000	Luijten (1960)
M	i.p.	acetate	7.9 (6.6–9.7)	Stoner (1966)
—	i.p.	hydroxide	10; 8; 17; 1	Pieters (1961)
M	sub-cutan.	acetate	44.0 (28.0–69.1)	Ueda et al. (1961)

^{a} 95% confidence limits in parentheses.
^{b} Quoted by FAO/WHO—evaluations (1971).

the bromide. The "tolerated dose" for mice, administered in the same way as the tetraphenyltin, was only 0.5 mg/20 g of body wt. Later, numerous other experiments were conducted, mainly with triphenyltin, acetate and hydroxide in mice (Table XVII), rats (Table XVIII), guinea pigs and

Table XVIII. *Acute toxicity of triphenyltin compounds on rats.*

Sex	Route	Compound	LD$_{50}$ (mg/kg)^a	References
M	oral	acetate	136	Klimmer (1964)
F	oral	acetate	140 (125–160)	Scholz (1964 d)
F	oral	acetate	491 (322–750)	Stoner (1966)
F	oral	acetate	429 (150–1,222)	Stoner (1966)
—	oral	acetate	150	Stoner et al. (1955)
—	oral	chloride	125	Brueckner et al. (1959)
F	oral	chloride	130–135	Scholz (1963)
F	oral	hydroxide	110 (90–140)	Scholz (1965 b)
M	oral	hydroxide	240 (189–305)	Gaines & Kimbrough (1968)
F	oral	hydroxide	360 (283–457)	Gaines & Kimbrough (1968)
M	oral	hydroxide	171 (100–295)	Marks et al. (1969)
F	oral	hydroxide	268 (205–344)	Marks et al. (1969)
—	oral	hydroxide	120	Scholz (1969 a)
—	oral	oxide	155	Scholz (1969 b)
M	i.p.	acetate	13.2	Klimmer (1963, 1964)
M	i.p.	acetate	8.5 (7.5–9.1)	Stoner (1966)
F	i.p.	acetate	11.9 (9.1–15.7)	Stoner (1966)
—	i.v.	acetate	18.0	Klimmer & Schulemann (1957)

^{a} 95% confidence limits in parentheses.

rabbits (Table XIX), but also with other triphenyltin compounds (Table XX).

α) *Acute oral toxicity of pure triphenyltin compounds.*—The LD$_{50}$ results, found by oral administration of triphenyltin acetate to rats, are—with few exceptions—in the range of 125 to 150 mg/kg. Similar results might be expected for the hydroxide, since it is certain that in the intestines, the acetate and the chloride are converted to the hydroxide (GOR-BACH 1968). Thus, the sometimes much higher results (Table XVII) are difficult to explain. It is possible that, in administering oil solutions of the hydroxide, the less toxic oxide was formed. The occurrence of such more or less complete reaction would also explain the wide deviation of results.

For rabbits, the acute oral LD$_{50}$ of triphenyltin acetate is about 30 to 50 mg/kg. When intoxicated, they show the same symptoms as rats (KLIMMER 1964).

It was impossible to determine the acute oral LD$_{50}$ for dogs since they vomited shortly after a dose of 10 mg/kg or more of triphenyltin acetate was given.

Cats can tolerate an oral dose of 32 mg/kg without showing any reactions. Higher doses led to vomiting (SCHOLZ 1960 a). KLIMMER (1964) reported the same result. The symptoms of intoxication were as uncharacteristic as in other animals. Of those animals tested, rats, mice, and hens were the least sensitive; oral doses of 0.5 and 1.0 g triphenyltin

Table XIX. *Acute toxicity of triphenyltin compounds on guinea pigs and rabbits.*

Sex	Route	Compound	LD$_{50}$ (mg/kg)[a]	References
			Guinea pig	
—	oral	acetate	25	WEIGAND (1962)
M	oral	acetate	21	KLIMMER (1964)
M	oral	acetate	23.5 (15.8–34.9)	STONER (1966)
M	oral	acetate	41.2 (19.0–89.5)	STONER (1966)
—	oral	acetate	10	STONER et al. (1955)
—	oral	acetate	25 (24–27)	SCHOLZ (1964)
M	oral	hydroxide	27.1 (20–36.9)	*Philips-Duphar* (1964 a)[b]
F	oral	hydroxide	31.1 (21.5–46.4)	*Philips-Duphar* (1964 a)[b]
—	oral	hydroxide	25	WEIGAND (1962)
M	i.p.	hydroxide	5.3	KLIMMER (1964)
M	i.p.	hydroxide	3.74 (3.1–4.6)	STONER (1966)
			Rabbit	
M	oral	acetate	30–50	KLIMMER (1964)
—	oral	acetate	40	STONER et al. (1955)
—	oral	acetate	140	STONER (1966)
M	i.p.	acetate	±10	KLIMMER (1964)
—	i.p.	acetate	±16	STONER (1966)

[a] 95% confidence limits in parentheses.
[b] Quoted by FAO/WHO—evaluations (1971).

Table XX. *Acute toxicity of various triphenyltin compounds on mice.*

Compound	Appli-cation	LD_{50} (mg/kg)	References
$(C_6H_5)_3Sn-S-Sn(C_6H_5)_3$	p.o.	$\geqslant$1,470	Van Der Kerk (1961)
$(C_6H_5)_3Sn-S-Sn(C_6H_5)_3$	p.o.	$\geqslant$680	Van Der Kerk (1961)
$(C_6H_5)_3Sn-S-Sn(C_6H_5)_3$	p.o.	$\geqslant$700	Van Der Kerk et al. (1962)
$(C_6H_5)_3Sn-S-Sn(C_6H_5)_3$	i.p.	$\geqslant$680	Van Der Kerk (1961)
$(C_6H_5)_3Sn-S-Sn(C_6H_5)_3$	i.p.	$\geqslant$1,470	Van Der Kerk (1961)
$(C_6H_5)_3Sn-S-S-Sn(C_6H_5)_3$	p.o.	$\geqslant$1,470	Van Der Kerk (1961)
$(C_6H_5)_3Sn-S-S-Sn(C_6H_5)_3$	p.o.	$>$700	Van Der Kerk et al. (1962)
$(C_6H_5)_3Sn-S-S-Sn(C_6H_5)_3$	i.p.	200	Van Der Kerk (1961)
$(C_6H_5)_3Sn-S-S-Sn(C_6H_5)_3$	i.p.	$\geqslant$580	Van Der Kerk (1961)
$(C_6H_5)_3SnCl \cdot$ Quinoline N-Oxide	p.o.	376	Schicke et al. (1968)
$(C_6H_5)_3SnCl \cdot$ Dimethyl-sulfoxide	p.o.	325	Schicke et al. (1968)
$(C_6H_5)_3SnCl \cdot$ Decyltriphenyl-phosphonium bromide	p.o.	400	Schicke et al. (1968)

chloride proved to be fatal for young chickens (Guthrie and Harwood 1941).

By far the lowest LD_{50}, between 10 and 41.2 mg/kg of body wt, was found for guinea pigs. The differences between reactions of the tested animals have been explained to result from varying resorption rates of active ingredient in the intestines. Also, the vital centers may show different sensitivities toward metalorganic compounds (Klimmer 1963, Stoner 1966).

Toxic effects of triphenyltin compounds in oral administration appeared only slowly; sometimes it took several days for the animals to die. Symptoms of poisoning were weakness, anorexia, rough coat, watery diarrhea, staggering, reddish lachrymal fluid, and death in coma (Klimmer 1964, Gaines and Kimbrough 1968). "In all species, the main action of these organo-tin compounds is thought to be on the central nervous system, but in contrast to triethyltin compounds, cerebral oedema did not occur" (Cahen et al. 1970).

The complexes of triphenyltin chloride show—even if the higher molecular weight is taken into consideration—a slightly lower toxicity than triphenyltin compounds.

β) *Acute intraperitoneal toxicity.*—Intraperitoneal as well as intravenous injections show the direct course of effects avoiding the gastrointestinal passage. For rats, the LD_{50} of triphenyltin acetate with intraperitoneal administration was found to be 8.5 to 13.2 mg/kg (Klimmer 1963 and 1964, Stoner 1966). Mice showed a similar sensitivity. The LD_{50} of triphenyltin acetate or hydroxide for mice is 7.9 to 17.1 mg/kg with intraperitoneal application (Stoner 1966, Pieters 1961).

Guinea pigs showed an LD_{50} of 3.74 to 5.3 mg/kg for triphenyltin acetate in intraperitoneal application. The symptoms of poisoning are the same as in oral intoxication. The test animals die after 1 to 2 days, lying on their sides (KLIMMER 1964, STONER 1966).

The intraperitoneal LD_{50} for rabbits is 10 to 16 mg/kg. Fatal doses of triphenyltin acetate were accompanied by beginning muscular dystrophy and declining rate of breathing (KLIMMER 1964, STONER 1966).

γ) *Acute intravenous toxicity.*—With intravenous injection, the LD_{50} of triphenyltin acetate for rats was determined as 18 mg/kg. The active ingredient, dissolved in diglycolmonomethylether (0.8%), was injected into the vein at the tail. After a few min the rats showed weakness, after 10 to 15 min inability to run, prone position, cyanosis, and dyspnea. They died within 24 to 48 hr (KLIMMER and SCHULEMANN 1957).

In cats, intravenous injection of triphenyltin acetate or chloride in 0.5% solution (1 mg/kg body wt) caused short respiratory arrests. Repeated doses of 1 to 2 mg/kg at intervals of 20 to 60 min, totaling 6 to 14 mg triphenyltin acetate led to death due to paralysis of the respiratory center (TAUBERGER 1963).

δ) *Acute dermal toxicity.*—The physical form of the active ingredient is most important in determining the dermal toxicity. In oil solutions or oil in water emulsions, triphenyltin compounds are—like many other active ingredients—better absorbed by the skin than in the aqueous form of wettable powder formulations commonly used for triphenyltin compounds.

ε) *Skin tests with pure triphenyltin compounds.*—

αα) *Tests on rats:*

A 10% solution of triphenyltin acetate, dissolved in Ol. Arachidis, was rubbed into the depilated back skin of Wistar-CFN-rats in several doses of 150 mg/kg body wt. During the test period, the animals appeared weak, thirsty, and sensitive to touch. The test area of the skin became severely inflamed. After high doses, the rats began to stagger. They died after 3 to 7 days, independent of the dose. The dermal LD_{50} of triphenyltin acetate for rats was found to be about 450 mg/kg body wt (KLIMMER 1964).

SCHOLZ (1964) reported similar results. After application of an oil-in-water emulsion of triphenyltin acetate 5 times with 100 mg active ingredient (=500 mg/kg body wt) to the depilated nape of the neck, the rats showed only minor local irritation and minor loss of wt. After the rats had been treated 5 times with 200 mg active ingredient/kg body wt (=1,000 mg/kg) in oil in water emulsion, 4 of 5 animals died.

ββ) *Tests on mice:*

Groups of 6 male mice with a body wt of 18 to 22 g were selected for the test. An emulsion of triphenyltin acetate was applied in increasing concentration to the depilated back between the shoulder blades. The treated animals showed no specific symptoms. They died within 2 to 5 days with no apparent relation to the dose. The dermal LD_{50} for mice was 350 mg/ky body wt (UEDA et al. 1961).

γγ) *Tests on rabbits:*

The reaction of rabbits, treated 5 times with 6.25 mg triphenyltin acetate in oil in water emulsion on the depilated skin around the flanks (=31.25 mg/kg), was unchanged. At this amount, the gain in body wt was not different from that of the control animals. After application of altogether 62.5 mg active ingredient/kg body wt, 2 of the 5 tested animals died after 11 days. At 125.0 mg/kg, 3 of 5 treated animals died, showing considerable loss of weight (SCHOLZ 1964).

δδ) *Tests on guinea pigs:*

In guinea pigs, only a high dose of 180 mg/kg body wt of triphenyltin acetate in Ol. Arachidis caused the death of several animals (STONER 1966; Table XXI).

Table XXI. *Dermal toxicity of triphenyltin acetate in guinea pigs* (after STONER 1966).

Dose (mg/kg)	No. of animals	No. that died	Survival time
30	2	0	—
60	2	0	—
90	2	0	—
180	8	3	22 hr, 22 hr, 11 days

ζ) *Acute oral toxicity of formulated products.*—

According to SCHOLZ (1962), the acute oral LD_{50} for a formulated product (Brestan Super® with 60% triphenyltin acetate + 20% maneb) is 215 mg/kg for female rats; later, results of 398 mg/kg for female and 252 mg/kg for male rats were found (SCHOLZ and HOLLANDER 1973).

This means that formulation of triphenyltin acetate did not result in higher toxicity.

η) *Skin tests with formulated triphenyltin compounds.*—

αα) *Tests on rats:*

To check for compatibility and resorption effects on the skin, 1 ml of 15% freshly prepared suspension of a 20% triphenyltin acetate product (Brestan®) was rubbed into the abdominal skin of 10 rats (220 to 250 g body wt) once daily for 10 consecutive days. The total amount applied was 300 mg active ingredient/rat, which equals 1,200 to 1,300 mg triphenyltin acetate/kg of rat. After 5 to 6 applications, the skin showed very slight superficial inflammation at the site of test. The effects subsided after termination of the experiment. General symptoms, as noticed after application of triphenyltin in Ol. Arachidis could not be found in this test. All animals survived (KLIMMER and SCHULEMANN 1957, KLIMMER 1964).

θ) *Compatibility with skin and mucous membranes.*—Following the intracutaneous injection procedure of Barail, 0.02 ml of a digested and filtrated solution of triphenyltin acetate were injected into the depilated skin of the flanks of rabbits. This did not cause any irritation, neither with the undiluted nor with the diluted filtrate. Application of the undiluted filtrate onto the depilated, intact flank skin of rabbits caused only slight local irritation. Injection of 0.1 ml of undiluted filtrate into the conjunctiva of rabbits did not cause any irritation or other changes (SCHOLZ and WEIGAND 1964).

ι) *Inhalation toxicity.*—In a chamber with a volume of 200 L, 5 rats, 2 rabbits, and 1 cat were exposed to a continuous flow of air containing triphenyltin acetate dust 4 times for 1 hr on 4 days. During the experiment, the animals showed distinct signs of irritation in the mucous membranes, the eyes, and the respiratory tract, together with restlessness, grooming, and coughing. No general or nervous signs of poisoning were registered, neither during the test nor during the following observation period. All animals survived.

After the experiment, the rats were killed and dissected. They showed signs of irritation in the respiratory tract and in the connective membranes of the eyes. The lungs were strongly supplied with blood. The other organs showed no changes (KLIMMER and SCHULEMANN 1957).

κ) *Acute subcutaneous toxicity.*—A triphenyltin acetate emulsion was injected in 6 increasing doses into the backs of mice, at a rate of 0.1 ml per 20 g of body wt.

Intoxication was manifested by unsteady walk, followed by light tremor and symptoms such as shortness of breath before dying. As Table XXII shows, the subcutaneous LD_{50} is about 44 mg/kg body wt (UEDA *et al.* 1961).

Table XXII. *Subcutaneous toxicity of triphenyltin acetate*[a] (after UEDA *et al.* 1961).

No.	Dosage (mg/kg)	No. of mice dead/tested after days							Mortality (%)	LD_{50} (mg/kg)
		1	2	3	4	5	6	7		
1	90	3/6		5/6				5/6	83.3	
2	60	1/6	3/6		4/6			4/6	66.7	44.0
3	40	0/6		1/6		2/6		2/6	33.3	(28.0–69.1)
4	26.7	0/6						0/6	0	
5	17.8	0/6			1/6			1/6	16.7	
6	11.9	0/6						0/6	0	

[a] Materials used: triphenyltin acetate emulsion containing 10% of acetone and Sorpol 2170; animals tested: male D-D mice, weighing between 16 and 20 g each.

c) Chronic toxicity of triphenyltin compounds

1. General.—Chronic toxicities of mono- and diphenyltin compounds have not yet been determined, but thorough short- and long-term studies with triphenyltin compounds have been undertaken on rats, guinea pigs, and dogs. In feeding tests, the results depend partly on the mode of feeding, *i.e.*, by probang or *ad libitum*. At voluntary feeding and with higher concentrations of active ingredient, one observes reluctancy to eat, leading to slower growth and general deterioration of health. Small doses cause no detectable histological changes in organs or tissue. From these tests, no-effect levels of 1.0 to 5.0 ppm could be derived for rats, 5.0 ppm for guinea pigs, and 2.0 ppm for dogs.

Based on their evaluation of short- and long-term experiments, the FAO/WHO (1971) determined a nontoxic effect level of triphenyltin compounds for rats at levels of 2.0 ppm in the food, which equals 0.1 mg/kg body weight. This results in an Acceptable Daily Intake for man (ADI) of 0.0005 mg/kg body wt for triphenyltin acetate, hydroxide, and chloride.

2. Short-term studies.—

α) *Experiments on rats.—*

αα) *Triphenyltin acetate:*

Rats, fed by probang for up to 170 days, did not show any negative effects as long as the food contained no more than 25 ppm of triphenyltin acetate. On the contrary, they showed a slightly increased gain in wt over the controls. If the food contained 50 ppm triphenyltin acetate, the animals had signs of weakness and slower increase in wt, and within the test period of 105 days, 14 of 20 animals died (KLIMMER 1963 and 1964).

In experiments by STONER (1966), rats were able to tolerate triphenyltin acetate much better. At first, they were hesitant to accept food with 200 ppm of active ingredient, and their increase in wt was lower than that of the control animals; however, after several wk, the rats ate normally and gained wt like the controls, though they never caught up with them.

After 10 wk, the content of triphenyltin acetate in the food was raised to 300 ppm. This resulted in wt loss after about 3 wk, and in the course of 117 to 168 days, 5 out of 6 animals died.

Studies by VERSCHUUREN et al. (1966) on rats with 2.5 to 50 ppm triphenyltin acetate or hydroxide in the food showed no significantly higher mortality within 12 wk compared with the control group. Out of 20 animals, at 25 ppm of active ingredient none died during the test period; at 50 ppm, one each died from the test group and from the control group.

KLOTZSCHE (1960) reported results from a different set of experiments. The rats did not get food with a certain concentration of active ingredient, but rather a fixed amount of triphenyltin acetate/day. At 20 mg/kg/rat/day, 4 out of 5 animals died within 1 mon, at 10 mg/kg/rat/day, all 4 tested animals survived more than 4 wk.

SCHOLZ and BAEDER (1968) used groups of 20 rats, 10 male and 10 female, for their feed tests with triphenyltin acetate. During a period of 4 mon the groups received 0, 15, 40, 100, 250, and 600 ppm of active ingredient. Up to 100 ppm of triphenyltin acetate in the food, no abnormalities could be noticed compared with untreated animals. Only at more than 100 ppm, was growth slowed down, depending on the dose. At 600 ppm, 30% of the test animals died.

ββ) *Triphenyltin hydroxide:*

Esch and Arnoldussen (1962) treated groups of 10 male and 20 female rats for 20 days with 0, 5, 20, 50, and 100 ppm of triphenyltin hydroxide in the food. At doses of 20 ppm and more, food intake and body wt decreased. At 50 ppm, 9 out of 20 animals died, at 100 ppm 9 out of 10 animals died.

Feeding doses of 0, 100, 200, and 400 ppm of triphenyltin hydroxide, Gaines and Kimbrough (1968) observed that only at 400 ppm all animals died. At 100 and 200 ppm all animals (10 each in a test group) survived for 14 wk. Rats of the 200 ppm group lost wt compared with the control animals. Rats of the 100 ppm group showed normal behavior, except for a lower food intake.

γγ) *Triphenyltin chloride:*

Groups of 10 male and female rats were fed by probang during a period of 41 days with 28 doses of 0, 6.25, 12.5, 25.0, and 50 mg triphenyltin chloride/kg body wt. Except for a slower gain in body wt of the female animals at 12.5 mg/kg, doses of 6.25 and 12.5 mg/kg did not result in any abnormalities (growth, behavior, blood and urine composition, macroscopic or histopathologic appearance of organs). Higher doses, depending on the amount of active ingredient, resulted in increased mortality compared with untreated animals (Scholz 1963).

β) *Experiments on guinea pigs.—*

αα) *Triphenyltin acetate:*

Guinea pigs showed higher sensitivity also in long-term tests. At a concentration of 12 ppm of triphenyltin acetate in the food, 5 out 6 animals died within 1 yr; at 25 ppm, 5 out of 6 animals died within 77 days; at 50 ppm, all animals died within 31 days (after an intake of 54 ± 14 mg of active ingredient). In another experiment, groups of 5 guinea pigs were fed 0, 1.0, and 5.0 ppm of triphenyltin acetate over a period of 392 days with no higher rate of mortality, but food intake was reduced even at 1 ppm. The histopathological examination showed that none of these experiments with triphenyltin compounds led to a significant increase of water content in the central nervous system nor to the typical histological damage of the white substance, as it is caused by triethyltin compounds (Magee *et al.* 1957, Stoner 1966, Stoner and Heath 1967).

In a 30-day experiment, 4 groups of 10 male and 10 female guinea pigs were fed with 0, 7.5, 30, and 125 ppm of triphenyltin acetate daily. The behavior, the wt gain, and the general health of the test groups with 7.5 and 30 ppm were the same as those of the controls. One male each of the groups with 7.5 and 30 ppm died during the experiment. The animals in the group with 125 ppm developed rough coat and showed deteriorating health. Right after the start, their body wt decreased. Eight male and 5 female animals died during this experiment (Scholz 1964).

In another experiment with doses of 0, 1.0, 5.0, 10, 50, and 100 ppm of triphenyltin acetate over a period of 4 mon, the groups of 10 male and 10 female guinea pigs tolerated amounts of 1.0 and 5.0 ppm without any change in health, food intake, wt gain, or behavior. Ten ppm merely resulted in a slightly lower wt gain. At a dose of 100 ppm, all animals died. At a dose of 50 ppm, 30% of the animals died between the 36th and the 56th day after start of the test. All animals in the two highest dose groups showed bad health and high wt loss (SCHOLZ and WEIGAND 1968).

ββ) Triphenyltin hydroxide:
Groups of 10 male and 10 female guinea pigs were treated for 12 wk with 0, 2.5, 5.0, 10, 20, and 50 ppm of triphenyltin hydroxide in the food. In the groups with the two highest doses the growth rate was slowed down, while the animals in the 50 ppm groups died within the first half of the experiment. One animal each of the 10 and 20 ppm groups died during the experiment (VERSCHUUREN et al. 1966).

In another experiment, groups of 20 male and female rats were treated for 30 days with 0, 7.5, 30, and 125 ppm of triphenyltin hydroxide in their daily food. During the experiment, 9 male and 9 female animals in the 125 ppm group and 1 male animal in the 30 ppm group died of heart failure. The animals in the 125 ppm group showed rough coat, deteriorating health, and wt loss (SCHOLZ 1965 b).

γγ) Triphenyltin chloride:
Feeding experiments with triphenyltin chloride in various concentrations resulted in average life spans as shown in Table XXIII (SCHOLZ 1965).

Table XXIII. *Average life expectancy of guinea pigs at different concentrations of $(C_6H_5)_3SnCl$ in food* (according to a graphic representation by SCHOLZ 1965).

$(C_6H_5)_3SnCl$ (ppm)[a]	Average life expectancy (days)
380	Approx. 12
190	Approx. 28
90	Approx. 55
45	Approx. 70

[a] Approximate values.

γ) Experiments on dogs.—
αα) Triphenyltin acetate:
Four groups of beagles with 4 male and 4 female animals each were treated for 120 days with amounts of 0, 1.0, 5.0, and 20 ppm of triphenyltin acetate in their food. At 20 ppm, the food intake decreased and resulted in wt loss in half of the animals. No abnormalities could be noticed in the behavior of the animals nor in the macro- and microscopic examination of organs and tissues (SCHOLZ and BRUNK 1968).

ββ) Triphenyltin hydroxide:
Four groups of 6 male and female dogs were fed with 0, 10, 25, and 37 ppm of triphenyltin hydroxide for 100 days; except for darkening of the skin in all groups, only the group with 37 ppm showed signs of lethargy (Til and Feron 1965, quoted by FAO/WHO Evaluations 1971).
   δ) *Experiments on cattle.*—
αα) Triphenyltin acetate:
Experiments on cattle were conducted by feeding sugarbeet leaves from a field treated with triphenyltin acetate. For 46 days the animals received 42 kg fresh leaves daily with an average of 0.76 ppm of triphenyltin acetate, *i.e.*, each animal received approximately 32 mg of active ingredient/day. Neither the clinical examination nor the milk yield showed any significant differences compared with untreated animals (Brueggemann *et al.* 1964 a).
   3. Long-term studies.—
   α) *Experiments on rats.*—
αα) Triphenyltin hydroxide:
Six groups of 25 male and female rats each were treated during a period of 2 yr with 0, 0.5, 1.0, 2.0, 5.0, and 10 ppm of triphenyltin hydroxide in the food. The group with the highest dose showed a slightly lowered wt of the thyroid gland, a somewhat lower number of leucocytes in the first yr and, in the females, a slightly higher mortality rate. Otherwise the animals of all groups showed no differences compared with the controls. General health, wt gain, food intake, growth rate, appearance, blood-pressure, and urine were normal. The macro- and microscopic examinations of all organs and tissues were not different from the controls (Til *et al.* 1970).
   β) *Experiments on guinea pigs.*—
αα) Triphenyltin acetate:
Seven groups of guinea pigs, consisting of 10 male and 10 female each, were treated with 0, 1.0, 5.0, 10, 50, 100 and 200 ppm of triphenyltin acetate in the food for 2 yr. At the 3 highest doses, all animals died within 16 wk. The mortality rate at the other doses was not different from the control groups. Health, food intake, behavior, appearance, growth, blood-picture, and urine at 1.0, 5.0, and 10 ppm were the same as that of the control animals. The groups with 50 ppm and more showed wt loss. The histopathological examination of organs and tissues in the 1.0 and 5.0 ppm groups revealed no differences compared with the controls. At 10, 50, 100, and 200 ppm, fat deposits appeared in the cells of the liver and heart (Weigand and Kief 1965, Weigand 1975).
   γ) *Experiments on dogs.*—
αα) Triphenyltin acetate:
Groups of 3 male and 3 female beagles were fed daily with doses of 0.5, 1.0, and 5.0 ppm for 2 yr. Besides a slightly lower wt gain in the 5.0 ppm group caused by decreased food intake, there was no abnormality in appearance, behavior, health, blood-picture, or urine of the test ani-

mals compared with the controls. The histopathological examinations of organs and tissues showed no difference either (SCHOLZ and BRUNK 1968).

$\beta\beta$) *Triphenyltin hydroxide·*

Four groups of 3 male and 3 female beagles each were fed for 2 yr with 0.5, 2.5, 5.0, and 10 ppm of triphenyltin hydroxide. One control group consisted of 5 male and 5 female beagles. The general behavior, appearance, growth, food intake, and health, as well as blood and urine tests, showed no differences compared with the controls. In the 5.0 and 10 ppm group, the relative wt of liver and kidneys and the water content of the brain were slightly higher than those of the control animals, but there was no histological change. The macro- and microscopic examinations showed no abnormalities attributable to the intake of triphenyltin hydroxide (TIL and FERON 1968).

4. Special studies on reproduction.—

α) *Triphenyltin acetate and triphenyltin chloride.*—During a reproduction experiment, SPF-Wistar-K-rats were treated with triphenyltin acetate in their food in concentrations of 25, 80, and 250 ppm. The test served to examine the influence of this compound on the fertility and on the intrauterine and postnatal development of the offspring.

Continued treatment with 25 ppm of triphenyltin acetate did not adversely effect general health, fertility, and postnatal development of the rats in the examined generations.

At the 80 ppm dose, the numbers in the litters decreased slightly. Feeding of 250 ppm of triphenyltin acetate caused distinct reduction of the food intake, lowered growth rates, and disorders in the female cycles. The female animals were difficult to mate, produced fewer offspring, and raised fewer young. The poor raising result in this group was due to lower viability of the young and to lactation disorders (insufficient development of the mammary glands) in the mothers (SCHOLZ and BAEDER 1970).

PATE and HAYS (1968) fed 35- to 40-day old male rats for 19 days with triphenyltin acetate or chloride in a dose of 20 mg/kg body wt. Each group consisted of 20 animals; they were killed on the 20th day, and the testes were examined histologically. In this unusually high dose, both compounds caused changes in the tissue and size of the testes; however, later studies have shown spontaneous recovery of normal spermatogenesis and cogenesis occurs after cessation of treatment (SNOW 1970).

Similarly, a group of female rats (15 animals aged 42 to 46 days) was treated with 20 mg of triphenyltin acetate or chloride/kg body wt/ day. On the 4th, 9th, 14th, and 19th day after beginning the experiment, 3 animals each were killed and the ovaries examined histologically. At this high dose both compounds caused distinct changes in the tissue of the ovaries as reduced numbers of matured follicles and a decrease in corpora lutea; this explains the lowered ovulation and fertility rate (NEWTON and HAYS 1968).

BROWN (1972 and 1973) administered triphenyltin acetate and chloride orally at a dosage rate of 10 mg/kg body wt/day to 7-mon old mice for the first 6 wk of treatment and then increased the dosage to 15 mg/kg body wt for the rest of the treatment of 14 mon. The results indicated that administration of triphenyltin acetate and chloride to adult mice at a slightly lower dosage schedule as it was used by NEWTON and HAYS (1968) does not cause the decrease in fertility that is seen when the compounds are given to immature females. The reproductive tissue of immature animals is apparently much more sensitive to the inhibitory effects of the two organotin compounds. This may be due to the rapid proliferation rate that occurs in reproductive tissue as adolescence is reached.

Mice, after obtaining 15 mg/kg of triphenyltin acetate or chloride for 60 days, showed at the end of the experiment no changes in the ductular development, in the density of alveoli in the mammary glands, or in the number of follicles, corpora lutea, or concretions in the gonads (BROWN 1973).

β) *Triphenyltin hydroxide.*—Extensive reproduction experiments on rats were also conducted with triphenyltin hydroxide. One, 2, or 5 ppm of active ingredient in the food caused no negative influence on fertility, number of young in the litter, vitality, body wt, and mortality of the offspring in three consecutive generations. A 90-day feeding experiment with the F_1b- and F_2b- generations produced no negative change in the blood or urine and no histopathologic abnormalities (TIL et al. 1967).

In three experiments with male rats (10 and 20 in each group), which were fed 0 to 25 ppm of triphenyltin hydroxide, the development of the testicle tissue was examined under the microscope. After 2 and 4 wk, no differences to the control groups were visible (TIL and FERON 1968).

In another experiment, 4 groups of 10 weaned, young male rats were fed with high amounts of triphenyltin hydroxide; the animals received 50, 100, or 200 ppm of the tin compound during a period of 296 days and were then mated with untreated females at different intervals. In the groups that had been fed 100 or 200 ppm of the active ingredient, the number of offspring was approximately 27% lower than that of the untreated animals. Since the food intake of these groups was significantly lower during the first wk, this effect could also be attributed to malnutrition. In the seventh wk, the animals had become used to the triphenyltin compound.

Sixty-four days after the feeding of active ingredient had been discontinued, the decrease in offspring was about 18%, and when the males were mated thereafter their fertility had improved further. One-hundred-and-thirteen days after the rats had returned to normal food, their fertility was comparable with the untreated control animals. Examination of organs and tissue showed no abnormalities; there was no evidence of the offspring being harmed (GAINES and KIMBROUGH 1968).

5. **Special studies on influence on lymphatic tissue and immune re-**

sponses.—For testing the influence of triphenyltin compounds on lymphatic tissue, VERSCHUUREN *et al.* (1970) conducted 3 experiments with guinea pigs. During the first 2 experiments, female animals were treated with 15 ppm of triphenyltin acetate in their daily food for 49 and 77 days, respectively. During the third experiment, the same amount was fed for 104 days. Twenty-one and again 7 days before the feeding of the active ingredient was discontinued, the animals received an intramuscular tetanus-toxoid injection. For later examination, part of the treated animals received normal food for an additional 14 days, that is to the 118th day of the experiment. A group of control animals was fed with normal food without any active ingredient.

The blood test showed no confirmed difference in the number of leukocytes and lymphocytes between the treated and untreated test groups. The wt of the organs was not different either, except that the thymus glands of the treated animals in the 49-day test weighed less than those of the control group.

Also, the 3 immune globulin fractions of the treated animals in the 49- and 77-day test and of those in the 104-day test up to the date of the tetanus-toxoid injection showed no difference compared with the controls.

However, based on the histological examinations and with the aid of the fluorescent antibody test technique, a decrease of plasma-cells in all lymphatic tissues could be shown in the triphenyltin acetate group of the 77-day experiment. This was especially apparent in the mesenteric lymph nodes. Here, the number of immunoblasts had also decreased distinctly (Table XXIV). In the 49-day test, though, similar changes could not be noticed.

By serological analysis (Table XXV) and with the aid of the fluorescent antibody test technique, it was found that after the tetanus-toxoid injection of the 104-day test group the number of immunologically active cells in the popliteal lymph nodes had decreased. These results, however, could not be found in other lymphatic tissue. It can be concluded that treatment of guinea pigs with 15 ppm triphenyltin acetate over a longer period results in a decrease of immunologic resistance. This effect, even after a recovery period of 14 days, had not disappeared, as was proven in immunochemical and serological tests (Table XXVI).

d) Other physiologic effects of triphenyltin compounds

1. Effects on mitochondria, chloroplasts, erythrocytes, and ferments.— Mitochondria and erythrocytes, suspended in isotonic NaCl- or NH_4Cl-solutions, swell after triphenyltin chloride or hydroxide has been added, and the permeability of the mitochondria-membranes to Na-, K- and Cl-ions as well as to fumarate, malate, and citrate increases (SYROWATKA 1969 and 1970, SELWYN *et al.* 1970). The swelling of mitochondria can be prevented by adding sucrose or sulfide; the latter probably causes formation of bis-(triphenyltin)sulfide, which, contrary to other triphenyltin

Table XXIV. *Semiquantitative record of plasma cells and immunoblasts in lymphatic tissues of control and TPTA-fed guinea pigs after 77 days (experiment 1)* (after Verschuuren et al. 1970).

Tissue and group	No. of animals	No. of cells[a] +	++	+++
Spleen		*Plasma cells*		
Control	8	1[b]	4	3
TPTA, 15 ppm	7	4	3	0
Cervical lymph node				
Control	8	1	1	5
TPTA, 15 ppm	8	2	2	4
Mesenteric lymph node				
Control	7	2	2	3
TPTA, 15 ppm	8	6	1	1
Axillary lymph node				
Control	8	2	2	3
TPTA, 15 ppm	8	3	4	1
Right popliteal lymph node				
Control	8	4	1	2
TPTA, 15 ppm	7	3	4	0
Left popliteal lymph node				
Control	8	4	1	2
TPTA, 15 ppm	7	5	2	0
Mesenteric lymph nodes		*Immunoblasts*		
Control	8	1	3	3
TPTA, 15 ppm	8	5	2	1

[a] + to +++ = a comparative estimate of the number of cells.
[b] Values are number of animals in cell number category.

compounds, causes no swelling. Tetraphenyltin as well as diphenyl- and monophenyltin compounds are also practically ineffective (Wulf and Byington 1975). Triphenyltin compounds (just as trialkyltin compounds) aid the exchange of F^-, Cl^-, Br^-, I^-, and SCN^- with OH^-; the Cl^-/OH^- exchange in particular was investigated in different membrane systems: in liposomes (Selwyn et al. 1970), in chloroplasts (Watling-Payne and Selwyn 1974), and in mitochondria (Selwyn et al. 1970, Stockdale et al. 1970).

On various occasions an influence of triphenyltin compounds on ferment systems was observed. Some authors (Pieper and Casida 1965, Stoner 1966) described inhibition of adenosinetriphosphatase in the brain microsomes, caused by triphenyltin acetate and chloride, and also (Ascher and Ishaaya 1973, Ishaaya and Casida 1975) of proteases and amylase by the acetate. According to El-Sebae and Ahmed (1972/73) the activity of monoamino-oxidase is inhibited by triphenyltin acetate but not by the hydroxide. Both compounds reduce the activity of aldehydedehydrogenase.

Table XXV. *Semiquantitative record of tetanus antitoxin-producing cells in popliteal lymph nodes of control and TPTA-fed guinea pigs after 104 days and after 2 weeks recovery (118 days) (experiment 2)* (after VER-SCHUUREN *et al.* 1970).

Group	No. of animals	Tetanus antitoxin-producing cells[a]		
		+	++	+++
104 days				
Control + tetanus ppt.	6	0[b]	3	3
TPTA + tetanus ppt.	6	3	2	1
118 days				
Control + tetanus ppt.	3	0	1	2
TPTA + tetanus ppt.	3	2	1	0

[a] + to +++ = a comparative estimate of the number of cells.
[b] Values are number of animals in cell number category.

The repeatedly observed inhibition of oxydative phosphorylation of ADP to ATP in mitochondria (SELWYN *et al.* 1970, STOCKDALE *et al.* 1970) and chloroplasts (WATLING-PAYNE and SELWYN 1974) through trialkyl- and triphenyltin compounds is important for cell-respiration and photosynthesis. Also, inhibition of the Na^+-K^+ATPase in cell membranes and of ATPase for Ca^{2+}-transfer in sarcoplasmic reticulum has been found. Results of these studies have been summarized by SELWYN (1976).

KIMMEL *et al.* (1977) tested the monooxygenase attack on trialkyltin compounds and on triphenyltin acetate; the acetate is more resistant, though destannylation occurs in rats.

2. Effects on rats.—When >20 mg of triphenyltin acetate/kg body wt is fed to rats, the first symptoms are reluctance to move and eat; weakness of the muscles, general weakening, and staggering appear at a later point (KLOTZSCHE 1960).

The same symptoms were reported by KLIMMER (1964), who, in addition, noticed diarrhea, rough coat, slower breathing, and drop of temperature. There were no reactions to stimulation. After 4 to 6 days the rats died in coma, sometimes with tremor and convulsions. Dissection

Table XXVI. *Antitoxin titer in A-E/ml, 5 Lf in the sera of control and TPTA-fed guinea pigs after tetanus-toxoid injection at 83 and 104 days[a]* (after VERSCHUUREN *et al.* 1970).

Group	Day 97	Day 104	Day 118
Control + tetanus ppt.	0.032	0.064	6.00
TPTA, 15 ppm + tetanus ppt.	0.016	0.032	1.81

[a] The TPTA feeding was discontinued at 104 days. The titers were obtained from pooled samples (experiment 2).

showed no abnormal or pathological condition in the organs. The histo-logical examination of organs and tissue showed no degenerative changes.

After intraperitoneal injection of >6 mg of triphenyltin acetate/kg body wt, the rats became listless; higher doses led to increasing weakness. Female animals became agitated after injection of 6 to 8 mg/kg (STONER 1966). With intraperitoneal application, the symptoms were practically the same as with feeding of the active ingredient. The rats died after 2 to 6 days in coma, sometimes accompanied by convulsions. Dissection of the organs showed the same results as in animals that died after ingesting triphenyltin acetate with their food (KLIMMER 1964).

Feeding experiments by VERSCHUUREN et al. (1966) proved that the growth rate of female rats is not noticeably reduced when 50 ppm of triphenyltin acetate is added to their food; male rats showed the same results at 10 ppm. The corresponding data for triphenyltin hydroxide were 50 and 25 ppm. Higher concentrations of the active ingredient led to diminished food intake, and the body wt of the rats was lower than that of the control animals.

After 12 wk some of the animals showed significant changes in the blood. The number of leukocytes and lymphocytes had decreased; the effect was more obvious with triphenyltin hydroxide than with the acetate and was more frequent in female than in male animals; however, no relation to the doses was apparent. At 50 ppm triphenyltin acetate in the food, the blood of the male rats showed a lowered hemoglobin content.

Further, the wt of various organs (thyroid gland, hypophysis, pan-creas, uterus, ovaries) was lower, that of the brain higher, but these symptoms appeared only at high doses of triphenyltin acetate or hy-droxide. In female rats that had been fed with 50 ppm of a triphenyltin compound, the water content of brain and spinal cord was slightly increased.

In rats that had died after treatment with 150 to 400 mg of triphenyltin hydroxide/kg, GAINES and KIMBROUGH (1968) found extended intra-alveolar hemorrhages and pulmonar edema, sometimes also inflamed spots in the lungs. A diet with 200 ppm of triphenyltin hydroxide led to severe wt loss and a decreased number of leukocytes compared with the control animals.

3. Effects on mice.—Mice react to triphenyltin acetate much the same as rats (STONER 1966). Daily feedings of 15 mg of triphenyltin chloride or acetate/kg over a period of 60 days resulted in reduced wt of the kidneys (BROWN 1973). The chloride led to decreased wt of the spleen and increased wt of the liver. Chloride and acetate did not appear in significantly higher concentrations in the brain, spleen, or fatty tissue; however, the tin content of kidneys and the feces was higher and the amount of acetate (not chloride) in the liver had increased. During a 14-mon feeding experiment with 10 mg of triphenyltin acetate or hydroxide/kg/day for the first 8 wk, and 15 mg/kg/day for the remainder of the experiment, wt loss was observed only after 8 or 10 wk, respectively.

Triphenyltin acetate (not the chloride), when administered orally over 14 mon or by injection over 3 wk, resulted in decreased growth of tumors in mice (BROWN 1973).

4. Effects on guinea pigs.—When guinea pigs were fed with triphenyltin compounds, they showed reluctance to eat. At a concentration of only 1 ppm (C_6H_5)$_3$SnOCOCH$_3$ (equivalent to an intake of about 0.1 mg/kg/day of active ingredient) there was significant loss of wt compared with the control animals, though no other visible symptoms appeared during the 95 days of the experiment (STONER 1966).

Higher doses led to unsteady movements, tremor, paralyzation of the hind extremities, slowed respiration, and lowered temperature after 1 to 2 days. Usually the animals died in coma within 1 to 4 days after administration of lethal doses (KLIMMER 1964, STONER 1966). The organs showed no significant changes; sporadically, there were congestions and oedemas in the lungs. The histological results were similar to those of rats.

Intraperitoneal injection of triphenyltin acetate caused basically the same symptoms; the animals died after 1 to 2 days, lying on their side. Organs and histological diagnosis were similar to the findings in the above feeding tests. There was an irritation in the abdominal cavity (KLIMMER 1964).

Twelve-wk feed tests with triphenyltin acetate by VERSCHUUREN et al. (1966), contrary to the findings of STONER (1966), showed significant reduction of growth rate only at triphenyltin concentrations of 5 ppm (female animals) or 10 ppm (male animals) in the food; triphenyltin hydroxide was effective only at a concentration of 20 ppm.

Both triphenyltin compounds caused reduction of the number of leukocytes and lymphocytes. At 10 and 20 ppm of triphenyltin acetate, the wt of the hypophysis, and at 20 ppm also the wt of the kidneys was increased in the female animals. The wt of the liver in male animals increased at 20 ppm, that of the brain in both sexes at 5 to 20 ppm. Uterus and testicles decreased in size at 20 ppm triphenyltin acetate in the food, and the water content of brain and spinal cord increased, but the organs showed no histological changes.

In concentrations of 10 and 20 ppm, triphenyltin hydroxide caused wt loss of the spleen in female animals; also at 20 ppm the wt of the thyroid glands in both sexes diminished as well as that of testicles and uterus. The histological results are shown in Table XXVII.

5. Effects on rabbits.—After oral administration of 140 mg of triphenyltin acetate/kg, rabbits showed symptoms of weakness and unsteady gait. The symptoms became stronger and, finally, the animals lay down on their sides, breathing slowly. The muscles of the hind legs were not relaxed, the reflexes being stronger than normal. At this dose, 2 out of 3 animals died within 1 to 4 days. One animal which survived a dose of 160 mg/kg showed the above mentioned symptoms for 3 days and then recovered slowly. Dissection brought about the same results as in rats. Intraperitoneal administration of triphenyltin acetate led almost instantly to vasodilation in the ears and hyperpnea. Sublethal doses did not cause

Table XXVII. *Histopathological findings in the lymphopoietic system of guinea pigs receiving triphenyltin acetate (TPTA) or triphenyltin hydroxide (TPTH) at dietary levels of 0 to 50 ppm for 12 weeks* (after VERSCHUUREN *et al.* 1966).

| | No. of animals displaying lesions in group | | | | | | | |
| | TPTA (ppm) | | | | | TPTH (ppm) | | |
Histopathological findings	0	5	10	20	50	0	20	50
Atrophic white pulp of spleen	–	–	–	1	5	–	5	15
Mycotic necrotizing inflammation of mesentric lymph glands	–	1	1	2	12	–	–	8
Mycotic enteritis	–	–	1	–	5	–	–	3

any further symptoms, and the animals recovered. Lethal doses caused general weakness of the muscles after several hr; the animals laid listlessly on their sides and died after 24 to 48 hr under decreased frequency of respiration and increasing weakness (KLIMMER 1964, STONER 1966).

6. **Effects on cats.**—After intravenous injection of 1 mg/kg of triphenyltin acetate (dissolved in 20% diethyleneglycol monomethylether + 10% Tween 80 in physiological sodium chloride solution), the blood pressure increased sharply and breathing stopped for a short period; then, breathing was stimulated. Repeated injections of the same amount of organotin compound resulted in gradually diminishing increases of the blood pressure until finally a drop was observed. At this point, the pressoric effect of noradrenaline was also reduced. Death was caused by paralysis of the respiratory center. Some of the animals probably also showed circulatory damage. There were no significant differences between the effects of triphenyltin acetate and chloride (TAUBERGER 1963).

7. **Effects on chickens.**—Chickens showed no reaction to oral administration of triphenyltin acetate, even at 30 mg/kg. Twenty-four hr after intraperitoneal doses of 20 mg/kg, their gait became unsteady and they seemed to have landing difficulties (STONER 1966).

8. **Effects on human beings.**—Some reports state that handling of triphenyltin compounds caused no harm or poisoning to the persons involved (BRUEGGEMANN *et al.* 1964 b, BARNES and MAGOS 1968, KLIMMER 1969, PLUM 1972). However, in isolated instances skin irritations and other reactions were observed.

Inhalation of triphenyltin acetate, for example, led to irritation of the mucus membranes and to inflamed eyes (HAERTEL 1962, KLIMMER 1968). In another case inflammation of the skin has occurred (SEIFFER and SCHOOF 1967). Triphenyltin iodide in concentrations of $1:10^6$ was intolerable (McCOMBIE and SAUNDERS 1947). Further, 3 cases of poisoning were reported when triphenyltin acetate solutions were carelessly prepared and then applied by spraying (GUARDASCIONE and MARZELLA DI BOSCO 1967). The symptoms were headache, nausea, and vomiting.

To date, the most serious accidents happened when a combined preparation of triphenyltin acetate and manganous-ethylenebisdithiocarbamate was sprayed from an airplane (HORÁČEK and DEMEČIK 1970). Two pilots exhibited rather strong symptoms; 3 mechanics were harmed less. The latter complained about heartburn and thirst which could not be quenched. When contact with the active ingredient stopped, the symptoms disappeared quickly; all laboratory tests were negative.

The more seriously affected pilots suffered from heartburn, thirst, stomachache, and diarrhea. In addition, one had impaired vision. The laboratory tests showed hyperglycemia and glycosuria; a strong diffuse steatosis was found in the parenchyma of the liver. In one case the symptoms lasted for several mon. Eleven mon after beginning of the illness the damaged liver parenchym started to regenerate.

The accidents were caused by improper handling of the active ingredient (which evidently had contaminated food consumed in the field), untight cockpits of the airplanes, and masks which did not function correctly.

The following treatment has been recommended by SCHOLZ (1959) in cases of poisoning with triphenyltin compounds: Induce vomiting, evacuation of the intestines with salinic laxatives (no drastica!), plenty of carbo medicinalis, in serious cases BAL (2,3-mercapto-1-propanol, dimercaprol) or EDTA (ethylenediaminetetraacetic acid, disodium salt, Versenate) (cp. also FAO/WHO 1976).

e) Absorption, accumulation, distribution, and secretion of triphenyltin compounds

1. Experiments on rats.—In experiments by HEATH (1963 a and b), a solution of radioactively labeled triphenyltin chloride in Ol. Arachidis was administered to rats, both orally and intraperitoneally. The amount excreted with feces and urine was checked. The result showed that depending on the quantity given by oral administration, 80 to 88% of the ^{113}Sn was eliminated within 7 days (Table XXVIII).

The amount of tin retained in various organs was measured after oral (Table XXIX) and intraperitoneal (Table XXX) treatment and found to decrease rapidly.

Table XXVIII. *Excretion of tin from rats; total amount excreted as percentage of the dose given (after HEATH 1963 a and b).*

Treatment[a]	Time after treatment (days)								
	1	2	3	4	5	6	7	8	9
A	10	25	41	55	73	86	88	88	88
B	17	36	62	78	79	80	80	—	—

[a] A = 185 mg/kg (C_6H_5)$_3$SnCl in arachis oil to 2 rats; B = 50 mg/kg (C_6H_5)$_3$SnCl in arachis oil to 3 rats.

Table XXIX. *Tin concentrations in rats treated orally with triphenyltin chloride* (after HEATH 1966).

Tissue	Concentrations[a] [μg (C_6H_5)$_3$SnCl/g tissue]		
	A	B	C
Brain	11	3.3	7.1
Liver	25	3.1	21
Kidney	22	7.5	26
Fat	9.2	0.7	7.6
Muscle	3.9	1.1	—
Stomach	331	0.6	1,230
Small intestine	331	1.5	35
Large intestine	542	2.3	61

[a] A = given 185 mg/kg in arachis oil and killed 2 days after injection; average of 2 rats; B = same, but killed 9 days after treatment; and C = given 67 mg/kg in dimethylformamide, average of 3 rats, which died 2 days after treatment.

Table XXX. *Concentrations[a] of tin in the tissues of rats treated with triphenyltin chloride intraperitoneally* (after HEATH 1963 a and b).

Tissue	Times after injection[b]						
	2 hr	1 day		2 days	4 days		10 days
	A	A	B	A	A	B	C
Brain	3.2	5.8	2.6	3.6	4.4	1.2	0.7
Liver	41	28	8.7	10	13	2.2	1.1
Kidney	15	20	8.0	19	14	6.8	4.4
Fat	16	10	3.0	21	—	1.2	0.7
Muscle	2.2	2.7	1.2	2.1	—	0.6	0.4
Heart	8.7	2.2	2.2	3.3	—	—	—
Blood	2.3	1.1	0.5	0.8	—	0.14	0.05

[a] Concentrations are expressed as μg triphenyltin chloride/g tissue.
[b] A = rats given 8.8 mg/kg, B = rats given 3.5 mg/kg, and C = rats given 4.4 mg/kg.

In another experiment the decrease of tin content in the brain of rats was determined depending on time. Initially, all [113]Sn was calculated as triphenyltin chloride (Fig. 9b; see p. 29). With increasing time, however, the percentage of total [113]Sn found which could be extracted with $CHCl_3$ appeared to be decreasing (Table XXXI). Only triphenyltin compounds are soluble in chloroform. It was concluded, therefore, that triphenyltin chloride partly had been converted to other tin derivatives.

According to HAERTEL (1964 a), a dietary level of 5 to 25 ppm triphenyltin acetate does not lead to accumulation of the active ingredient in rats. Excretion experiments with triphenyltin chloride led to basically

Table XXXI. *Percentages of ¹¹³Sn appearing in the chloroform layers of brain extracts of rats* (after HEATH 1966).

i.p.		Oral	
Day	%	Day	%
1	94	2	89
2	88	9	68
4	79	18	43
10	73	25	25
18	49		

Table XXXII. *Excretion of ¹¹³Sn from rats after single oral administration of 3 mg triphenyltin chloride* (FREITAG and BOCK 1974 b).

Days after administration	Tin eliminated (%)			
	In urine		In feces	
	Male	Female	Male	Female
1	0.2	0.3	16.9	11.8
2	1.0	0.5	21.6	21.7
3	0.9	0.9	31.5	28.5
4	0.5	0.7	11.4	18.4
5	0.2	0.4	4.4	5.5
6	0.1	0.2	1.6	1.4
7	0.1	0.1	0.8	0.6
Total	3.0	3.1	88.2	87.6

the same results (FREITAG and BOCK 1974 b). After a single oral administration of 3 mg of active ingredient (¹¹³Sn-labeled) the animals excreted within 7 days an average of 88% of the administered tin with the feces and 3% with the urine (Table XXXII).

Only 0.5% was found in the organs (Table XXXIII); the difference towards 100% may be explained by administration and analytical errors and by the fact that only parts of the animals' bodies were examined.

Experiments with repeated administration of triphenyltin chloride (5 × 3 mg in 5 days) after 1 wk showed excretion of 81.5% of the tin in the feces and 3.3% in the urine.

KIMMEL et al. (1977) conducted experiments on rats with oral administration of 1.6 mg of triphenyltin acetate/kg body wt (the active ingredient was dissolved in methoxy-triglycol). Within 10 days 85.9% of the ¹¹³Sn was excreted with urine and feces; the major amount was found unchanged in the feces. After intraperitoneal administration of the same amount, only 64.5% of the tin was excreted within 10 days.

2. **Experiments on guinea pigs.**—Guinea pigs, according to HEATH

Table XXXIII. *Distribution of the remaining activity in the organs of rats; average of 5 males and 5 females* (after FREITAG and BOCK 1974 b).

Organ	Total amount of organ (g)		% of the activity given	
	Males	Females	Males	Females
Stomach and intestines	43.07	40.14	0.079	0.077
Kidneys	7.43	6.61	0.055	0.097
Gonads	12.36	0.39	0.117	0.002
Liver	34.02	29.30	0.164	0.212
Heart	3.28	3.02	0.008	0.010
Muscles	23.71	18.68	0.045	0.018
Fat (skin)	5.83	6.96	0.004	0.004
Fat (kidneys)	3.75	4.08	0.003	0.005
Brain	7.25	7.16	0.025	0.030
Total	141.30	116.84	0.50	0.46
$=$	18%	16.7%		
	of the total wt			

(1966), after oral administration of $(C_6H_5)_3SnCl$ eliminate 80% of the tin within 1 wk; thereafter excretion slows down (Fig. 10).

The elimination from the brain proceeds with a "half-life period" of several days (Fig. 11); similar to the previously described experiments on rats, part of the triphenyltin compound is metabolized.

Contrary to the observations cited, several aforementioned authors claim that the triphenyltin acetate accumulates in the organs of guinea pigs. BARNES and MAGOS (1968) reported this finding without describing any experiments; STONER (1966), through theoretical calculations based on results of feeding experiments, came to the same conclusion. After further feed studies on guinea pigs, however, STONER and HEATH (1967) (Table XXXIV) somewhat altered their position. Based again on calcu-

Table XXXIV. *Survival time and dose rate in guinea pigs given a diet containing 12 ppm triphenyltin acetate for up to 1 yr* (after STONER and HEATH 1967).

Animal no.	Survival time (days)	Approx. average daily dose (mg/kg body wt/day)			Total dose consumed (mg/kg body wt)
		1–28 days	200–228 days	Overall	
29	22	—	—	0.89	19
22	248	0.83	0.45	0.81	196
27	259	1.15	0.73	0.87	247
24	294	1.10	0.76	0.96	277
32	306	1.22	0.93	0.80	279
28	S[a]	1.23	0.77	0.93	352

[a] S = alive at end of experiment.

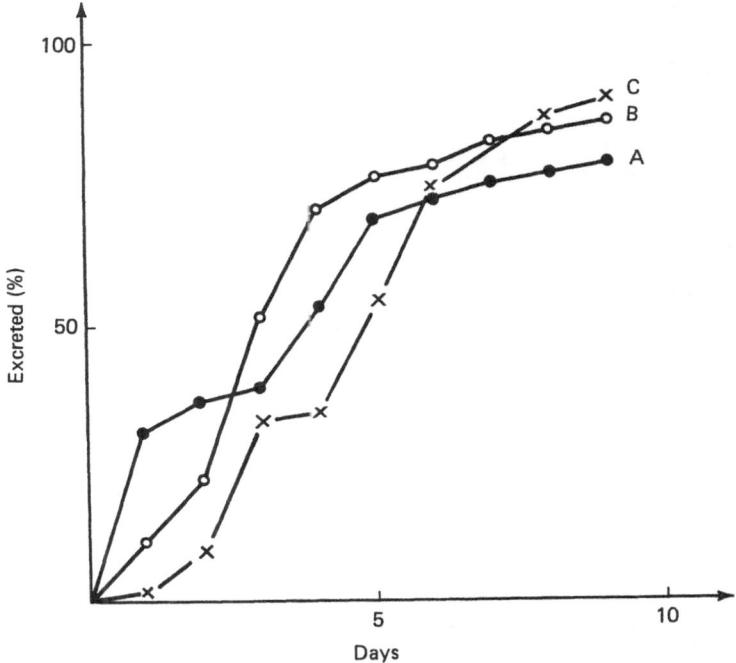

Fig. 10. Excretion of 113Sn from guinea pigs given triphenyltin chloride (after
HEATH 1966): A = 20 mg/kg, B = 5 mg/kg in arachis oil, and C =
5 mg/kg in DMF.

lations, they concluded that triphenyltin acetate is not accumulated altogether, but may be excreted under certain conditions with a "half-life
period" of about 50 days.

The above, as mentioned before, are theories which lack supportive
evidence from well defined analytical or radio-chemical experiments.

3. Experiments on dogs.—Feed studies on dogs, over a period of 2 yr,
did not show any accumulation of triphenyltin acetate (HAERTEL 1964,
a and b).

4. Experiments on sheep.—Sheep were fed daily with 10 mg of radioactively labeled triphenyltin acetate for 25 days; the concentration of the
active ingredient in the food was 3.2 ppm. The excretion was monitored
by measuring the 113Sn-activity in urine, milk, and blood (HEROK and
GOETTE 1961, 1963, and 1964). During feeding, the 113Sn-concentrations
in blood and milk remained almost unchanged at 2 to 3 μg/L, in the urine
at about 10 μg/L. They gradually subsided when the triphenyltin acetate
doses had been stopped (Fig. 12).

Most of the 113Sn was discharged with the feces (Table XXXV). Time-

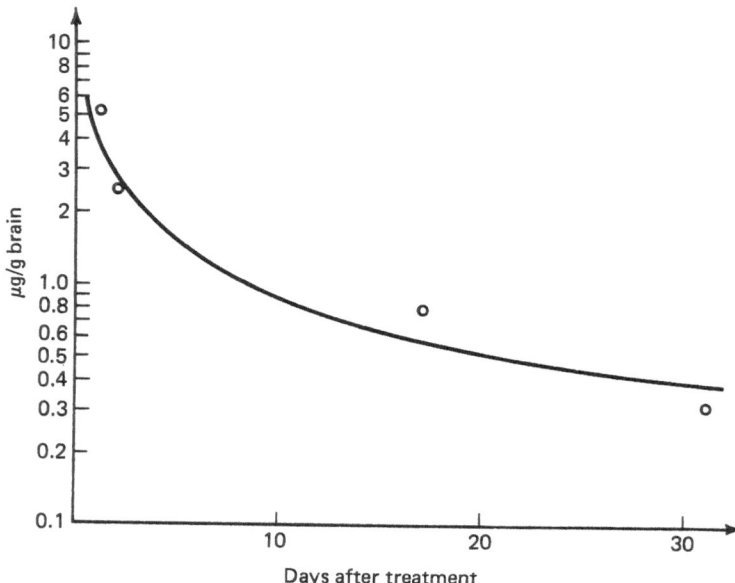

Fig. 11.   Excretion of [113]Sn from the brain of rats after intraperitoneal administration of 3 mg of triphenyltin chloride/kg (after HEATH 1966).

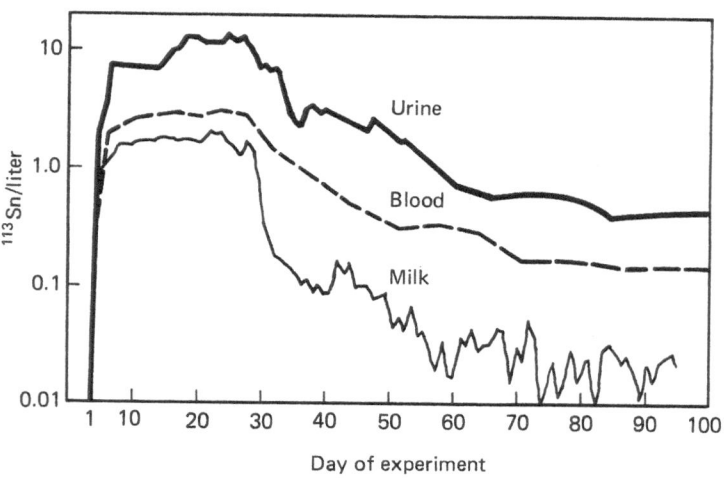

Fig. 12.   [113]Sn concentrations in milk, blood, and urine; first to 25th day of experiment 10 mg of [[113]Sn] triphenyltin acetate fed/animal/day (after HEROK and GOETTE 1961).

Table XXXV. *Total amount of* ^{113}*Sn excreted until the animals were killed* (after HEROK and GOETTE 1961).

Sheep	% of given ^{113}Sn excreted after		
	28 days	52 days	218 days
1	89.9	—	—
2	91.8	94.4	—
3	93.0	94.5	95.2

wise, the discharge was similar to the one through milk; however, the decline in concentration was steeper after feeding of active ingredient had been stopped (Fig. 13).

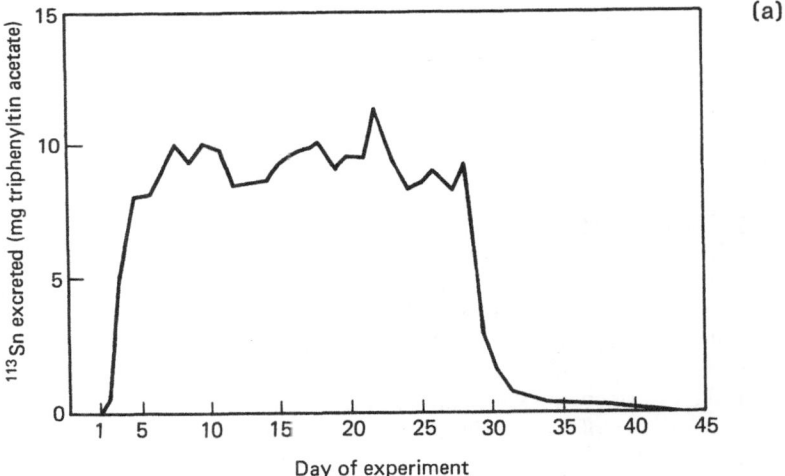

Fig. 13.   (a) Amounts of ^{113}Sn excreted daily with feces and urine; first to 25th day of experiment 10 mg of [^{113}Sn] triphenyltin acetate fed/animal/day (after HEROK and GOETTE 1961); (b) degradation of triphenyltin chloride depending on time, discharge of metabolites in feces of rats (after FREITAG and BOCK 1974 b): A = $(C_6H_5)_3Sn^+$, B = $(C_6H_5)_2Sn^{2+}$, C = $(C_6H_5)Sn^{3+}$, and D = Sn^{4+}.

Twenty-eight, 52, and 218 days after start of experiment, one animal each was killed and the distribution of ^{113}Sn was checked. As can be seen from the recovery, the tin is being discharged almost completely after feeding of the active ingredient has been discontinued (Table XXXVI). No detailed analysis for individual structure has been undertaken; however, it is known that a large part of the material fed had been converted into inorganic tin. Presumably, it is this conversion that is responsible for the slow secretion after prolonged feeding.

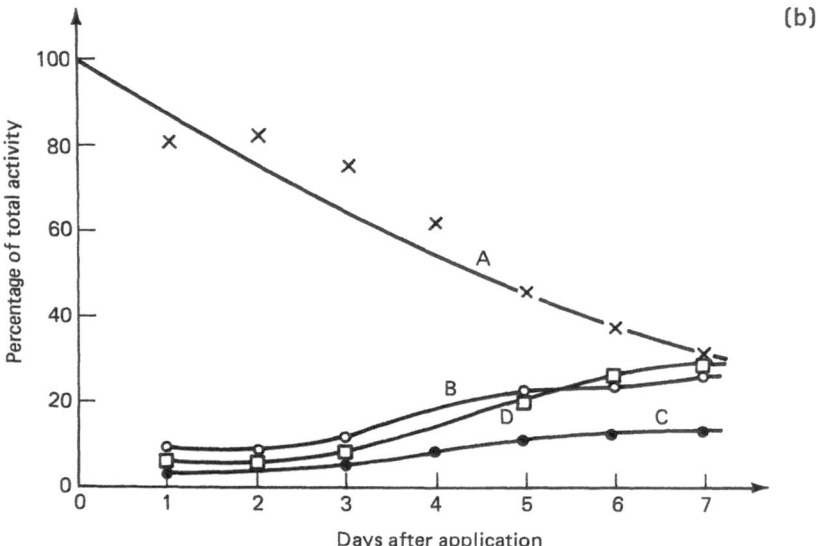

Days after application

5. Experiments on cows.—Cows fed over a period of 46 days with sugarbeet leaves containing an average daily dose of 32 mg of triphenyltin acetate discharged most of the tin with the feces. Already 2 wk after feeding was discontinued, the tin contents had dropped to the level of the control animals. Traces of phenyltin compounds (tri-, di-, and mono-phenyltin compounds, calculated as triphenyltin acetate) were found in the milk (BRUEGGEMANN *et al.* 1964 a and b).

The organs of two animals killed at the end of the 46-day feeding period contained—especially in the liver and kidneys—slightly increased overall tin concentrations (Table XXXVII), a considerable part of it appearing, however, in inorganic form. In one animal, killed 8 wk after the end of the experiment, the tin had been discharged almost completely (Table XXXVIII).

f) Metabolism of triphenyltin compounds in warm-blooded animals and degradation on plants

The metabolism of triphenyltin chloride in the organs of rats was examined by quantitative analysis of the degradation products in feces and urine (FREITAG and BOCK 1974 b). After a single oral dose of 3 mg of $(C_6H_5)_3$ $^{113}SnCl$, its degradation was measured as a function of time. It appeared that the concentration of di- and monophenyltin compounds in the feces increased with a corresponding decrease of $(C_6H_5)_3Sn^+$ (Fig. 13 b; see this page).

Phenyltin compound concentrations in the urine were too low for daily

Table XXXVI. *^{113}Sn content in μg of Sn/100 g of fresh substance of parts of the body after applying 10 mg of [^{113}Sn] triphenyltin acetate/animal/day for 25 days (1st to 25th day of experiment)* (after HEROK and GOETTE 1961).

Organ or tissue, etc.	μg of ^{113}Sn/100 g of fresh substance		
	Sheep 1 28th day	Sheep 2 52nd day	Sheep 3 218th day
Liver	89.6	92.9	13.1
Kidney	34.0	11.8	0.92
Adrenal gland	2.03	0.52	0.33
Lungs	6.14	0.95	0.60
Spleen	2.42	0.56	0.36
Pancreas	3.12	0.45	1.18
Gall	19.27	—	0.15
Mammary gland	1.49	0.43	0.36
Thyroid gland	2.02	0.63	0.16
Lymph node	1.43	1.68	3.44
Eyes	0.22	0.20	0.10
Brain	3.95	0.13	0.32
Spinal marrow	2.21	0.33	0.13
Bone marrow	—	0.36	0.13
Omental fat	1.01	0.64	0.04
Depot fat and kidney fat	1.13	0.70	0.04
Bladder	1.45	0.45	0.29
Uterus	1.02	0.31	0.20
Oesophagus	1.45	0.35	0.14
Wall of rumen	0.83	0.17	0.12
abomasum	1.05	0.39	0.17
large intestine	3.99	0.35	0.12
small intestine	4.82	0.44	0.23
Heart	2.77	0.29	0.23
Tongue and trachea	2.24	0.41	0.15
Muscle psoas	1.95	0.27	0.03
quadriceps	2.11	0.29	0.05
brachiocephalicus	2.02	0.33	0.03
Skin	2.59	1.00	0.32
Long bone	—	0.36	0.11
Wool, degreased	2.80	2.36	0.22

analysis. Therefore, a composite 7-day sample was analyzed several times showing that di- and monotin compounds and inorganic tin had formed in corresponding quantities (Table XXXIX).

FISH et al. (1976) and KIMMEL et al. (1977) also studied the breakdown of ^{113}Sn-triphenyltin acetate in rats after oral and intraperitoneal administration. Confirming the above mentioned results, both test series showed that mainly unchanged active ingredient and polar tin compounds ($C_6H_5Sn^{3+}$ as well as inorganic tin) were discharged in urine and feces, apart from small amounts of diphenyltin compounds. Also, traces of an unidentified nonpolar substance were found, containing tin.

In several *in vitro* tests, the same authors subjected triphenyltin

Table XXXVII. *Total tin content of cattle organs after feeding 32 mg of triphenyltin acetate/day/animal for 46 days* (after BRUEGGEMANN et al. 1964 a).

Organ	Analyzed amount (g)	Tin content (μg) in		
		Animal 1	Animal 2	Control
Muscle	100	< 5	11	9
Udder	100	10	10	< 5
Fat	100	16	15	< 5
Spleen	100	15	54	< 5
Liver	100	152	82	6
Heart	100	23	22	< 5
Kidney	100	236	49	< 5
Lung	100	30	19	< 5
Intestine	100	12	276	10
Parotic gland	100	18	16	< 5
Pancreas	100	26	22	8
Brain	100	34	33	< 5
Ovary	16	—	< 5	< 5
Spinal marrow	70	20	15	6
Thyroid gland	22	—	< 5	< 5
Adrenal gland	30	—	< 5	< 5

Table XXXVIII. *Tin content of cattle organs 8 wk after feeding of triphenyltin acetate had been discontinued* (after BRUEGGEMANN et al. 1964 a).

Organ	μg Sn/100 g	Organ	μg Sn/100 g
Muscle	< 5	Intestine	< 5
Udder	< 5	Parotic gland	11
Fat	< 5	Pancreas	10
Spleen	< 5	Brain	< 5
Heart	< 5	Ovary	< 5
Liver	21	Spinal marrow	< 5
Kidney	< 5	Thyroid gland	—
Lung	< 5	Adrenal gland	< 5
Gall (100 ml)	< 5	Blood (500 ml)	13

Table XXXIX. *Concentration of tin compounds in rat urine* (after FREITAG and BOCK 1974 b).

Compound	Radioactivity[a] (%)
$(C_6H_5)_3Sn^+$	25
$(C_6H_5)_2Sn^{2+}$	17–19
$(C_6H_5)Sn^{3+}$	20–29
Inorganic tin (IV)	29–33
Not extracted from the aqueous solution	0.6– 5

[a] Percentage of total.

Table XL. *Degradation of triphenyltin chloride on sugarbeet leaves in the greenhouse*[a] (after FREITAG and BOCK 1974 b).

Days after application	Percentage of total radioactivity applied				
	$(C_6H_5)_3Sn^+$	$(C_6H_5)_2Sn^{2+}$	$(C_6H_5)Sn^{3+}$	Sn^{4+}	Not extracted[b]
0	100	—	—	—	—
3	86	9.8	0.6	2.7	1.3
7	67	13	1.0	16	3.0
14	47	8.5	1.0	38	5.0
21	33	6.5	1.0	55	4.0
28	26	4.3	1.2	60	9.0
35	22	2.7	1.0	65	10
42	19	1.6	1.0	67	12

[a] 288 μg of $(C_6H_5)_3SnCl$/leaf.
[b] Hydrolyzed and aged $(C_6H_5)Sn^{3+}$ and Sn^{4+}.

acetate to conditions of biological oxidation (microsomal monooxygenase-reaction). Surprisingly—and contrary to the behavior of the tributyltin compound—there was no significant degradation. Only traces of possible oxydation products (their nature could not be determined) were found by qualitative thin-layer chromatography. Apparently, an H-atom in 1-position is necessary for oxidation (CASIDA *et al.* 1971).

This result leads to the conclusion that the degradation of triphenyltin compounds *in vivo* is not caused by cytochrome-P-450-dependent mono-oxygenase, but rather through another enzyme system or entirely without such.

There are several reports indicating the formation of diphenyltin compounds on plants (KROELLER 1960, ANONYMOUS 1965). Table XL shows the exact course of the degradation of triphenyltin chloride on sugarbeet leaves during a daylight experiment in a greenhouse (BOCK and FREITAG 1972, FREITAG and BOCK 1974 b). Within 10 days the amount of triphenyltin compound had decreased to about one-half. There were only small amounts of di- and monophenyltin compounds, since they decompose to inorganic tin rather rapidly.

Since benzene has also been found as a breakdown component of triphenyltin compounds, it can be concluded that the degradation in warm-blooded animals as well as on plants in daylight proceeds along the following path:

$$(C_6H_5)_3Sn^+ \rightarrow (C_6H_5)_2Sn^{2+} \rightarrow (C_6H_5)Sn^{3+} \rightarrow Sn^{4+} \qquad (32)$$

V. Biological properties of triphenyltin compounds; application

a) Introduction

In systematic experiments VAN DER KERK and coworkers have studied the efficiency of organotin compounds of the type R_nSnX_{4-n} on various species of fungi. The number and kind of the organic groups R were varied, also the anionic residue X. Subsequently, the influence of the especially efficient compounds R_3SnX on a number of other organisms was tested. At first, alkyltin compounds were used for these experiments, later and with increasing frequency also aryl compounds (with R=phenyl).

The fungitoxicity of alkyltin compounds R_nSnX_{4-n} depends on the number of R. The first result of experiments with the following series $(C_2H_5)_4Sn - (C_2H_5)_3SnCl - (C_2H_5)_2SnCl_2 - (C_2H_5)SnCl_3$ was that the triethyl compound is by far the most efficient in inhibiting the growth of various fungi (VAN DER KERK 1952, VAN DER KERK and LUIJTEN 1954; Table XLI).

Table XLI. *Fungitoxic properties of ethyltin chlorides and of tin chlorides; maltagar, pH approximately 6, at 24°C* (after VAN DER KERK and LUIJTEN 1954).

Compound	Fungistatic concentration (mg/L)			
	Botrytis allii Munn.	Penicillium italicum Wehm.	Aspergillus niger v. Tigh.	Rhizopus nigricans Ehrenb.
$(C_2H_5)_4Sn$	50	>1,000	100	100
$(C_2H_5)_3SnCl$	0.5	2	5	2
$(C_2H_5)_2SnCl_2$	100	100	500	200
$(C_2H_5)SnCl_3$	>1,000	>1,000	>1,000	>1,000
$SnCl_2$	>1,000	>1,000	>1,000	>1,000
$SnCl_4$	>1,000	>1,000	>1,000	>1,000

The fungitoxicity of alkyltin compounds R_nSnX_{4-n} depends on the nature of aliphatic R. When varying the group R in a series of trialkyltin acetates $(R_3SnOCOCH_3)$ maximum efficiency was found for the tri-*n*-propyl- and tri-*n*-butyltin compounds (VAN DER KERK and LUIJTEN 1954; Table XLII).

Mixed tinorganic compounds showed no distinct advantages, though efficient fungicides were also found among them (VAN DER KERK and LUIJTEN 1956 c). By introducing functional groups into R (for example, R=CH_2CH_2COONa) the fungitoxicity is almost lost (NOLTES et al. 1961).

Contrary to the results described so far, the chemical nature of X did not seem to have particular influence on the fungitoxicity of alkyltin compounds. This was shown through experiments with triethyltin compounds (VAN DER KERK and LUIJTEN 1954, LUIJTEN and VAN DER KERK 1955; Table XLIII) and with tri-*n*-propyltin compounds (CZERWIŃSKA et al. 1967).

The reason for this behavior was seen in their hydrolysis to the hydroxide in very diluted solutions (KAARS SIJPESTEIJN et al. 1969).

The tributyltin compounds, however, showed partly different results. Depending on the nature of X the fungistatic concentrations differed by

Table XLII. *Fungitoxicity of triorganotin compounds $R_3SnOCOCH_3$; peptone-glucose-agar, pH 6.4 at 24°C* (after VAN DER KERK and LUIJTEN 1954, VAN DER KERK *et al.* 1962).

	Fungistatic concentration (mg/L)			
R	*Botrytis allii* Munn.	*Penicillium italicum* Wehm.	*Aspergillus niger* v. Tigh.	*Rhizopus nigricans* Ehrenb.
Methyl	200	500	200	500
Ethyl	1	10	2	2
n-Propyl	0.5	0.5	0.5	0.5
i-Propyl	0.1	0.5	1	1
n-Butyl	0.5	0.5	1	1
i-Butyl	1	1	10	1
Pentyl	5	2	5	5
Hexyl	>500	>500	>500	>500
Octyl	>500	>500	>500	>500
Cyclopentyl	0.5	0.5	5	0.5
Cyclohexyl	20	20	50	20
Phenyl	10	1	0.5	5

Table XLIII. *Fungitoxic properties of triethyltin compounds $(C_2H_5)_3SnX$ with varying X* (excerpt from LUIJTEN and VAN DER KERK 1955).

	Fungistatic concentration (mg/L)			
X	*Botrytis allii* Munn.	*Penicillium italicum* Wehm.	*Aspergillus niger* v. Tigh.	*Rhizopus nigricans* Ehrenb.
Hydroxide	0.2	5	0.5	0.5
Chloride	0.5	2	5	2
Bromide	0.5	2	1	1
Sulfide	0.2	1	1	1
Sulfate	0.2	0.2	5	5
Acetate	1	2	5	2
Laurate	0.2	0.2	5	5
Benzoate	2	10	5	5
Phenolate	0.5	1	2	1

several orders of magnitude. Also, some of these compounds showed a certain selectivity towards one or the other of the three fungi tested (Table XLIV). Experiments with another fungus *(Coriolellus palustris)* also produced these differences (FUSE and NISHIMOTO 1964, NISHIMOTO and FUSE 1966).

Field experiments showed that the tripropyl- and tributyltin compounds which *in vitro* proved so effective are not only very phytotoxic, but also (probably due to high volatility) not well suited to control fungi. Triphenyltin compounds, however, were found to be tolerated much better by the plants and to be highly fungistatic (BRUECKNER and HAERTEL

Table XLIV. *Fungitoxic properties of tri-n-butyltin compounds* $(C_4H_9)_3SnX$ *with varying X; agar-culture* (excerpt from CZERWIŃSKA *et al.* 1967).

	Fungistatic concentration (wt %)		
X	*Pusarium culmorum* Sacc.	*Alternaria tenuis* Nees	*Rhizoctonia solani* Kuehn
F	0.001	0.001	0.001
Cl	0.5	0.5	0.5
OSn(C_4H_9)$_3$	0.05	0.05	0.0005
OCH$_3$	0.005	0.005	0.005
OCOCH$_3$	0.001	0.001	0.001
OCOCH$_2$OC$_6$Cl$_5$	0.05	0.0005	0.0005
OCOCH$_2$SC$_6$Cl$_5$	0.1	0.1	0.1

1955, 1956, and 1959; HAERTEL and BAUMANN 1956; HAERTEL 1958 a and b, and 1962).

Systematic experiments were also conducted with triaryltin compounds, varying the nature of R. However, the fungitoxic efficiency could not be significantly improved compared with the phenyl compound (Table XLV).

Table XLV. *Fungitoxic properties of various triaryltin acetates; maltagar, pH ca. 6, at 24°C* (after VAN DER KERK *et al.* 1962).

	Fungistatic concentration (mg/L)			
R	*Botrytis allii* Munn.	*Penicillium italicum* Wehm.	*Aspergillus niger* v. Tigh.	*Rhizopus nigricans* Ehrenb.
Phenyl	5	5	1	10
m-Tolyl	5	5	2	10
p-Tolyl	5	5	2	5
Benzyl	2	20	50	20
2-Phenylethyl	1	5	50	20
α-Naphthyl	>50	>50	>50	>50

The theory that for these triphenyltin compounds, too, the influence of X on the toxicity is rather low, was proven in laboratory and field tests with several simple anions (X = Cl⁻, OH⁻, acetate, and others) and for the complex of $(C_6H_5)_3SnCl$ with decyltriphenyl-phosphonium bromide (SRIVASTAVA and TANDON 1964, BONGIOVANNI 1968 a, SCHICKE *et al.* 1968). Other authors, however, found marked differences by varying the anion X (JEN-HSI CHO *et al.* 1966). In experiments by BYRDY *et al.* (1966 a) with spores of *Venturia inaequalis* (Cooke) and *Alternaria tenuis* Nees, namely the triphenyltin acetate and methoxy compound $(C_6H_5)_3SnOCH_3$, were far more efficient than, for example, hydroxide,

oxide, or phthalate. CZERWIŃSKA *et al.* (1967) too, found very distinct differences in the efficiency of triphenyltin compounds (Table XLVI).

Table XLVI. *Fungitoxic properties of triphenyltin compounds* $(C_6H_5)_3SnX$ *depending on the nature of X* (excerpt from CZERWIŃSKA *et al.* 1967).

	Fungistatic concentration (wt %)		
X	*Fusarium culmorum* Sacc.	*Alternaria tenuis* Nees	*Rhizoctonia solani* Kuehn
Cl	0.5	0.05	0.005
CN	0.1	0.01	0.05
SCN	0.05	0.005	0.01
OH	0.5	0.05	0.05
$OSn(C_6H_5)_3$	0.005	0.005	0.005
$SSn(C_6H_5)_3$	>0.1	>0.1	>0.1
$OCOCH_3$	0.5	0.05	0.005
$OCOCH_2CH_3$	0.05	0.05	0.1
$OCOC_6H_5$	0.05	0.01	0.01

b) Control of microorganisms

1. **Efficiency spectrum of triphenyltin compounds.**—The first quantitative observations about the biocidic effects of triphenyltin compounds were made by VAN DER KERK (1952): hydroxide in concentrations of 5.0, 0.5, 0.5 and 2.0 in maltagar-cultures prevents the growth of *Botrytis cinerea* Pers., *Penicillium italicum* Wehm., *Aspergillus niger* v. Tigh and *Rhizopus nigricans* Ehr. Similar results were published later for triphenyltin acetate; however, the experimental conditions had been somewhat different (see Tables XLII and XLV and KAARS SIJPESTEIJN 1959, KAARS SIJPESTEIJN *et al.* 1962). BRUECKNER and HAERTEL (1955 and 1956) and HAERTEL (1962 and 1964 a) described the efficiency of the acetate towards *Peronospora*, *Phytophthora infestans* de B., *Fusicladium*, *Botrytis allii* Munn., *Septoria apii* Chester, *Helminthosporium*, *Fusarium nivale*, *Tilletia*, *Colletotrichum lindemuthianum* Bri. et Cav., and *Cercospora beticola* Sacc. The experiments were conducted partly in the field, partly in greenhouses.

BAUMANN (1958) studied the effects of triphenyltin chloride, acetate, and hydroxide on several types of fungi through spore germination tests. The concentrations which prevented germination of 90% of the spores ("LD_{90}") were 6 to 10 ppm for *Botrytis cinerea* Pers., 20 ppm for *Alternaria tenuis* Nees, and 30 to 50 ppm for *Sclerotinia fructicola* (Wint.) Rehm.

More extensive findings about the efficiency spectrum of triphenyltin acetate were published by HAERTEL (1962, 1963 c, and 1964 a) (Table XLVII) and POLSTER and HALAČKA (1971) (Table XLVIII).

Table XLVII. *Action spectrum in field tests of triphenyltin acetate and doses compared to other fungicides*[a] (after HAERTEL. 1962 and 1964 a).

Fungus disease	Triphenyltin acetate	Copper oxychloride (copper content)	Zincethylene bis-dithiocarbamate	N-Trichloromethyl-thiotetrahydroph-thalimide
Venturia inaequalis (Cooke) Aderh	20–40 g/100 L	100–200 g/100 L	160 g/100 L	100–150 g/100 L
Plasmopara viticola (Berk. et Curt) Berk. et de Toni	20–30 g/100 L	250–375 g/100 L	160 g/100 L	100–150 g/100 L'
Colletotrichum lindemuthianum Sacc. et Magn.	20–30 g/100 L	250–375 g/100 L	260–200 g/100 L	100–150 g/100 L
Dothichiza populea Sacc. et Briard.	40–60 g/100 L	275–500 g/100 L	—	—
Cercospora beticola Sacc.	150–240 g/ha	3,000 g/ha	>2,400 g/ha	—
Septoria apii (Br. et Cav.) Chester	120–200 g/ha	>3,000 g/ha	>2,400 g/ha	—
Phytophthora infestans de By.	200–360 g/ha	2,500–3,000 g/ha	1,200–1,400 g/ha	600–800 g/ha
Lophodermium pinastri (Schrad.) Chev.	360 g/ha	2,000–3,000 g/ha	1,200–1,400 g/ha	

[a] 1 g/100 L = 20 oz/gal, 100 g/ha = 1.42 oz./A.

Table XLVIII. *Efficiency spectrum of triphenyltin acetate*
(after POLSTER and HALAČKA 1971).

Microorganism	Concentration (ppm)	
	Microbiostatic	Microbiocidic
Bacteria		
Staphylococcus aureus	0.5	5
Spacrotilus natans	10	10
Escherichia coli	>10	>10
Proteus mirabilis	>10	>10
Pseudomonas aeroginosa	10	>10
Pseudomonas indoloxydans	>10	>10
Mycobacterium phlei	5	5
Bac. cereus var. *mycoides*	1	1
Serrata marcescens	>10	>10
Xanthomonas begoniae	1	1
Yeast plants		
Candida albicans	0.05	0.5
Candida tropicalis	0.5	0.5
Candida pseudotropicalis	0.5	1
Candida kruzei	0.1	0.5
Candida parapsilosis	>10	>10
Candida guillermondii	0.1	0.5
Geotrichum candidum	0.5	>10
Pulluaria pulluans	0.1	1
Saccharomyces carlsbergensis	0.1	0.5
Torulopsis magnoliae	5	5
Fungi		
Aspergillus flavus	>10	>10
Aspergillus niger	1	>10
Penicillium expansum	5	5
Mucor racemosus	0.05	0.05
Trichoderma viridiae	10	10
Cladosporium cladosporoides	10	10
Botrytis allii	0.5	0.5
Myrothrecium verrucaria	0.5	0.5
Paccilomyces varioti	5	10
Chaetomium globosum	10	10

Their results showed that low concentrations of this compound prevent not only the growth of fungi, but also of yeasts and of various kinds of bacteria.

The efficiency spectrum of the complex triphenyltin chloride/decyl-triphenylphosphonium bromide is shown in Table IL and the effectiveness of several triphenyltin compounds of phosphoric acid esters is shown in Table L.

Besides these experiments, many others were conducted to test the usefulness of triphenyltin compounds against different microorganisms under varying conditions (Table LI).

Table XLIX. *Efficiency spectrum of the complex triphenyltin chloride decyltriphenylphosphonium bromide; 500–750 g/ha* (after *CELA/ Landwirschaftliche Chemikalien GmbH*, Ingelheim; CA 6830).

Culture	Pathogenic agent
Corn	*Helminthosporicum turcicum* Pass.
Onion	*Alternaria porri* (Ell.) Cif.
Carrot	*Alternaria dauci* Kuehn.
	Cercospora carotae Pass.
Cucumber	*Colletotrichum lagenarium* (Pass.) Ell. et Halst.
	Mycosphaerella citrullina c.o.Sm.
Fruit tree	*Nectria galligena* Bres.
Tobacco	*Alternaria tabacina* Ell. et Evh.
	Cercospora nicotianae Ell. et Evh.
Asparagus	*Fusarium* sp.
	Puccinia asparagi D.C.
Tomato	*Alternaria solani* Sorauer
	Phytophthora infestans de Bary
Pecan nut	*Fusicladium effusum* Wint.
Peanut	*Mycosphaerella personata* Higgins
	Mycosphaerella arachidicola W. A. Jenkins
Coffee	*Hemileia vastatrix* Berk. et Br.
	Colletotrichum coffeanum Noack
Peppermint	*Puccinia menthae* Pers.
Gum tree (Hevea)	*Phytophthora palmivora* Butl.
	Melanopsammopsis ulei (P. Henn.) Stahel
	Pellicularia salmonicolor
Cacao	*Phytophthora palmivora* Butl.
Citrus	*Phytophthora* sp.
	Gloeosporium sp.
Almond	*Cercospora circumscissa* Sacc.
Cotton	*Alternaria* sp.
	Rhizoctonia solani Kuehn
Olive	*Cycloconium oleaginum* Cast.
Rice	*Corticium sasakii* Shiraÿ
	Sphaerulina oryzina Hara
	Helminthosporium oryzae B. de Haan
	Piricularia oryzae Cav.
	Algae
Sugarcane	*Ceratostomella paradoxa* Pade

With regard to the effectiveness of triphenyltin compounds against bacteria, in low concentrations, triphenyltin compounds inhibit the growth of gram-positive species of bacteria, whereas gram-negative bacteria are relatively insensitive (KAARS SIJPESTEIJN *et al.* 1962, KAARS SIJPESTEIJN *et al.* 1969; Table LII).

These results, confirmed by SRIVASTAVA and TANDON (1964) may be important for fermentation processes from which, if gram-positive bacteria are involved, triphenyltin compounds should be kept away. Within a group of various triphenyltin compounds, the hydroxide and acetate were found to be most efficient (CZERWIŃSKA *et al.* 1967; Table LIII).

Table L. *Fungicide activity of triphenyltin-phosphorus compounds* (after Kubo 1965).

Compound	R	Cercospora beticola[a]	Piricularia oryzae[a]	Cochliobolus miyabeanus[a]	Botrytis allii[b]	Penicillium italicum[b]	Aspergillus niger[b]	Rhizopus nigricans[b]
$RO\!\!-\!\!\overset{\displaystyle S}{\underset{}{\underset{RO}{\|}}}P\!-\!S\!-\!Sn(C_6H_5)_3$	CH_3	0.5	0.3	2	0.5	0.1	0.2	0.2
	C_2H_5	0.5	0.2	2	0.5	0.5	0.2	1
	$n\text{-}C_3H_7$	0.3	0.2	2	0.5	0.2	0.2	0.2
	$n\text{-}C_4H_9$	0.3	0.2	2	0.5	0.2	0.2	0.2
$RO\!\!-\!\!\overset{\displaystyle O}{\underset{RO}{\|}}P\!-\!O\!-\!Sn(C_6H_5)_3$	C_2H_5	0.5	0.3	1	2	0.5	0.5	1
	C_6H_5	—	—	—	1	0.2	0.2	0.2
$R\!\!-\!\!\overset{\displaystyle O}{\underset{R}{\|}}P\!-\!O\!-\!Sn(C_6H_5)_3$	C_6H_5	0.5	0.3	1	2	0.5	0.5	1
$R_2N\!\!-\!\!\overset{\displaystyle O}{\underset{R_2N}{\|}}P\!-\!O\!-\!Sn(C_6H_5)_3$	CH_3	0.5	0.3	2	—	—	—	—
	C_2H_5	0.5	0.3	2	—	—	—	—
$RNH\!\!-\!\!\overset{\displaystyle O}{\underset{RNH}{\|}}P\!-\!O\!-\!Sn(C_6H_5)_3$	C_2H_5	0.5	0.3	1	—	—	—	—
	C_6H_5	0.5	0.5	1	—	—	—	—
$(C_6H_5)_3SnOCOCH_3$	—	0.3	0.3	1	2	1	0.5	10

[a] Paper disk agar method; concentration (μg) causing complete inhibition of growth.
[b] Roll culture method; concentration (ppm) causing complete inhibition of growth.

Table LI. *Effectiveness of triphenyltin compounds against different microorganisms.*

Microorganism	Triphenyltin compound	Concentration	References
Physalospora miyabeana Fukush.	Acetate	0.4%	DIERCKS (1957)
Ceratostomella paradoxa Pade	Acetate (20%)	0.5–2.5%	ANTOINE (1957)
Botrytis cinera Pers.	Acetate (20%)	20 ppm	ESTIENNE & HENNEBERT (1959)
Rhizoctonia solani Kuehn	Acetate (20%)	4–20 ppm	ESTIENNE & HENNEBERT (1959)
Pestalozzia sp.	Acetate (20%)	4–20 ppm	ESTIENNE & HENNEBERT (1959)
Monilia fructigena Aderh. et Ruhl.	Acetate (20%)	4–20 ppm	ESTIENNE & HENNEBERT (1959)
Corticium sp.	Acetate (20%)	40–80 ppm	ESTIENNE & HENNEBERT (1959)
Verticillium dahliae Kleb.	Hydroxide	40 ppm	HAERTEL (1962)
Verticillium dahliae Kleb.	Acetate	20 ppm	HAERTEL (1962)
Verticillium dahliae Kleb.	Chloride	10 ppm	HAERTEL (1962)
Botrytis cinerea Pers.	Hydroxide	80 ppm	HAERTEL (1962)
Botrytis cinerea Pers.	Acetate	60 ppm	HAERTEL (1962)
Botrytis cinerea Pers.	Chloride	40 ppm	HAERTEL (1962)
Alternaria tenuis Nees	Hydroxide	80 ppm	HAERTEL (1962)
Alternaria tenuis Nees	Acetate	80 ppm	HAERTEL (1962)
Alternaria tenuis Nees	Chloride	40 ppm	HAERTEL (1962)
Penicillium sp.	Acetate	63–4,000 ppm	MILLER & GOULD (1963)
Colletotrichum obiculare	Hydroxide	0.1%	HORN (1963)
Pseudoperonospora cubensis Berk. et Curt.	Hydroxide	0.1%	NUGENT (1963)
Erysiphe cichoracearum DC	Hydroxide	0.1%	NUGENT (1963)
Mycosphaerella melonis	Hydroxide	0.4 kg/ha	SITTERLY (1963 a)
Pythium and Rhizoctonia spp.	Hydroxide	0.35 kg/ha	SITTERLY (1963 b)
Cercospora arachidicola Hori	Hydroxide	0.07 kg/ha	BOYLE (1963)
Cercospora personata Ellis	Hydroxide	0.07 kg/ha	BOYLE (1963)
Aspergillus niger v. Tigh.	8-Hydroxy-quinoline-compound	0.01–1%	FOELDESI & STRÁNER (1965)

Table LI. (*continued*)

Microorganism	Triphenyltin compound	Concentration	References
Chaetomium globosum	8-Hydroxy-quinoline-compound	0.01–1%	FOELDESI & STRÁNER (1965)
Penicillium cyclopium	8-Hydroxy-quinoline-compound	0.01–1%	FOELDESI & STRÁNER (1965)
Penicillium brevi compactum	8-Hydroxy-quinoline-compound	0.01–1%	FOELDESI & STRÁNER (1965)
Paecilomyces varioti	8-Hydroxy-quinoline-compound	0.01–1%	FOELDESI & STRÁNER (1965)
Stachybotris astra	8-Hydroxy-quinoline-compound	0.01–1%	FOELDESI & STRÁNER (1965)
Aspergillus amstelodami	8-Hydroxy-quinoline-compound	0.01–1%	FOELDESI & STRÁNER (1965)
Trichophyton rubrum	Hydroxide	45 ppm	KHOSA & DIXIT (1969)
Trichophyton rubrum	Acetate	100 ppm	KHOSA & DIXIT (1969)
Trichophyton rubrum	Benzoate	40 ppm	KHOSA & DIXIT (1969)
Pellicularia sasaki Shiray	Hydroxide	40 ppm	KHOSA & DIXIT (1969)
Pellicularia sasaki Shiray	Acetate	100 ppm	KHOSA & DIXIT (1969)
Pellicularia sasaki Shiray	Benzoate	45 ppm	KHOSA & DIXIT (1969)
Fusarium vasiinfectum Atk.	Hydroxide	45 ppm	KHOSA & DIXIT (1969)
Fusarium vasiinfectum Atk.	Acetate	1,000 ppm	KHOSA & DIXIT (1969)
Fusarium vasiinfectum Atk.	Benzoate	40 ppm	KHOSA & DIXIT (1969)
Epidermophyton floccosum	Hydroxide	40 ppm	KHOSA & DIXIT (1969)
Epidermophyton floccosum	Benzoate	40 ppm	KHOSA & DIXIT (1969)
Microsporum gypsum	Hydroxide	75 ppm	KHOSA & DIXIT (1969)
Microsporum gypsum	Benzoate	60 ppm	KHOSA & DIXIT (1969)
Aspergillus flavus	Hydroxide	50 ppm	KHOSA & DIXIT (1969)
Aspergillus flavus	Benzoate	45 ppm	KHOSA & DIXIT (1969)
Curvularia lunata	Hydroxide	45 ppm	KHOSA & DIXIT (1969)
Curvularia lunata	Benzoate	40 ppm	KHOSA & DIXIT (1969)
Absidia glauca	Acetate	2–50 ppm	TRINCI & GULL (1970)
Colletotrichum gloeosporioides	Acetate	2 ppm	BOLKAN *et al.* (1976)

Table LII. *Effectiveness of triphenyltin acetate against gram-positive and gram-negative bacteria* (after Kaars Sijpesteijn et al. 1962, Kaars Sijpesteijn et al. 1969).

Microorganism	Growth inhibiting concentration (ppm)
Gram-positive	
Bacillus subtilis	2; 0.5
Mycobacterium phlei	1; 0.1
Staphylococcus aureus	5
Streptococcus lactis	5
Gram-negative	
Escherichia coli	>100; >500
Pseudomonas fluorescens	>100; >500
Pseudomonas phaseolicola Burkh.	>100

Table LIII. *Bacteriostatic effect of various triphenyltin compounds; agar-culture* (after Czerwińska et al. 1967).

Triphenyltin compound	Growth inhibiting concentration (%)		
	Staphylococcus aureus	*Escherichia coli*	*Bacillus subtilis*
Hydroxide	0.0001	0.0001	0.0001
Fluoride	0.001	0.01	0.0001
Chloride	0.0001	0.01	0.0001
Acetate	0.0001	0.001	0.0001

The triethanolamine-complex of triphenyltin acetate also has bacteriostatic effects (inhibits the growth of *Flavobacterium peregrinum* and *Bac. subtilis*) (Inaba and Watanabe 1963).

With regard to the effectiveness of triphenyltin compounds against algae, Tisdale reported already in 1943 that growth of algae on ship hulls can be controlled by triphenyltin linoleate. Later, triphenyltin chloride (Sparmann 1957, Evans 1970) and hexaphenyldistannoxane (Brueckner et al. 1959) were proposed for the same purpose.

Baumann (1958) found that triphenyltin acetate inhibits the growth of algae in water cultures of celery. Green algae in fresh water are harmed by triphenyltin chloride in concentrations of 2 to 5 ppm (Floch and Deschiens 1962). Single-cell green algae are destroyed by 0.67 ppm of triphenyltin acetate (Chiapparini et al. 1964). The algae *Chlamydomonas* sp. dies already at a concentration of 1.6×10^{-2} ppm of hexaphenyldistannoxane (Rivett 1965). The complex triphenyltin acetate/triethanolamine is effective in concentration of several ppm against blue-, green-, and silico-algae (Inaba and Watanabe 1963).

2. Use of triphenyltin compounds in agriculture.—

α) *Survey.*—To control pests in agriculture, numerous experiments have been conducted with various triphenyltin compounds containing

different anionic residues X. Inorganic as well as organic acids were introduced to the molecule, and different complex compounds of triphenyltin
chloride prepared. Further, frequent attempts have been made to expand
the field of use for triphenyltin compounds or to intensify their efficiency
by using anions with fungicidal or insecticidal activity. As examples should
be mentioned 8-hydroxyquinoline (HAERTEL 1962, FOELDESI and STRÁNER
1965), sorbic acid (HAERTEL 1962), ethylenebisdithiocarbamic acid
(BRUECKNER and HAERTEL 1955), and several phosphoric acid esters
(KUBO 1965).

However, as far as known to date, these experiments were of no practical importance. The only compounds that have been used on a larger
scale to control fungi on plants are triphenyltin acetate, triphenyltin
hydroxide, and (on a smaller scale) triphenyltin chloride. Besides these,
a formula is used that combines triphenyltin acetate and manganous
ethylenebisdithiocarbamate and has a lower phytotoxicity.

Also, the broad application spectrum described in the last section is
only partly utilized in practice. The most important applications are the
control of *Phytophthora infestans* in potatoes, of *Cercospora beticola* in
sugar beets, and of *Septoria apii* in celery. A summary of these and some
other uses is shown in Table LIV.

Table LIV. *Use of triphenyltin compounds in agriculture*
(after *Codex Committee* 1970).

Microorganism	Plant
Phytophthora infestans de Bary	potato
Alternaria solani I. et Gr.	potato
Cercospora beticola Sacc.	sugarbeet
Septoria apii Chester	celery
Alternaria porri f. *dauci* Neerg.	carrot
Fusicladium effusum Wint.	pecan nut
Gnomonia spp.	pecan nut
Mycosphaerella caryigena	pecan nut
Fusarium spp.	hops
Verticillium alboatrum Reinke et Berth.	hops
Pseudoperonospora humili Wils.	hops
Algae	rice
Cercospora spp.	peanut
Colletotrichum coffeanum Noack	coffee
Hemileia vastatrix Berk. et Br.	coffee
Phytophthora palmivora Butl.	cacao

The phytotoxicity of triphenyltin compounds is for many plants too
high to allow a wider application.

β) *Use of triphenyltin compounds on potatoes.*—Triphenyltin compounds are mainly used to control leaf and tuber blight in potatoes. They
have also proved effective against alternaria (early blight) and, to a
certain extent, against powdery mildew.

αα) *Late blight:*

Late blight (leaf blight) caused by *Phytophthora infestans* de Bary is especially capable of epidemic appearance. The reason is a short incubation period of only 3 to 5 days, their great capability for reproduction, and the fact that the reproductive organs of the microorganism are immediately infectious in the presence of water. Under favorable weather conditions, a *Phytophthora* epidemic can proceed in a relatively short period of time. Within a few days, it can cause complete defoliation which leads to considerable yield loss mainly during the period of major growth.

BAUMANN (1957 and 1958) was first to report application of triphenyltin compounds against leaf blight. If used in protective application at the right time (just like other fungicides), they are about 10 times more efficient than copper oxychloride and 5 times more effective than zineb or maneb (HAERTEL 1962, BYRDY *et al.* 1966 a, also BAUMANN 1957, KREXNER and WENZL 1958). Compared with untreated crops seriously affected by leaf blight those treated show little defoliation and much higher yields.

The normal application rates of triphenyltin compounds are, according to attack conditions, 200 to 360 g/ha, at most 480 g/ha each time (HAERTEL 1962). The *Biologische Bundesanstalt,* Braunschweig, recommended increasing amounts with increasing threat of infection: 240 g for the first treatment, 300 g for the second, from then on 360 g of triphenyltin compound/ha in 400 to 600 L of spray liquid. This recommendation has been followed widely in Germany. However, in especially threatened areas spraying with 360 g/ha or more each time can be more advantageous (HOLMES and STOREY 1962).

The number of treatments depends on local weather and climatic conditions; usually there are 3 to 5 applications; however, in areas with high humidity, frequent precipitation, and mainly cool weather with temperatures between 13 and 18°C, 10 or more treatments may be necessary.

Since potato plants are more sensitive towards triphenyltin compounds during early growth, spraying should be done after flowering, if possible. However, the first treatment should be done before the attack of the fungus can be noticed visually (HUTCHINSON 1974 and 1975).

BAUMANN (1957 and 1958), for a series of 8 experiments, reported good effects and significantly higher yields compared with untreated fields and other fungicides after 3 applications with 360 g of triphenyltin acetate or chloride in 600 L of water (Table LV). Increasing the amount of triphenyltin acetate to 600 g/ha did not bring any advantage. An amount of 600 g of triphenyltin chloride/ha was phytotoxic, which was confirmed by lower crop yields.

In a year with late leaf blight appearance of medium degree, KREXNER and WENZL (1958) applied 360 g of triphenyltin acetate in 550 L of water 4 times. The crop yield increased by 19% compared with untreated controls, by 7.75% compared with copper oxychloride, and by 7.5% compared with manganous ethylenebisdithiocarbamate ("Maneb").

Table LV. *Leaf blight and tuber yield in experiments 1955–57*
(after BAUMANN 1958).

Formulation[a] (kg/ha)	Test no. 1	2	3	4	5	6	7	8	Yield (average)	P (%)	Leaf rot
Co –	100	100	100	100	100	100	100	100	100	—	4.4
Cu 4.0–6.0	101	107	122	117	109	112	122	132	117	0.1	3.0
Zn 1.8	105	103	103	115	106	108	124	—	109	1	2.8
Sn-Ac 1.8	104	111	132	128	115	113	125	164	124	0.1	2.5
Sn-Ac 3.0	116	118	135	130	113	114	125	146	125	0.1	2.1
Sn-Cl 1.8							126	172		1	
Sn-Cl 3.0							115	155		1	

[a] Co = control, untreated = 100, Cu = copper oxychloride (50% W.P.), Zn = zinc ethylenebisdithiocarbamate (78% W.P.), Sn-Ac = triphenyltin acetate (20% W.P.), Sn-Cl = triphenyltin chloride (20% W.P.).

During a severe blight attack, THURSTON et al. (1960) conducted experiments in Columbia, spraying 12 times in weekly intervals with 420 g of triphenyltin acetate in 1,180 L of water/ha. The increase in crop yields was 149% compared with the control field. The comparative preparations maneb and Bordeaux mixture were less effective (Table LVI).

Table LVI. *Effect of protective spraying on defoliation and yield of potatoes* (excerpt from THURSTON et al. 1960).

Fungicide and rate of usage	% Defoliation, days after first application 48 days	68 days	Order of Disease control	Safety to potato	Yield[a] (%)
Untreated	69.0	65.0	5	—	100
Bordeaux mixture 1,200 L/ha 3.6–3.6–380	12.0	12.0	4	safe	209
Folpet[b], 1.78 kg/ha	10.5	10.5	3	safe	230
Maneb[c], 1.68 kg/ha	12.0	6.0	2	safe	237
Triphenyltin acetate, 0.425 kg/ha	6.5	7.5	1	?	249

[a] Untreated = 100.
[b] N-(trichloromethanesulphenyl) phthalimide.
[c] Manganous ethylenebisdithiocarbamate.

An experiment conducted in England showed also favorable results (LLOYD et al. 1962); 675 g of triphenyltin acetate/ha in 1,125 L of water, applied 3 times, delayed the onset of blight (Fig. 14) and brought considerably higher crop yields.

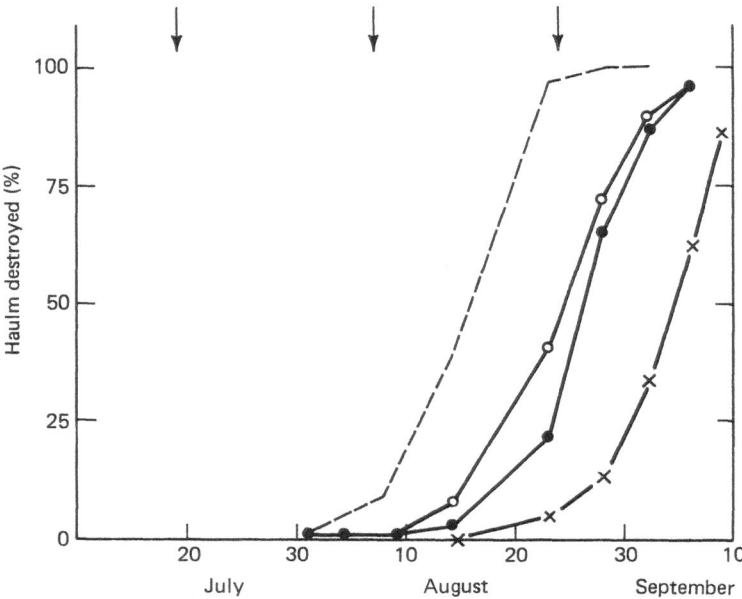

Fig. 14. Development of blight, measured as % foliage destroyed, on unsprayed
potatoes (----), and crops sprayed with zinc (O–O), copper (●–●)
and tin (×–×) fungicides (after LLOYD et al. 1962); arrows are spraying
dates.

The tin compound was distinctly superior to the other formulations
copper oxychloride and zinc ethylenebisdithiocarbamate (Table LVII).

Table LVII. *Effect of tin, copper, and zinc fungicides on the yield of potatoes in a
severe blight attack* (after LLOYD et al. 1962).

Fungicide used	Concentration[a] (% w/v a.i.)	Gross yield washed tubers (t/ha)
Unsprayed controls	—	33.75
Zinc ethylenebisdithiocarbamate	0.13	36.75
Copper oxychloride	0.25	37.5
Triphenyltin acetate	0.06	40.75

[a] Spray application: 1,125 L/ha.

After applying triphenyltin acetate, HOLMES and STOREY (1962) obtained increased yields which were especially apparent in years with
humid weather and strong attacks. After spraying 6 to 8 times at weekly
intervals (940 L/ha with 690 g of active ingredient) they obtained the
yields shown in Table LVIII.

Table LVIII. *Effect of protective sprayings on total yield at lifting 1959–1961* (after HOLMES and STOREY 1962).

| Year | Total yield (tons/ha) | | | | | Significant difference (p = 0.05) |
	Un-sprayed	Copper	Zineb	Maneb	Fentin acetate	
1959 Mean	43.0	40.75 −2.25[a]	43.50 +0.5[a]	44.25 +1.25[a]	43.25 +0.25[a]	2.25
1960 Mean	38.5	39.75 +1.25[a]	44.0 +5.5[a]	43.5 +5.0[a]	44.75 +6.25[a]	3.0
1961 Mean	40.5	43.75 +3.25[a]	44.5 +3.5[a]	43.5 +3.0[a]	45.25 +4.75[a]	2.5

[a] Gain (or loss) significant at p = 0.05.

After a severe blight attack ZAHIR (1968) sprayed the variety "Bintje" which is susceptible to *Phytophthora*, with increasing dosages of a combined formulation of triphenyltin acetate (60%) and manganous ethylenebisdithiocarbamate (20%). The result is shown in Table LIX. The dosages were 400 g/ha for the first time, then 500 g, and in the following 3 treatments 600 g each in 600 L of water. CAČA and JAŠEK (1975) also obtained excellent results by application of triphenyltin acetate and maneb or Liro-manzeb.

Table LIX. *Leaf blight-control in potatoes (variety Bintje)* (after ZAHIR 1968).

Fungicide	Amount (kg/ha)	Blight attack (%)	Yield (t/ha)	Increased yield (t/ha)
Untreated	—	100[a]	34.9	—
Copper oxychloride (50% Cu, W.P.)	2.0–3.0	50	36.7	1.8
Propineb[b] (80% W.P.)	1.8	14	41.7	6.8
Maneb (78% W.P.)	1.8	10	41.8	6.9
Triphenyltin acetate + maneb (60% + 20% W.P.)	0.4–0.6	3	42.8	7.9

[a] Complete defoliation.
[b] Zinc propylenebisdithiocarbamate.

In Ceylon, where weather conditions are extremely humid, the control of blight had to be started 25 to 35 days after planting the seed potatoes (SENEVIRATNE 1970). Triphenyltin compounds did control the disease under these conditions; however, the foliage was damaged. Higher yields could be obtained with other fungicide compounds. Therefore, a combined treatment with dithiocarbamates and triphenyltin compounds was considered.

Earlier, CAMPACCI and SANTOS (1959) had made similar observations with triphenyltin compounds in Brazil. They tried to lower the phytotoxicity by alternate sprayings with maneb and triphenyltin acetate (CAMPACCI 1960, CAMPACCI and SANTOS 1962). A combination of 94 g of triphenyltin with 1.25 kg of maneb was also well tolerated by the plants and yielded about the same amount as with alternate spraying (Table LX.) Triphenyltin acetate alone was effective against the blight; however, yields were lower.

Table LX. *Effect of protective spraying with different fungicides against Phytophthora on potatoes on yield and disease incident*[a] (excerpt from CAMPACCI 1960).

Fungicide used and dosage	Disease (%)	Yield (kg/plot)	Safety to plants[b]
Untreated control	69.3	11.2	−
Copper oxychloride, 3.0 kg/ha	16.8	14.2	1
Zineb, 2.25 kg/ha	13.9	15.7	1
Triphenyltin acetate, 0.25 kg/ha	9.6	16.3	3
Maneb, 1.75 kg/ha	9.1	18.3	1
Triphenyltin acetate, 0.094 kg/ha, + maneb, 1.25 kg/ha	8.7	19.8	1
Triphenyltin acetate, 0.25 kg/ha, alternating with maneb, 1.75 kg/ha	8.3	20.2	1

[a] Least significant difference at the 1% level: 4.1
Least significant difference at the 5% level: 3.1
[b] 1 = safe.

In the United States, MANZER and MERRIAM (1963) obtained 11.6% higher yields compared with untreated plots when they sprayed 156 g/ha of triphenyltin hydroxide against *Phytophthora* blight (in spite of low, but distinct phytotoxicity; compare section V g)1 β).

Experiments with warm and cold aerosols of different triphenyltin compounds (acetate, hydroxide, chloride, oxide, sulfide, oleate, carboxylate, benzoate) were conducted by KOULA (1971 a and b) in the greenhouse and in the field. The high fungitoxic effect of acetate in the greenhouse could not be reproduced using the same application method in the field.

BYRDY *et al.* (1966 a) also tested the effectiveness of various triphenyltin compounds against blight in the greenhouse. Acetate, benzoate, and the methoxy compound were quite effective in concentrations of 0.1 to 0.2%, whereas a copper oxychloride formulation had to be used in 10 times higher concentration.

It is hard to understand why in these experiments triphenyltin hydroxide was considerably less effective than acetate and more phytotoxic.

Finally, some results of sprayings with less common triphenyltin compounds should be mentioned. McIntosh (1971) tested the effectiveness of bis-(triphenyltin) sulfide, but this compound was distinctly inferior to acetate. Good results were obtained with hexaphenyldistannoxane (Luijten 1960), also with triphenyl-(phenylthio)tin and with triphenyl-(p-chlorophenylthio)tin (McIntosh 1971). Experiments with the complex triphenyltin chloride/decyltriphenylphosphonium bromide in the greenhouse were also successful (McIntosh 1966). Further data about increased yields will be presented in the next section ("tuber blight").

$\beta\beta$) *Tuber blight:*

Tuber blight in potatoes is caused by the same fungus as leaf blight (*Phytophthora infestans* de Bary). Infection of the potato tubers occurs when zoosporangia and zoospores, washed from the leaves and stems by rain, seep through fine cracks into the ground and get to the young tubers (McIntosh 1965).

Whereas the early destruction of potato leaves by blight leads to considerable yield losses, the tuber blight results in much lower quality of the crop. When the infected potatoes are stored, further losses occur.

The tuber blight, often hardly detectable from the outside, is especially damaging to the seed potatoes, because each infected tuber can cause the primary infection of the whole field. The fungus quickly spreads out to neighboring plants, and within a short period of time it can grow to epidemic levels, threatening the whole stock of a large area.

The best protection against tuber blight is control of the blight in the leaves (Zeck 1957), but this method does not always work. Sometimes the results are quite the opposite, and after spraying with copper formulations or organic fungicides the tuber rot attack has been observed to be even more severe (Brase 1956, Baumann 1957 and 1958). Stolze (1957) explained this with a "creeping" blight, remaining after application of the fungicide (see also Baumann 1958). Biedermann and Mueller (1952) and Zeck (1956 and 1957) showed that copper formulations, even in unusually high dosages (more than 200 kg/ha) are inactivated in the ground by complex formation; Haertel (1963 c) confirmed this with another fungus (*Alternaria solani*) (see Table LXI).

Baumann (1957 and 1958) was first to point out that triphenyltin compounds, contrary to copper compounds and organic fungicides, are able to reduce the tuber blight significantly. There had been some indications that organotin compounds affect the *Phytophthora* zoosporangia and zoospores also in the ground (Holmes and Storey 1962), and Haertel 1963 c) showed with *Alternaria solani* that triphenyltin acetate retains its fungicidic effects even in loamy soil (Table LXI). Treatments of the ground with triphenyltin compounds, however, were only partly successful.

It should be noted, however, that the danger of tuber infection increases in spite of treatment with triphenyltin compounds if the time intervals, especially between last spraying and harvest, are too long and

Table LXI. *Influence of loam on the effects of copper oxychloride and triphenyltin acetate against Alternaria solani* (after HAERTEL 1963 c).

Fungicide	Concentration	Nutritive substratum[a]	
		Water	Loam + Wort agar
Untreated	—	+++	+++
Copper oxy chloride	100 ppm	0	+++
Copper oxy chloride	1,000 ppm	0	+++
Triphenyltin acetate	50 ppm	0	0

[a] 0 = no growth, +++ = uninhibited growth.

the foliage has not been destroyed early enough (BAUMANN 1958) (Fig. 15; see also Table LXII).

In an experiment by BAUMANN (1957), 3 sprayings with triphenyltin acetate (360 g or 600 g/ha in 600 L of water) during the growth period produced significantly higher yields and reduction in tuber blight (Table LXIII). The higher amount of organotin compound did not show any additional advantages.

Based on their experiments over several years, HOLMES and STOREY

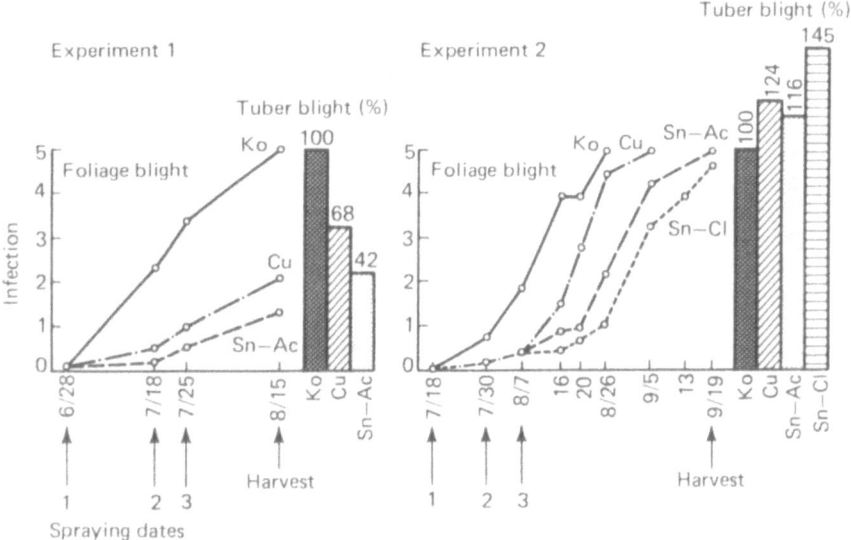

Fig. 15.   Increase of leaf blight in potatoes (variety Bona) up to harvest time and relation of tuber blight infection to harvest date (after BAUMANN 1958): experiment 1 = harvest 21 days after last spraying and experiment 2 = harvest 42 days after last spraying; Ko = untreated control, Cu = copper oxychloride, Sn-Ac = triphenyltin acetate, and Sn-Cl = triphenyltin chloride.

Table LXII. *Result of protective spraying against leaf and tuber blight with various fungicides (harvest date delayed and foliage not destroyed)* (excerpt from ANDRÉN and OLOFSSON 1962).

Fungicide and amount	Total yield t/ha	Total yield rel. (%)	Tuber blight (%)	Leaf blight (%) on 8/18	Leaf blight (%) on 8/24	Leaf blight (%) on 8/28
Untreated	41.0	100	37	58	79	88
Copper oxy chloride (50% W.P.) 6.0 kg/ha	45.2	110	48	5	8	25
Maneb (78% W.P.) 2.5 kg/ha	48.0	117	56	4	4	10
Zineb (78% W.P.) 2.5 kg/ha	43.3	106	46	4	4	13
Triphenyltin acetate (20% W.P.) 0.5–0.6 kg/ha	45.6	111	36	6	15	19
Triphenyltin acetate (20% W.P.) 1.5 kg/ha	47.6	116	35	7	9	14
Triphenyltin acetate (20% W.P.) 1.5–1.8 kg/ha	45.9	112	43	6	9	16
Triphenyltin acetate + maneb (10% + 62.5% W.P.) 1.5–1.8 kg/ha	48.1	117	33	4	8	9

Table LXIII. *Effect of protective spraying on yield, leaf blight, and tuber blight in potatoes* (after BAUMANN 1957).

Fungicide and dosage	Rating[a]	Yield t/ha	Yield d (%)	Infected tubers (per 1,000) No.	Infected tubers (per 1,000) d (%)	Infected tubers (per 1,000) kg	Infected tubers (per 1,000) d (%)
Untreated	4.7	22.9	—	50	—	65	—
Copper oxy chloride (50%) 3.0 kg/ha	2.1	24.9	+8.7	45	−11.4	44	−32.1
Zineb 1.44 kg/ha	1.9	24.2	+5.7	55	+8.7	55	−15.8
Triphenyltin acetate 0.36 kg/ha	1.6	26.3	+14.8	27	−43.7	27	−57.8
Triphenyltin acetate 0.60 kg/ha	1.3	25.8	+12.7	27	−43.7	30	−54.0

[a] 5 = complete infection.

(1962) confirmed these results. Besides the increased yield already mentioned (Table LVIII), the percentage of infected tubers had gone down from 15.8 to 8.7% (average 1960), respectively, from 7.9 to 1.5% (average 1961). Copper oxy chloride, zineb, and maneb had almost no effect. The authors concluded, "Triphenyltin acetate was the only fungicide to have a direct effect on reducing tuber blight."

LLOYD *et al.* (1962) also obtained significant decrease of tuber blight in their experiments mentioned earlier (Table LVII). After treatment with triphenyltin acetate, the number of infected tubers went from 26.5 (untreated control) down to 15.7%. Treatment with copper oxy chloride, however, resulted in 27.5%, with zineb in 29.7% infected tubers.

Table LXIV sums up test results with 4 different triphenyltin compounds (after PIETERS 1962). It shows that acetate, oxalate, and hydroxide were quite effective against tuber blight, whereas the *p*-toluene sulfonamide compound was much inferior. Yield and percentage of uninfected tubers had increased significantly compared with untreated controls and others treated with a copper oxychloride formulation.

CETAS (1962 b and 1963) obtained, after spraying 168 to 338 g of triphenyltin hydroxide 4 to 5 times, increased yields in tubers without or with only very minor infection.

Good results were also reported by CALLBECK (1963) with 7 applications of 562 g of triphenyltin hydroxide in 380 L of water/ha. The yield had increased considerably, and the percentage of infected tubers was reduced from 34.6 (untreated) to 7.6%.

In experiments by OLOFSSON and ANDRÉN (1963), potatoes were treated 3 times with increasing amounts of triphenyltin acetate (240 to 360 g of active ingredient in 750 L of water/ha). High yields with low infection of the tubers were obtained, while comparative formulations had distinctly lower effects (Table LXV).

Figure 16 shows median results from 98 experiments with different fungicides under varying conditions (after HAERTEL 1963 c); triphenyltin formulations proved to be the most effective.

Unusually high increases in yield, together with lowered infection rate of tuber blight, were obtained in a year with exceptionally humid weather and extremely strong *Phytophthora* infection (FRASELLE 1967). Eight to 10 protective sprayings with triphenyltin compounds proved to be more effective than treatment with comparative formulations (copper oxy chloride and maneb) (Table LXVI).

JARVIS *et al.* (1967) reported excellent effects of triphenyltin acetate and hydroxide applied in very high dosages of about 840 g/ha during experiments in 1962 and 1965. Apart from a significant yield increase, the tuber blight was reduced. They recommended application of tin compounds, especially with potato varieties that are sensitive to tuber blight.

By application of triphenyltin acetate + maneb (0.1 to 0.2% of the 60% product) or triphenyltin hydroxide (0.1 to 0.2% of the 47.5% product), RUDKIEWICZ (1973) reduced tuber blight from 10.7% (untreated) to 1.4 to 4.5%. Good results, especially with triphenyltin acetate + maneb, were reported also by SOSNA and RUDNA (1972), MACKIEWICZ and SOSNA (1974), and CAČA and JAŠEK (1975). CALLBECK (1974) mentioned some phytotoxicity for potatoes and recommended re-examination.

Soil treatment experiments were first conducted by BRUIN (1961). During the first yr, the application of triphenyltin acetate resulted in

Table LXIV. *Effect of protective spraying against late blight on yield and tuber blight; variety: Bintje* (after PIETERS 1962).

Treatment	Spraying dates and dosages (kg/ha)							Seasonal averages leaf infestation[a]		Average injury figures[c]		Average yield of 50 plants			
	24/6	3/7	15/7	25/7	6/8	18/8	27/8		Transf.[b]	18/8	Transf.[b]	Healthy in kg	Copper oxychl. =100	% diseased	Transf.[b]
1. Untreated								2.4	2.75	10	5.74	40.2	81	19.9	26.41
2. Copper oxychloride 50%	7	8	9	10	10	10	10	5.6	4.29	8.1	5.15	49.7	=100	8.6	16.80
3. TPT-acetate 20% w.p.	1.8	2.0	2.2	2.4	2.4	2.4	2.4	7.4	4.94	6.6	4.64	53.1	107	2.6	8.75
4. TPT-oxalate 20% w.p.	1.8	2.0	2.2	2.4	2.4	2.4	2.4	6.1	4.46	8.3	5.20	53.4	107	4.9	12.67
5. TPT-hydroxide 20% w.p.	1.8	2.0	2.2	2.4	2.4	2.4	2.4	7.1	4.84	9.4	5.55	52.4	105	3.1	10.03
6. TPT-p-toluene sulfonamide 20% w.p.	1.8	2.0	2.2	2.4	2.4	2.4	2.4	4.4	3.76	9.6	5.61	49.4	99	11.9	19.95
P 0.05									0.49		0.31	4.7			4.64
P 0.01									0.74		0.47	7.2			7.03

[a] 10 = free from Phytophthora; 0 = completely dead.

[b] Transformation applied: lg sin $\sqrt{\dfrac{x}{10}}$.

[c] 10 = no injury.

Table LXV. *Effect of different fungicides on leaf and tuber blight and on the yield of potatoes* (excerpt from OLOFSSON and ANDRÉN 1963)

Fungicide and dosage	Yield[a]		Tuber blight (%)	Foliage blight on		
	t/ha	Rel.		2/21	8/28	9/4
Untreated	25.2	100	30.0	3.0	49	81
Bordeaux mixture (1.2%)	36.0	143	6.8	0	1.4	1.8
Maneb (80% W.P.) 2.5 kg/ha	33.5	133	16.2	0	0.3	1.4
Zineb (78% W.P.) 2.5 kg/ha	28.4	113	26.8	0.1	1.5	3.5
Triphenyltin acetate (60% W.P.) 0.4–0.6 kg/ha[b]	38.3	152	3.5	0	0.1	0.8
Triphenyltin acetate + maneb (60% + 20% W.P.) 0.4–0.6 kg/ha	39.6	157	3.4	0	0	1.0

[a] Without infected tubers.
[b] Dosage increased gradually.

Table LXVI. *Tuber blight experiments with different fungicides* (after FRASELLE 1967).

Fungicide and dosage	Experiments in Gembloux			Experiments in Sauvenière	
	Healthy tubers	% Tuber blight		Healthy tubers	% Tuber blight
		At lifting time	After 1 mon storage		At lifting time
Untreated	100	30.6	41.1	100	40.1
Copper oxy chloride (50% Cu) 8 × 2.5 kg/ha	196.8	14.6	19.7	417	15.9
Copper oxy chloride (50% Cu) 8 × 1.25 kg/ha	206.8	14.9	21.6	320	24.4
Copper oxy chloride (50% Cu) 8 × 1.25 kg/ha	185.1	17.8	26.4	271	31.5
Maneb 8 × 1.5 kg/ha	236.6	9.0	11.9	481	21.9
Triphenyltin acetate 2 × 0.3 kg/ha, 8 × 0.35 kg/ha	224.7	7.2	10.5	458	15.2
Triphenyltin hydroxide 1 × 0.3 kg/ha, 8 × 0.36 kg/ha	239.1	0.6	2.5	604	5.3
Triphenyltin acetate + maneb 8 × 0.36 + 0.12 kg/ha	233.8	3.7	9.3	603	7.9

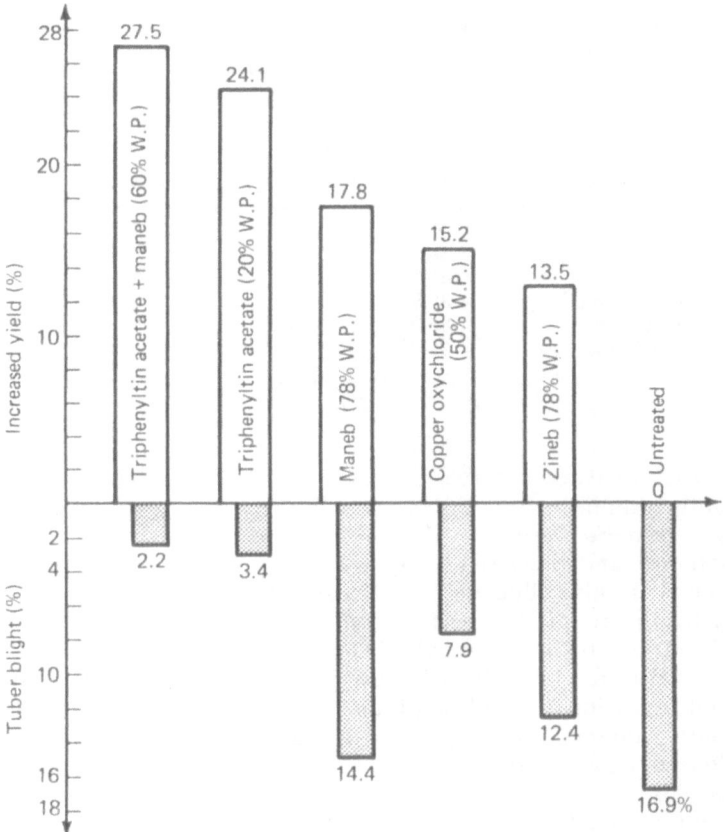

Fig. 16. Effect of different fungicides on leaf and tuber blight in potatoes; medians from 98 experiments (after HAERTEL 1963 c).

reduced tuber blight infection. However, out of 2 experiments conducted the following yr, only 1 was successful.

By treatments of the soil with triphenyltin compounds, McINTOSH (1965) obtained significant reduction of tuber blight infection (Table LXVII). However, a second experiment on a larger area could not confirm this result (McINTOSH 1966).

Another experiment of treating soil with triphenyltin acetate combined with maneb (ANONYMOUS 1968) had positive results (Table LXVIII). The formulation was sprayed onto the ridge before the last earthing up. Later, the leaves were treated in the usual way with Propineb. As a result, the percentage of infected tubers was much lower

Table LXVII. *Effect of soil treatment on yield and tuber blight*
(after McINTOSH 1965).

Fungicide used	Total yield (t/ha) and treatment date		% of blighted tubers (w/w) at lifting time and treatment date	
	6/28–6/30	7/27–7/29	6/28–6/30	7/27–7/29
None	36.75	36.75	11.0	11.0
Tetrachloroisophthalonitrile	38.5	33.75	6.9	6.3
Tributyltin acetate	39.5	36.5	6.9	7.6
Bis-(triphenyltin)sulfide	36.5	34.25	9.9	11.3
Triphenyltin acetate	35.0	37.0	5.0	5.1
Triphenyltin chloride	37.25	39.5	5.3	3.4
	Least sig. dif. (5%): 5.75		Least sig. dif. (5%): 5.6	

than in the untreated controls and in normal leaf spraying without treatment of the ground.

γγ) *Alternaria disease:*

CAMPACCI and SANTOS (1959) were first to notice that triphenyltin compounds are also effective against early blight on potato (*Alternaria*), caused by the fungi *Alternaria solani* J. and Gr. and *Alternaria tenuis* Nees. Contrary to the late blight (*Phytophthora*), the fungus *Alternaria* needs a warm, dry climate with few rainfalls, as it prevails in many of the potato growing areas of South and Southeast Europe, North, Central, and South America, and in parts of India. During warm and dry years, the *Alternaria* disease becomes more important in Germany, too (BACHTHALER 1962).

Even though *Alternaria* blight does not have the economic importance of the *Phytophthora* blight, the damages caused are frequently underestimated. Often the disease, which starts very early, is not recognized

Table LXVIII. *Effect of soil treatment with triphenyltin acetate (followed by regular leaf treatment with Propineb) on tuber blight in potatoes*
(after STADER SAATZUCHT 1968).

Soil treatment	Foliage treatment	No. of tubers (total)	Wet rot		Dry rot	
			No.	%	No.	%
—	—	715	91	12.7	133	18.6
—	Propineb 1.8 kg/ha	797	66	8.3	123	15.4
Triphenyltin acetate + maneb 1.2 + 0.32 kg/ha	Propineb 1.8 kg/ha	941	9.5	1.0	50.5	5.7

in time and then masked by the later appearing *Phytophthora* infection. The *Alternaria* fungus attacks leaves and stems as well as tubers. The resulting tuber blight can be recognized by small, indented spots. The rot does not penetrate deeply into the tuber, since it is soon encased by a cork layer, but it causes diminished quality.

In the potato field, the *Alternaria* attack usually sets in before the blossom. Therefore, control has to start early, especially if the weather is dry and warm (BACHTHALER 1962 and 1964). On the other hand, at this stage of plant development triphenyltin compounds can be phytotoxic, which leads to lower yields in spite of good fungicide effects. ABDEL-RAHMAN (1977) recommended weekly applications of small amounts of triphenyltin hydroxide in preference to higher amounts in intervals of 2 to 4 wk.

To control *Alternaria,* 240 to 360 g of triphenyltin compound/ha in 400 to 600 L of water is used.

In experiments by CAMPACCI and SANTOS (1959), 12 doses of 0.03% triphenyltin acetate suspension in 1,180 L of water ($=ca.$ 360 g/ha) were used at intervals of 11 or 7 days. The tin compound, of all the preparations, was the most effective against *Alternaria* blight, but because of its phytotoxic properties it did not produce the highest yields (Table LXIX).

Good fungicidic effects of triphenyltin acetate were confirmed in further experiments (CAMPACCI 1960, CAMPACCI and SANTOS 1962); here,

Table LXIX. *Effect of fungicides on Alternaria blight of potatoes[a]* (after CAMPACCI and SANTOS 1959, CAMPACCI 1960).

Fungicide used and concentration	Field I			Field II		
	Yield (kg)	Dis-ease (%)	Safety to potatoes	Yield (kg)	Dis-ease (%)	Safety to potatoes
Untreated	65.74	99.0	—	114.29	95.4	—
Copper oxychloride (50% W.P.) 0.5% = 0.25% a.i.	86.76	94.7	1	119.59	75.2	1
Bordeaux mixture (3.6–3.6–380) 1%	94.63	95.3	2	123.82	75.8	2
Triphenyltin acetate (20% W.P.) 0.15% = 0.03% a.i.	101.07	79.0	3	117.34	30.3	3
Zineb (78% W.P.) 0.25% = 0.195% a.i.	102.62	91.6	1	125.84	79.4	1
Maneb (78% W.P.) 0.2% = 0.15% a.i.	114.02	86.8	1	141.26	83.4	1

[a] 1,180 L/ha spray solution. Figures rounded off.

phytotoxic side effects were eliminated by combination with maneb or by alternating sprayings, starting with maneb (Table LXX).

ANDRADE (1960 a) conducted comparative experiments with triphenyltin acetate and numerous other formulations. Most successful was alternating spraying with triphenyltin acetate (20%, 600 g/380 L) and zineb (680 g/380 L).

Table LXX. *Effect of fungicides on yield and disease incidence in Alternaria blight on potatoes* (after CAMPACCI and SANTOS 1962).

	Variety					
	Aquila		Heimkehr		Marita	
Fungicide used and amount	Yield (kg)	Dis- ease (%)	Yield (kg)	Dis- ease (%)	Yield (kg)	Dis- ease (%)
Untreated	72.0	69.3	117.7	64.8	143.4	32.9
Copper oxy chloride 3.0 kg/ha a.i.	97.3	16.8	179.0	15.2	150.2	10.4
Maneb 1.70 kg/ha a.i.	120.0	9.1	183.6	3.5	159.4	7.7
Triphenyltin acetate 0.25 kg/ha a.i.	99.9	9.6	184.8	9.1	155.8	3.7
Triphenyltin acetate alternate maneb 0.25 + 1.70 kg/ha a.i.	125.7	8.3	207.3	4.4	175.8	5.4
Triphenyltin acetate + maneb 0.094 + 1.25 kg/ha a.i.	131.5	8.7	231.5	2.9	179.8	7.7

BACHTHALER (1964) was able to control *Alternaria* attack in potatoes during years with mainly warm, dry weather by using triphenyltin acetate alone and in formulations combined with maneb (Table LXXI). However, in these experiments the blight had progressed quite far at the beginning of the treatment, which prevented optimal results.

Early use of triphenyltin compounds produced much better results (BACHTHALER 1967). The total as well as the starch yield were better than average during years with strong *Alternaria* infection (Table LXXII).

In India, ADDY and DASH (1966) tested a combined formulation of triphenyltin acetate and maneb during a year with conditions favorable for infection. They applied 600 g of triphenyltin acetate and 200 g of maneb in 625 L of water/ha (5 treatments at intervals of 14 days each). The combined preparation was much better and resulted in higher yields than the comparative fungicides (Table LXXIII).

Under similar conditions of climate and *Alternaria* attack in India, DUTT *et al.* (1972) obtained good results against *Alternaria* blight, using

Table LXXI. *Effect of fungicides on yield and starch content in potatoes during Alternaria control 1961–1963* (excerpt from BACHTHALER 1964).

Fungicide used and dosage	Tuber yield[a]			Starch yield[a]		
	1961	1962	1963	1961	1962	1963
Untreated	100	100	100	100	100	100
Maneb 1.6 kg/ha a.i.	103.3	91.8	116.2	106.2	98.2	101.2
Triphenyltin acetate 0.3 kg/ha a.i.	114.4	—	—	106.2	—	—
Triphenyltin acetate + maneb, 0.1 + 1.5 kg/ha a.i.	—	95.1	113.6	—	100.6	104.9
Triphenyltin acetate + maneb, 0.36 + 0.12 kg/ha a.i.	—	101.8	125.6	—	97.0	101.2

[a] In % of untreated; mean values.

a "low volume sprayer" with 0.36 kg triphenyltin acetate/ha. The yield was increased from 12.6 t/ha (untreated) to 17.6 t/ha.

PLAMADEALA (1972) also recommended triphenyltin compounds against *Alternaria* blight in Romania. Concentrations used were 0.03% active ingredient combined with maneb and zineb (the latter two in higher amounts of 0.18 and 0.32%, respectively.

δδ) *Powdery mildew:*

Finally, it should be mentioned that triphenyltin compounds were found to be effective against powdery mildew in potatoes (VENTURA and HERVÉ 1962). After treatment with triphenyltin acetate, the yield from heavily infected fields increased significantly.

γ) *Use of triphenyltin compounds on sugar beets.*—Triphenyltin compounds are used to control leaf spot on sugar beets (caused by *Cercospora beticula* Sacc.). They also diminish the infection of seeds. Further, certain effects were noticed against powdery mildew in sugar beets caused by *Erysiphe betae* Van., *Erysiphe polygoni DC*, or *Erysiphe communis*.

Table LXXII. *Effect of fungicides on tuber and starch yield in potatoes during Alternaria control with 8 varieties of potatoes 1964–1966* (excerpt from BACHTHALER 1967).

Fungicide and dosage	Tuber yield[a] (t/ha)	Starch yield[a] (t/ha)
Untreated	40.4	6.67
Zineb 3.4 kg/ha	36.4	6.04
Propineb 1.4 kg/ha	42.9	7.09
Triphenyltin acetate + maneb 0.30–0.36 kg/ha	44.1	6.69
Triphenyltin chloride 0.60 kg/ha	51.8	8.36

[a] Median yields.

Table LXXIII. *Effect of fungicides on early blight severity in 1965–1966*
(after ADDY and DASH 1966).

Fungicide used and amount	Leaf area diseased (%)[a]			Av. yield (kg/plot)	Increased yield over untreated (%)
	Jan. 15	Jan. 30	Feb. 14		
Untreated	7.17 (15.52)	22.27 (28.15)	40.74 (39.66)	10.00	—
Bordeaux mixture 2.25–2.25–225	4.6 (12.37)	11.01 (19.49)	18.85 (25.73)	10.76	7.6
Zineb (78% W.P.) 2.5 kg/ha	2.35 (8.80)	7.27 (15.64)	15.37 (23.08)	11.79	17.9
Triphenyltin acetate (60% W.P.) + maneb 0.5 kg/ha	1.11 (6.09)	5.57 (13.65)	12.90 (21.03)	12.62	26.2
C.D. for ang. val. at 5%:	0.665	0.446	0.667		
C.D. for yield at 5%				1.000	

[a] Figures within parentheses indicate angular values.

αα) *Leaf spot:*

BAUMANN (1957 and 1958) was first to describe the effects of triphenyltin acetate against *Cercospora* leaf spot. Considerably higher yields in beets, leaves, and sugar were obtained, and a marked decrease in the content of harmful nitrogen and the percentages of soluble ash was noticed (see BAUMANN 1958, KOCH et al. 1959, 1960, and 1961, GUTSHTEIN 1963, KISS and HETZER 1967). Of course, the strongest effects were visible where plants were severely attacked; however, even with medium infections the yields could be improved considerably (BAUMANN 1957). In a light infestation during a dry year, the effectiveness of triphenyltin acetate (400 or 600 g/ha) was only similar to copper oxy chloride (MARIĆ and ČAMPRAG 1959). In fields without any infection, even repeated sprayings with triphenyltin acetate did not result in higher yields. The harvest was in the average range of untreated plants.

The disease can still be controlled some time after infection. Contrary to copper oxy chloride and to other contact fungicides, triphenyltin compounds show certain penetration action and, therefore, also limited curative action (DARPOUX et al. 1966, RAPPARINI and BENEVELLI 1970) (Table LXXIV). But this effect can be used successfully only during 3 to 4 days after infection. In field tests the curative effects of the acetate were diminished considerably 4 days after incubation (VETTER and VOELKER 1965) (Table LXXV).

It follows that the earlier the control of *Cercospora* leaf spot begins the more efficient it will be (SOLEL 1970 a). It should start no later than when the first spots of the infection become visible on the leaves. With

Table LXXIV. *Curative effects of triphenyltin acetate* (after DARPOUX *et al.* 1966).

Delay of treatment after inoculation	Spots on leaves (av.)
Untreated	14.8
24 hr	3.4
2 days	1.4
3 days	4.3
5 days	13.0
7 days	14.7
8 days	12.9

Table LXXV. *Curative effects of triphenyltin acetate in the field[a]* (after VETTER and VOELKER 1965).

No.	Date of treat-ment	Evaluation of infection[b] on							
		7/20	7/7	8/5	8/12	8/20	8/25	9/17	10/6
1[c]	7/9	0.0	0.0	0.0	0.0	0.5	0.5	1.0	2.5
2	7/9	0.0	0.0	0.0	0.0	0.0	0.0	1.0	2.8
3	7/13	0.0	0.5	1.0	1.0	1.0	1.5	2.5	4.2
4	7/20	0.0	0.8	2.0	2.5	2.5	3.0	3.0	5.0
5	7/27	0.0	1.0	2.3	2.5	2.8	3.0	3.5	5.8
6	8/5	0.0	0.8	2.0	2.5	2.8	3.2	4.0	6.0
7	8/12	0.0	1.0	2.3	3.0	3.0	3.5	4.2	7.0
8	8/20	0.0	1.0	2.3	3.0	3.2	4.0	5.0	8.5

[a] Artificial infection with *Cercospora beticola* on 7/9; single treatment with 0.03 kg of triphenyltin acetate/ha.
[b] 0 = no damage, 9 = total damage.
[c] Not infected.

optimal selection of spraying dates, triphenyltin compounds should be superior to all other fungicides. However, visible results can be obtained even after slightly delayed application (DONÀ DALLE ROSE 1958). A warning service has been proposed which would be able to predict the first optimal spraying date at least 10 days in advance (WELLNER and VETTER 1962).

When wetting and adhering agents are employed (for example Triton B-1956), even the relatively hydrophobic triphenyltin hydroxide can form a film of active ingredient on the surface of sugarbeet leaves. This film is not destroyed by repeated rain (STALLKNECHT and CALPOUZOS 1968). After 10, 30, and 90 mm of precipitation, the fungicide effects of the film were found not to be diminished (KOVÀCS 1962; Fig. 17). FOSCHI and RAPPARINI (1963) also confirmed the adhesiveness of triphenyltin compounds against rain.

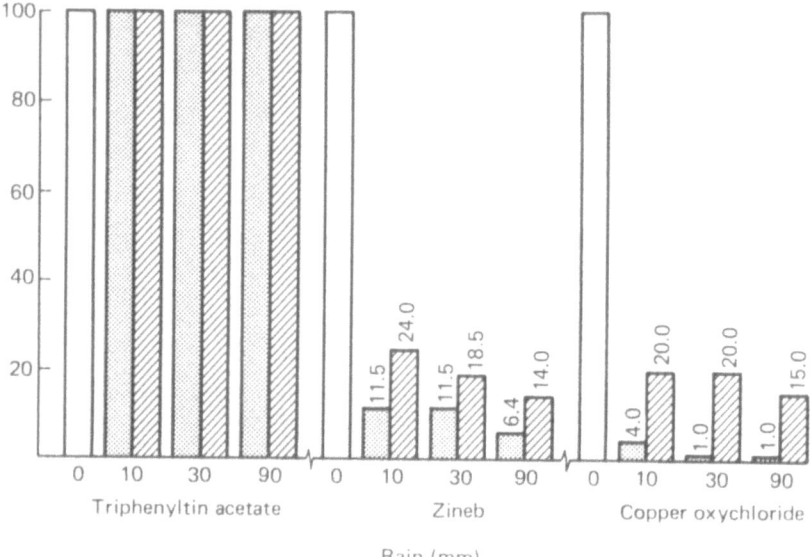

Fig. 17. Adhesion of different fungicides after rain (excerpt from KovÀcs 1962): □ = without rain, ■ = smooth surface, and ▨ = rough surface.

It should be noticed that *Cercospora*-strains do not seem to develop resistance against triphenyltin compounds as is the case with organic systemic fungicides (VAN DER KERK 1975). Strains that are completely resistant against benzimidazole and thiophanate fungicides can be effectively controlled by triphenyltin compounds at normal rates just as sensitive strains (GEORGOPOULOS and DOVAS 1973). Alternating or combined sprayings of benzimidazole or thiophanate fungicides with triphenyltin compounds are of no major advantage (DOVAS *et al.* 1976).

Even when combined with triphenyltin compounds, each application of benzimidazole-formulations causes the number of resistant *Cercospora* strains to increase selectively until total resistance to benzimidazole is reached (Fig. 18). A benzimidazole-resistant *Cercospora*-population can be controlled effectively with triphenyltin compounds. However, the resistance to benzimidazole cannot be eliminated any more, even with repeated application of triphenyltin compounds (DOVAS *et al.* 1976).

To control *Cercospora*-leaf spot in sugarbeets, triphenyltin compounds are applied in amounts and concentrations similar to those for potatoes. Depending on climate and attack conditions, 2 to 4, in severe cases 6 sprayings are necessary with amounts of 240 to 360 g of triphenyltin compound/ha in 200 to 600 L of water. If applied by air, 360 g of active ingredient/ha in 30 L of water produces the same film on the plants and an equal degree of control as application of the same amount of active

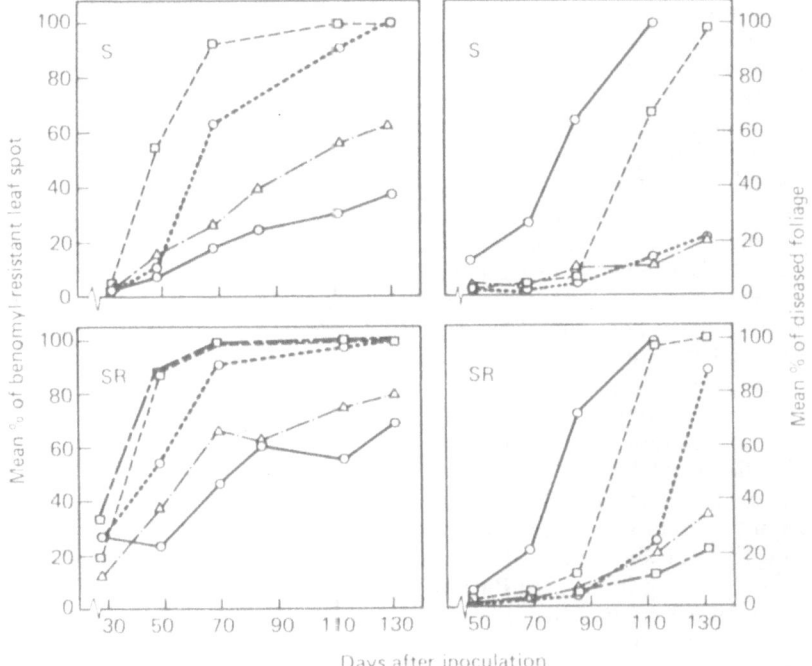

Fig. 18.   Effect of fungicide applications on the frequency of benomyl resistant strains of *Cercospora beticola* (left) and on the severity of disease (right) in plots artificially inoculated with benomyl-sensitive (S) or a mixture (SR) of sensitive and resistant conidia (after Dovas *et al.* 1976): □---□ = benomyl, △---△ = triphenyltin hydroxide, ○······○ = alternation of benomyl and triphenyltin hydroxide, □-×-×-□ = tank mix of 300 g triphenyltin hydroxide and 150 g benomyl/ha, and ○——○ = control.

ingredient in 50 or 100 L of water. Therefore, efficient spraying equipment provided, the results depend entirely on the amount of active ingredient/area unit (Hoesslin 1960; Anonymous, test report *Staatsgut Pik "Belje"* 1966 a). With 350 and 700 g of triphenyltin hydroxide/ha in 95 L of water, Wysong *et al.* (1968) obtained the same or even better results than if the same amounts of active ingredient had been suspended in 950 L of water. In any case, the tin compound was more efficient than the comparative compound tested.

In European areas with severe infestations (Greece, Italy, Yugoslavia, Hungary), the usual application rate was 300 to 360 g of triphenyltin compound/ha (Bongiovanni 1965, 1969 a and b, Kiss and Hetzer 1967, Schloesser *et al.* 1957, Picco 1957, Toderi and Viroli 1970). In Austria, however, 280 g of triphenyltin acetate/ha, applied 2 times, was usually sufficient (KREXNER 1960). In the United States, the recommendation for severe *Cercospora* attack is 4 applications of 168 g of triphenyltin compound in 450 L of water/ha; 675 g/ha did not show any significant advantages, and 2 treatments were not sufficient (FINKNER *et al.* 1966). In addition, triphenyltin compounds seem to have good effects on *Rhizoctonia* root rot and on army worm infection.

Similar amounts were used by CARLSON (1966) during his 4-year experiments in South Dakota. At 338 g/ha, triphenyltin hydroxide was superior to all other tested fungicides.

SCHNEIDER (1968), also in the United States, obtained very good results with 280 g of triphenyltin hydroxide. FROYD *et al.* (1967) used triphenyltin hydroxide and chloride (213 or 426 g/ha each) and triphenyltin acetate (340 g/ha) at spraying intervals of 14 and 28 days.

PAULUS *et al.* (1970 and 1971) came to the conclusion that triphenyltin compounds, applied at intervals of 11 and 22 days, are among the most effective fungicides for severe attacks.

Earlier experiments had already established the effect of dosage and spraying sequence on the control of *Cercospora* leaf spot disease. Decreasing or increasing amounts of active ingredient during the growth period did not show any advantages compared with uniform amounts at regular spraying intervals (BAUMANN 1958). KOCH *et al.* (1960) explored the question whether the number of sprayings can be reduced with increased dosages and changed intervals. They concluded that 4 sprayings with 360 g of triphenyltin acetate/ha in fairly regular intervals is most efficient (Table LXXVI).

KISS and HETZER (1967) also reported that early and continued sprayings (3 × 360 g of triphenyltin acetate/ha at intervals of 3 wk) produce optimal results (Table LXXVII).

During years with low to medium strong *Cercospora*-infestation, BAUMANN (1957) was still able to obtain considerably higher yields of beets, leaves, and sugar (Table LXXVIII). In severe attacks, advancing the spraying intervals and using 4 × 360 g of triphenyltin acetate/ha resulted in higher sugar yields of 63%, whereas 4 × 600 g/ha produced higher yields of 79% (BAUMANN 1958) (Table LXXIX).

In an area of northern Italy with severe infection, SCHLOESSER *et al.* (1957) obtained, depending on harvest time, increases in sugarbeet yields of 4.3 to 8.1 t/ha with 4 × 360 g of triphenyltin acetate, and 4.7 to 10.0 t/ha with 4 × 600 g of active ingredient/ha. Besides volume, the sugar content of the beets (mainly during late harvest) had improved considerably (Table LXXX).

PICCO (1957) used 2 × 600 g of triphenyltin acetate/ha for his experiments. The yield in sugarbeets increased by 12.4%, the sugar content from 16.05% (untreated controls) to 17.9%.

KOCH *et al.* (1959, 1960, and 1961), using 3 × 360 g of triphenyltin acetate/ha, obtained a significant increase in sugar, accompanied by a distinct reduction of harmful nitrogen and soluble ash (Table LXXXI).

Table LXXVI. *Effect of experimental spraying with triphenyltin acetate against Cercospora in sugarbeets. Refining: Laboratorium der Zuckerfabrik Regensburg (after Koch et al. 1960).*

Spraying date and dosage	Leaf infestation[a] 10/25	Polarization	Harmful nitrogen (mg/100 g)	Soluble ash (%)	Purified sugar (dz/ha rel.)		Theor. sugar	Yields relative to untreated	
								Beet	Leaf
Untreated	3.7	15.2	44.0	0.62	73.3	100.0	100.0	100	100
7/15; 8/4; 8/17; 9/8; 1/8 kg/ha Brestan[b]	0.3	15.7	34.0	0.56	88.4	120.6	113.5	110	114
7/27–1.8 kg/ha Brestan and 9/8–4.0 kg/ha OB 21[b]	1.1	15.6	35.0	0.55	85.7	116.9	110.0	107	112
7/15–3.0 kg/ha Brestan[b]	3.5	16.0	40.0	0.57	84.9	115.8	110.4	105	107
7/15–1.2 kg/ha Brestan and 8/17–1.8 kg/ha Brestan[b]	1.6	16.1	36.0	0.54	86.0	117.3	100.4	103	110
7/15–1.2 kg/ha Brestan and 8/17–2.4 kg/ha Brestan[b]	1.2	15.8	38.0	0.57	83.1	113.4	107.9	103	113
8/4–2.4 kg/ha Brestan[b]	1.6	15.8	38.0	0.56	85.5	116.6	110.3	106	119
GD = 5%							9.4	8	10
GD = 1%							12.6	11	13
GD = 0.1%							16.7	14	17

[a] Infestation scale: 0 = nil and 5 = 100%.

[b] Brestan = triphenyltin acetate (20% W.P.) and OB = copper oxy chloride (50% Cu W.P.).

Table LXXVII. *Effect of application date of triphenyltin acetate against leaf spot in sugarbeets (Cercospora) on infestation, yield, polarization, percentage of soluble ash, and harmful nitrogen* (after Kiss and Hetzer 1967).

Date of application (0.6 kg triphenyltin acetate/ha)	Cerco-spora infesta-tion[a]	Leaf yield (rel.)	Beet yield (rel.)	Polari-zation (rel.)	Sugar yield (rel.)	Sol. ash (rel.)	Harm-ful nitro-gen (rel.)
Untreated	2.81	100	100	100	100	100	100
6/17; 7/8; 7/31; 8/22; 9/14	0.81	280.5	108.6	101.4	111.1	94.4	74.0
6/17; 7/8; 7/31	0.75	265.2	100.9	103.4	114.2	93.1	69.3
7/8; 7/31; 8/22	1.19	274.8	107.0	104.7	114.9	94.4	62.0
7/31; 8/22; 9/14	2.38	230.7	105.7	98.6	104.3	98.6	75.0

[a] Rating: 0 = nil and 5 = 100%.

KREXNER and WENZL (1958), who applied 360 g of triphenyltin acetate/ha 3 times, reported a yield increase in sugarbeets from 42.6 to 46.4 t/ha; since the sugar content was above that of the untreated beets, the yield of sugar had improved even more (from 8.0 to 9.0 t/ha).

On crops in northern Italy severely attacked by *Cercospora*, ALDRO-VANDI (1958) also obtained much higher yields in beets, sugar, and leaves by applying triphenyltin acetate. After 2 sprayings with 360 respectively 600 g/ha of active ingredient, the sugar yield had increased by 48 to 61%, depending on the amount of active ingredient and harvest time.

Experiments by FORSYTH and BROADWELL (1962), who sprayed 6 times with a 0.3% suspension of triphenyltin acetate, using Triton X-100 as wetting and dispersing agent, showed an increase of sugar contents in beets from 14.2% (untreated) to 16.0%.

FOSCHI and RAPPARINI (1962) also noticed higher yields in beets and sugar. After 3 applications of 400 g of triphenyltin acetate/ha at intervals of approximately 4 wk, the rate of infection was reduced from 44 (un-treated) to 20%; the yield in beets increased from 45.0 to 53.2 kg/plot, the amount of leaves from 7.3 to 10.0 kg, and the sugar yield from 7.43 to 9.51 kg/plot.

COZZANI et al. (1963) reported an increase in beets of 20.7% after applying 2 × 300 g of triphenyltin acetate/ha; in addition, the sugar content of the beets had improved by 12.4% compared with the untreated crops.

Varying the harvest dates after applying triphenyltin acetate, GUTSH-TEIN (1963) also found increased sugar content and noticeably reduced soluble ash besides higher yields in beets and leaves (Table LXXXII).

A comparison of different fungicides used against leaf spot showed triphenyltin compounds to be the most effective (CARLSON 1963). Three

Table LXXVIII. *Effect of triphenyltin acetate on yield with severe and light Cercospora infestation* (after BAUMANN 1957).

Fungicide	Active ingredient (kg/ha)	Yield in % of untreated in severe infestation					Yield in % of untreated in light infestation				
		Leaf infestation[a]	Beets	Leaf	Purified sugar	Polarization	Leaf infestation[a]	Beets	Leaf	Purified sugar	Polarization
Untreated	—	4.6	100	100	100	100	1.85	100	100	100	100
Copper oxychloride	2.0	3.8	121.9	130.4	128.0	102.1	0.7	104.3	100.4	105.0	100.1
Triphenyltin acetate	0.36	2.5	140.1	148.3	168.0	112.5	0.2	104.9	103.1	112.0	100.1
Triphenyltin acetate	0.6	2.0	156.7	197.7	191.0	114.1	—	—	—	—	—

[a] Infestation scale: 0 = nil and 5 = 100%.

Table LXXIX. *Yield of sugarbeets in severe Cercospora infestation with advanced spraying dates* (after Baumann 1958).

Fungicide	Active ingred. (kg/ha)	Leaf infes-tation[a]	Yield in % of untreated			Polari-zation (%)	Solu-ble ash con-tent	Harm-ful nitro-gen
			Beets	Leaf	Sugar			
Untreated	—	4.7	100	100	100	15.4	0.42	44.4
Copper oxy chloride	2.0	4.4	110	112	114	16.1	0.39	39.5
Triphenyltin acetate	0.36	2.7	138	136	163	17.6	0.39	36.0
Triphenyltin acetate	0.60	2.3	147	203	179	17.8	0.38	35.9
GD 5%			18	36				
GD 1%			27	54				
GD 0.1%			44	86				

[a] Infestation scale: 0 = nil and 5 = 100%.

Table LXXX. *Yields in protective spraying[a] against Cercospora in Agna/Italy 1956* (excerpt from Schloesser et al. 1957).

Fungicide and dosage	Yield (t/ha)			Sugar content of beets (%)	Soluble ash (%)	Harmful nitrogen (mg/ 100 g)
	Beets	Sugar	Leaves			
Early harvest 8/13						
Untreated	55.8	8.39	35.8	15.03	0.68	0.27
Copper oxy chloride (45–50%); 5.0 kg/ha	56.2	8.97	44.2	15.96	0.69	0.23
Triphenyltin acetate (20% W.P.) 1.8 kg/ha	60.1	9.69	50.2	16.13	0.71	0.27
Triphenyltin acetate (20% W.P.) 3.0 kg/ha	60.5	9.74	52.0	16.10	0.68	0.23
Late harvest 8/29						
Untreated	58.4	8.38	21.5	14.38	0.66	0.33
Copper oxy chloride (40–50%); 5.0 kg/ha	61.4	9.39	27.3	15.32	0.64	0.31
Triphenyltin acetate (20% W.P.) 1.8 kg/ha	66.5	10.90	36.3	16.61	0.65	0.31
Triphenyltin acetate (20% W.P.) 3.0 kg/ha	68.4	11.18	40.5	16.33	0.65	0.31

[a] Variety: X KW-Zucca (Z); spraying dates: 6/16, 7/2, 7/17, 8/2.

Table LXXXI. *Effect of triphenyltin compounds on yield, polarization, soluble ash, and harmful nitrogen in sugarbeets after protective sprayings against Cercospora infection (after KOCH et al. 1961).*

Fungicide	Active ingredient (kg/ha)	Polarization (%)	Yield (t/ha)			Harmful nitrogen (mg/100 g)	Soluble ash (%)	Purified sugar (t/ha)	Purified sugar (%)	Yields of untreated		
			Leaf	Beets	Theor. sugar					Theor. sugar (%)	Beet (%)	Leaf (%)
Untreated	—	16.5	17.3	37.5	6.1	32.0	0.37	5.19	100	100	100	100
Copper oxy chloride (50% Cu W.P.)	2.0	17.6	21.3	43.8	7.71	30.0	0.35	6.61	127.4	124.6	117	123
Triphenyltin hydroxide (20% W.P.)	0.3	18.2	27.6	47.8	8.68	22.0	0.33	7.65	147.4	140.2	127	160
Triphenyltin acetate (20% W.P.)	0.3	18.2	25.6	47.7	8.68	21.0	0.34	7.61	146.6	140.2	127	148
Triphenyltin acetate + maneb	0.3 +0.1	18.3	29.0	49.8	9.10	21.0	0.33	8.03	154.7	147.0	133	168

Table LXXXII. *Effect of zineb and triphenyltin acetate against Cercospora leaf spot in sugarbeets with varying harvest dates* (after GUTSHTEIN 1963); *relative data (untreated = 100)*.

Fungicide and dosage	Harvest date May 6		Harvest date June 3		Harvest date July 2		Harvest date August 6		Mean data	
	Sugar yield	Ash content	Sugar yield	Ash content	Sugar yield	Ash content	Sugar yield	Ash content	Sugar yield	Ash content
Untreated	100	100	100	100	100	100	100	100	100	100
Zineb 4.68 kg/ha	102.7	110.0	110.3	93.4	121.5	90.3	113.4	85.6	112.4	93.1
Triphenyltin acetate 0.50 kg/ha	102.6	101.1	128.7	86.2	123.6	78.3	133.6	79.1	123.1	83.8

respectively 6 sprayings were applied with 1,700 g of a triphenyltin hydroxide formulation (content of active ingredient not mentioned) in 940 L of liquid. While 3 sprayings were of only limited success, yields had increased considerably after 6 applications (Table LXXXIII).

Table LXXXIII. *Control of Cercospora beticola in sugarbeets with triphenyltin hydroxide* (after CARLSON 1963).

Spraying dates	Leaves		Yield (t/ha)	
	Infected (%)	Decayed (%)	Beets	Sugar
Control[a]	93.6	73.2	39.9	3.32
8/5, 8/15, 8/26	52.6	45.6	45.1	4.78
7/15, 7/25, 8/5, 8/15, 8/26, 9/7	33.2	34.5	58.6	6.87

[a] Average from 4 fields.

SOLEL and MINZ (1963) compared the effects of triphenyltin acetate and hydroxide. After applying the active ingredient 5 times at intervals of 3 wk, the acetate resulted in increased sugar yield of 24%, the hydroxide of 8.8% compared with the controls. Later experiments (SOLEL 1970 b) showed that the attack had been controlled after 2 applications of 360 g of triphenyltin acetate; however, no yield data were reported.

DONÀ DALLE ROSE and OLIMPIERI (1967), using 2 × 350 g of triphenyltin acetate/ha, obtained a beet yield increased by 11%, while the sugar content had improved from 12.9% (control) to 13.3%.

BONGIOVANNI (1968 a and b; 1969 a, b, and c) described comparative experiments with various organic fungicides, the complex triphenyltin chloride/decyltriphenylphosphonium bromide as well as triphenyltin chloride, acetate, and hydroxide. The high efficiency of the triphenyltin compounds was confirmed, with acetate and hydroxide showing no significant differences. The chloride was somewhat less effective, and the complex showed no advantage (compare also BONGIOVANNI 1965). Later BONGIOVANNI (1975) found triphenyltin hydroxide (300 g/ha) more effective than benomyl (100 g/ha), and triphenyltin hydroxide + benomyl (150 + 50 g/ha) more effective than benomyl + maneb.

In India, MUKHOPADHYAY and RAO (1971) were able to increase the sugar yield by 34.8 respectively 28.0%, after applying 250 g of triphenyltin chloride or hydroxide 3 times at 10-day intervals. With 375 g/ha and 3 sprayings, the sugar yield was 80.1 respectively 47.3% higher than for the untreated controls. Similar results were obtained by other field experiments with 2 treatments applying triphenyltin chloride (360 g/ha, 20 day interval) (MUKHOPADHYAY and RAO 1974). Further reports on the effectiveness of triphenyltin compounds were given by KRASNOSHCHEKOV (1972), GALANDZOVSKAYA and GAMIN (1973), MOHIBULLA (1973), V'RBANOV (1974), and COMES et al. (1974).

One application of triphenyltin hydroxide (670 g/ha in 800 to 900 L of water) was sufficient to control attack of *Ramularia beticola* Faut. et Lamb. in sugarbeet seed crops (BYFORD 1976); the yield of seeds was increased by 4% (average of 11 trials).

ISAK (1962) conducted experiments with triphenyltin compounds on plants after flowering to prevent infection of sugarbeet seeds by *Cercospora*. After pretreating the sets by spraying about 300 g of triphenyltin compound/ha, 4 to 5 times, the seed-bearing plants, at a height of 15 to 20 cm, were sprayed during the following year only once with 300 g of active ingredient/ha. Areas threatened severely by *Cercospora* showed an average decrease in infestation of approximately 40%; in areas with medium infection threat, occurrence was reduced by 56%. Since the seed-bearing plants cannot be sprayed from the ground effectively, air application was recommended.

FOSCHI *et al.* (1969) showed that triphenyltin compounds can also be used to control seed infection by *Cercospora* while the beet plants are in blossom. Three to 4 treatments with 360 g of triphenyltin acetate/ha were used. Cytologic tests on progenies of a polyploid variety showed no anomalies, especially no increase in aneuploidies attributable to the treatment.

ββ) Powdery mildew:

It was noticed on various occasions that triphenyltin compounds also have distinct effects against powdery mildew in sugarbeets (caused by *Erysiphe betae* Van., *Erysiphe polygoni* DC, or *Erysiphe communis*) (PECHINEY PROGIL 1960, BONGIOVANNI 1963, SOLEL 1964, HILLS *et al.* 1975, BYFORD 1977 and 1978, and PAULUS *et al.* 1975). On the other hand, KREXNER and WENZL (1958) and KREXNER (1965) noticed that infection with powdery mildew (*Microsphaera betae* and *Erysiphe communis*) was intensified after application of the fungicide. The latter findings, however, were disproved by WELTZIEN (1968) after carefully evaluated greenhouse experiments. He showed that triphenyltin acetate has inhibitive effects on powdery mildew, especially when taking into account that the underside of the leaves is receiving little protection from normal sprayings.

Later, after evaluating all his experiments, KREXNER (1971) also came to the conclusion that "with the application of tin-containing fungicides usually positive side effects against powdery mildew can be expected." Therefore, triphenyltin compounds can be used without reservations in areas where powdery mildew appears in addition to *Cercospora*, since they will also work against the powdery mildew infection.

γγ) Rhizoctonia crown and root rot:

SCHNEIDER and POTTER (1974) reported some effect of triphenyltin hydroxide (*ca.* 425 g/ha) against *Rhizoctonia* attack of sugarbeet.

δδ) Phoma stem rot:

Three treatments with triphenyltin hydroxide (670 g/ha in 800 to 900 L of water) controlled stem rot of sugarbeets caused by *Phoma betae*;

in one trial, an increase of more than 10% in seed yield was observed (BYFORD 1976; see also HULL 1960 and 1962).

δ) *Application of triphenyltin compounds to celery.*[3]—Triphenyltin compounds are used to control leaf spot in celery (caused by *Septoria apii* Chester); their effects against celery root rot (caused by *Phoma apiicola* Kleb.) were noticed as a side result.

αα) *Leaf spot:*

Triphenyltin acetate was first used against *Septoria apii* by BAUMANN (1957). The compound is about 15 times more effective than copper oxychloride and zinc ethylenebisdithiocarbamate (HAERTEL 1962). Even in severely threatened areas, celery cultures can be kept free of the fungi by regular spraying, which results in much higher tuber yields; the tuber size and the yield in leaves is also increased (VON HOESSLIN 1960). According to LINDEMANN (1959), triphenyltin compounds are the most effective fungicides against *Septoria apii*.

Obviously, because of their penetrating effects, triphenyltin compounds also have a curative effect after infection has started (VON HOESSLIN 1960).

To control celery leaf spot, applications of 240 to 300 g of active ingredient in 600 L of water/ha were recommended (*Codex Committee* 1970). According to SCHUPP (1957), the disease can be controlled with 7 sprayings at intervals of 14 days. RYAN *et al.* (1972) showed that 4 to 6 sprayings of 900 to 1,800 L of liquid/ha at intervals of 14 days are sufficient. VON HOESSLIN (1960) recommended 4 to 6 sprayings, starting with 1,000 L of mixture/ha and increasing it to 2,000 L/ha for the last sprayings, since the amount of leaves increases, too. The best results were obtained by spraying a 0.2% dilution of the 20% formulation, which equals an amount of active ingredient of 400 to 800 g/ha. Whether the last spraying takes place 4 to 6 wk before the harvest is of no significance.

In the field tests with artificial inoculation, 6 sprayings with triphenyltin acetate (600 L of mixture each/ha) produced yields in tubers and leaves which were significantly higher than those obtained with any other fungicide (HAERTEL 1958 a; Table LXXXIV).

Similar results were obtained by BAUMANN (1957 and 1958). The plants produced particularly strong, dark green leaves and were almost free of disease up to the last spraying (Table LXXXV).

HOESSLIN (1960) obtained much increased yields in tubers and leaves (Table LXXXVI), and the tuber size distribution was also improved by the treatment with triphenyltin acetate (Table LXXXVII).

In comparative experiments with various fungicides to control *Septoria apii*, triphenyltin compounds were by far the best (RYAN and KAVANAGH 1971, RYAN *et al.* 1972). Table LXXXVIII shows results obtained with

---

[3] The name "celeriac" is also in use which actually stands for a variety of *Apium graveolens* that produces larger edible tubers.

Table LXXXIV. *Protective spraying against celery leaf spot (Septoria apii Chester) with various fungicides* (after Haertel 1958 a).

Fungicide and amount	Leaf infes-tation[a]	Tuber yield		Leaf yield	
		t/ha	Rel.	t/ha	Rel.
Untreated	4.5	16.95	100	9.75	100
Zineb (78% W.P.) 2.4 kg/ha a.i.	3.8	18.2	107.4	10.6	109.2
Copper oxy chloride (50% Cu, W.P.) 3.0 kg Cu/ha	3.0	21.7	128.0	12.9	132.8
Triphenyltin acetate (20% W.P.) 0.36 kg/ha a.i.	1.5	32.1	189.4	18.95	203.6
Triphenyltin acetate (20% W.P.) 0.60 kg/ha a.i.	1.0	36.15	213.2	21.65	222.1

[a] Rating: 0 = nil and 5 = 100%.

Table LXXXV. *Results from field tests with celery in severe Septoria apii-attack* (after Baumann 1958); *relative data, untreated = 100.*

Fungicide and amount	Leaf infes-tation[a]	Tuber yield	Leaf yield
Experiment 1			
Untreated	4.8	100	100
Copper oxy chloride (50% W.P.) 5 × 2.0 kg Cu/ha	4.0	186	288
Triphenyltin acetate (20% W.P.) 3 × 0.36 kg/ha a.i.	3.5	234	415
Triphenyltin acetate (20% W.P.) 5 × 0.36 kg/ha a.i.	0.6	249	557
GD 5%		21	71
1%		32	107
0.1%		51	172
Experiment 2			
Untreated	4.6	100	100
Copper oxy chloride (50% W.P.) 5 × 3.0 kg Cu/ha	3.8	123	200
Triphenyltin acetate (20% W.P.) 3 × 0.60 kg/ha a.i.	3.0	145	239
Triphenyltin acetate (20% W.P.) 5 × 0.6 kg/ha a.i.	0.8	188	328
GD 5%		21	61
1%		32	93
0.1%		51	149

[a] Rating: 0 = nil and 5 = 100%.

Table LXXXVI. *Protective spraying against celery leaf spot with various fungicides; 4–6 treatments, 1,000–2,000 L/ha each median from 2 experiments each (1957 and 1958) with 2 varieties* (after VON HOESSLIN 1960).

Fungicide and concentration	Yield (t/ha)			
	Variety: Magdeb. Markt.		Variety: Hilds Neckarland	
	Tubers	Leaf	Tubers	Leaf
Untreated	25.0	19.2	22.8	15.0
Copper oxy chloride (50% Cu W.P.) 0.25%	37.6	37.8	32.8	28.6
Triphenyltin acetate (20% W.P.) 0.06%	46.6	44.8	41.1	36.6
Triphenyltin acetate (20% W.P.) 0.04%	46.2	44.3	42.1	37.0
Triphenyltin acetate (20% W.P.)[a] 0.04%	39.0	37.7	32.8	27.8
Triphenyltin acetate (20% W.P.) 0.03%	44.6	43.2	40.4	35.8
Triphenyltin acetate (20% W.P.) 0.02%	45.4	42.1	38.1	34.3

[a] Half the no. of sprayings.

Table LXXXVII. *Tuber size distribution after treatment of celery with various fungicides; median from two years; conditions as in Table LXXXVI* (after VON HOESSLIN 1960).

Fungicide and concentration	Tuber size (%)				
	Over 15 cm	12–15 cm	9–12 cm	6–9 cm	Under 6 cm
Untreated	6.6%	25.8%	51.1%	15.8%	0.7%
Copper oxy chloride (50% Cu W.P.) 0.50%	17.4	47.0	29.9	5.6	0.1
Triphenyltin acetate (20% W.P.) 0.30%	28.4	51.6	17.8	2.2	0.0
Triphenyltin acetate (20% W.P.) 0.20%	26.8	52.5	18.5	2.1	0.1
Triphenyltin acetate (20% W.P.)[a] 0.20%	17.3	51.6	27.2	3.8	0.1
Triphenyltin acetate (20% W.P.) 0.15%	23.4	55.0	18.7	2.7	0.2
Triphenyltin acetate (20% W.P.) 0.10%	22.4	54.5	21.2	1.9	0.0

[a] Half the no. of sprayings.

Table LXXXVIII. *Application of triphenyltin compounds and of manganous ethylene-bisdithiocarbamate (maneb) to celery; 6 applications of 900–1,800 L/ha* (after RYAN et al. 1972).

Fungicide	g ingredient/ha/ application	Evaluation[a]	Marketable yield (kg/plot)
Control	None	4.2	1.5
Acetate + maneb	270 + 680	0.8	13.7
Acetate + maneb	400 + 135	0.3	13.0
Acetate + maneb	270 + 1,750	0.7	11.8
Maneb	3,240	2.8	4.1
Chloride	400	0.3	13.0
Hydroxide	400	0.3	12.9

[a] Rating: 0 = nil and 15 = maximum infestation.

acetate, chloride, and hydroxide; manganous ethylenebisdithiocarbamate, mentioned for comparative reasons, is much less effective.

Finally, experiments by PIETERS (1962) with 0.3% suspension of the hydroxide, and by SCHICKE et al. (1968) with complex triphenyltin chlorides should be mentioned. In all cases, the celery plants were effectively protected against *Septoria apii*.

ββ) *Celery root rot:*

SCHUPP (1957) already pointed out that triphenyltin compounds show fungicidic side effects against celery root rot. This was confirmed by SCHICKEDANZ and HOENICK (1967) (Table LXXXIX).

Table LXXXIX. *Protective spraying with triphenyltin acetate against celery root rot. Results after several years of field tests; treatment beginning 4 wk after planting, sprayings in 3-wk intervals* (after SCHICKEDANZ and HOENICK 1967).

Year	Fungicide and concentration	Yield[a] (%)	Infested with root rot (%)
1961	Untreated	100	91
1961	Triphenyltin acetate (20% W.P.) 0.2%	105	9
1962	Untreated	100	89
1962	Triphenyltin acetate (20% W.P.) 0.2%	117	33
1962	Triphenyltin acetate (20% W.P.) 0.3%	110	28
1963	Untreated	100	90
1963	Triphenyltin acetate (20% W.P.) 0.2%	98	33

[a] In % untreated.

ε) *Application of triphenyltin compounds to carrots and onions.—*
αα) *Carrots:*

Infection with *Alternaria porri f. dauci* Neerg. lowers the yield of carrots and makes mechanized harvesting more difficult. The fungus can be controlled by 1 to 3 sprayings with 240 to 300 g of triphenyltin compound/ha in 600 to 1,000 L of liquid (*Codex Committee* 1970).

SCHMIDT (1965) and FRANZ (1972) recommended triphenyltin compounds in amounts of 240 to 360 g/ha in 400 to 600 L of water against Alternaria blight (caused by *Alternaria dauci* Kühn).

NETZER and KATZIR (1966) were able to efficiently control not only the *Alternaria* blight, but also the powdery mildew in carrots (caused by *Erysiphe umbelliferarum* Lev. de By.), by using 300 g of triphenyltin acetate or hydroxide. The yields increased considerably (Table XC).

Table XC. *Incidence of Alternaria blight and powdery mildew in carrots sprayed with different compounds* (after NETZER and KATZIR 1966).

| | Diseased leaves (%) | | |
Fungicide used and amount	Alternaria blight	Powdery Mildew	Yield (kg/plot)
No treatment	80	62	28.2
Maneb (78% W.P.) 2.0 kg/ha a.i.	58	0	34.8
Triphenyltin hydroxide (20% W.P.) 0.3 kg/ha a.i.	28	0	35.3
Triphenyltin acetate (60% W.P.) 0.3 kg/ha a.i.	27	0	34.0
Triphenyltin acetate + maneb (60 + 20% W.P.) 0.3 + 0.1 kg/ha a.i.	38	0	37.3
LSD 0.01	13	37	

ββ) Onions:

In Brazil, COSTA DA LIMA et al. (1962) tested various fungicides against purple blotch in onions (caused by *Alternaria porri* Ell.) Triphenyltin acetate was superior to comparative fungicides both in disease control and in the effect on yields (Table XCI).

Table XCI. *Effect of different fungicides against Alternaria blight on onions in yield of plants and bulbs; 10 applications at intervals of 10 days* (after COSTA DA LIMA et al. 1962).

Fungicide and concentration	Plants (kg/plot; median)	Onions (kg/plot; median)
Zineb (78% W.P.) 0.35% a.i.	6.62	2.88
Copper oxy chloride (50% W.P.); 0.25% a.i.	6.79	2.91
Maneb (78% W.P.) 0.14% a.i.	7.32	3.20
Triphenyltin acetate (20% W.P.); 0.03% a.i.	8.40	3.59

ζ) Application of triphenyltin compounds to pecan nuts.—Triphenyltin compounds are used to control different diseases of pecan nuts. LARGE (1963) first used triphenyltin hydroxide against pecan scab (*Fusicladium effusum* Wint.). Later, also brown leaf spot (*Cercospora fusca* Rand), sooty mold (*Capnodium spp.*), nursery blight (*Elsinoe randii*), and powdery mildew (*Microsphaera alni* Wint.) could be controlled and the degree of attack by vein spot (*Gnomonia nerviseda*) and downy spot (*Mycosphaerella caryigeana*) was reduced. Finally, aphids and mites are also eliminated by organotin compounds.

The trees, after treatment with triphenyltin compounds, are conspicuous because of their healthy, dark green foliage (Barnes 1966); reduced infection of leaves and nuts results in higher yields and increased mean wt of the nuts.

Ninety to 135 g of triphenyltin hydroxide (or similar amounts of acetate) in 380 L of water were usually recommended (Large 1963 and 1965, Dodge 1966, Diener and Garrett 1965). There are no significant differences in the effects of hydroxide and acetate (Diener and Garrett 1967). In amounts of only 45 g, triphenyltin hydroxide proved to be insufficient.

The number of treatments depends on the conditions of the infection and the climate. While sometimes 3 sprayings are sufficient, 6 to 9 treatments may be required for severe infections and in humid weather.

Large (1963) was able to prevent infestation with *Fusicladium effusum* Wint. almost completely, spraying 3 times with 135 g of triphenyltin hydroxide in 380 L of water. During a severe infection of untreated trees, Cole (1965) observed reduced disease, much higher yields, and higher mean wt of the nuts after 6 sprayings with triphenyltin hydroxide (Table XCII).

Table XCII. *Results of spray treatments on yield and weight of nuts harvested from Schley pecan trees, 1964, Hurst orchard, Albany, Ga. (after Cole 1965).*

Treatment	Concentration (% a.i.)	No. of applications	Yield (kg/tree)	No. of nuts/kg
Untreated	—	–	0.45	254
Dodine[a]	0.04	6	6.8	194
Dodine[a]	0.04	2		
	0.08	4	9.9	166
Triphenyltin hydroxide	0.05	6	13.1	150

[a] *N*-Dodecylguanidine acetate.

Another series of experiments produced even higher yield gains (from 0.45 to 19.5 kg nuts/tree) after 5 sprayings.

During a wet year ("a wet season with 1,163 mm of rain, 162.8 mm above normal during the growing season, March 1 to October 1") with severe pecan scab infection, Large (1965) tested different fungicides. The disease could be well controlled with triphenyltin hydroxide only in concentrations of 90 to 135 g/380 L of water (a concentration of 45 g/380 L was not strong enough). The tin compound had also effects on aphids and mites (Table XCIII).

Nine treatments of pecan nut trees with 75 L of liquid each (135 g of triphenyltin hydroxide in 380 L of water) caused the average wt of the nuts to increase from 0.817 g/nut (untreated) to 2.449 g/nut (BARNES 1966).

Table XCIII. *Scab control on Mahan pecans in 1964* (after LARGE 1965).

Fungicide	Rate for 6 applications (g a.i./380 L)	Scab control (%)	Scab infection Clean	Light	Heavy	Yield (kg/ tree)
Not sprayed	—	0	0	0	100.0	0.45
Bordeaux	1,800–450, then 2,700–950	65.4	3.8	61.6	34.6	2.3
Dodine	146	84.3	37.5	46.8	15.7	4.66
Triphenyltin hydroxide	45	73.0	15.6	58.2	26.2	3.4
Triphenyltin hydroxide	90	98.5	50.8	47.8	1.5	5.5
Triphenyltin hydroxide	135	99.1	50.3	48.8	0.9	5.66

DIENER and GARRETT (1967) reported a reduction in pecan scab disease after using triphenyltin hydroxide or acetate (Table XCIV). The tin compounds also prevented other diseases (Table XCV). According to BARNES (1974), *Fusicladium* is better controlled by benomyl or thiophanate methyl than by triphenyltin hydroxide.

Table XCIV. *Control of scab with fungicides applied by air-blast sprayer to pecan trees in Mobile County, Alabama 1966; 7 treatments at intervals of 3 wk* (after DIENER and GARRETT 1967).

Fungicide	Rate (g active ingredient/ 380 L)	Scab infection Clean	Light	Heavy
Unsprayed check	—	0	0	100
Dodine (65% W.P.)	292	80.5	15.5	4.0
Triphenyltin acetate (60% W.P.)	108	73.7	23.3	3.0
Triphenyltin hydroxide (50% W.P.)	45	47.5	42.0	10.5
Triphenyltin hydroxide (50% W.P.)	90	88.3	11.7	0

Table XCV. *Control of brown leaf spot and other diseases with fungicides applied by mist blower to Lewis pecan trees in Montgomery County, Alabama, 1965; 5 treatments at intervals of 4 wk* (after DIENER and GARRETT 1967).

Fungicide	Rate (g a.i./ 380 L)	Brown leaf spot (%)	Scab on leaves	Nursery blight	Sooty mold	Defolia- tion
Unsprayed	—	85.4	+	+++	+++	25
Zineb (78% W.P.)	675	29.4	0	+	+++	4
Dodine (65% W.P.)	292	8.6	0	+	++	1
Triphenyltin hydrox- ide (50% W.P.)	135	1.2	0	0	0	0

LARGE (1965) and BARNES (1966) pointed out that trees treated with triphenyltin hydroxide showed no powdery mildew infection; however, the infestation had been rather low.

DODGE (1966) described good effects of triphenyltin hydroxide against several other diseases of pecan nuts. Three treatments with 90 g of the active ingredient in 380 L of water reduced infection with vein spot from 13.69 (untreated) to 4.1%, and with downy spot from 62.2 to 6.7% (Table XCVI).

Experiments to control the stem-end blight, a form of shuck disease, with triphenyltin hydroxide showed no clearly positive results. Prepollination sprays were effective in 1969, but not in 1970; later applications were all unsuccessful (SCHALLER and KENKNIGHT 1972).

η) *Application of triphenyltin compounds on rice.*—Triphenyltin compounds are used against rice blast ("Imochi-disease," caused by *Piricularia oryzae* Cav.) and against sheath blight (*Pellicularia sasakii* Shirai). Effects against brown spot disease (*Helminthosporium oryzae* Hori) have also been reported. Further, the development of algae in rice fields can be prevented with these fungicides.

αα) *Rice blast and sheath blight:*

HAERTEL (1962) was first to describe the effect of triphenyltin acetate against rice blast (*Piricularia oryzae* Cav.). TAMURA (1965 a and b) conducted extensive experiments with various organotin compounds to control rice blast and sheath blight. During the tests he developed a combined preparation from tributyltin oxide and triphenyltin acetate.

Greenhouse experiments with *Piricularia* and *Pellicularia* (TAMURA 1965 b) showed that of the 3 tributyltin compounds (chloride, acetate, oxide) the bis(tri-*n*-butyltin)-oxide was the most effective against *Piricularia oryzae*; however, each of these compounds caused considerable lesions in the leaves.

Triphenyltin chloride and acetate were much less phytotoxic but also not very effective, even in higher concentrations; therefore, mixtures of tributyl- and triphenyltin compounds were tested.

Table XCVI. *Mean number of lesions on leaf of Stuart-pecan when sprayed with indicated fungicides; 3 applications* (after Dodge 1966).

Fungicide	Rate (g active ingredient/ 380 L)	Vein spot	5%	1%	Downy spot	5%	1%	Liver spot	5%	1%	Foliage on tree 12/6/65 (%)	5%	1%
Untreated	—	13.69	ab[a]	—	62.2	a[a]	a	3.06	a[a]	a	1	a[a]	ab
Dodine (65% W.P.)	292	3.04	c	—	3.5	d	b	1.02	c	c	30	c	cd
Dodine (65% W.P.)	584	4.32	c	—	3.0	d	b	0.29	d	c	35	c	d
Triphenyltin hy-droxide (20% W.P.)	90	4.11	c	—	6.7	cd	b	0.34	d	c	29	c	cd

[a] For a given disease, means followed by the same letter are not significantly different at the level specified when compared with Duncan's Multiple Range.

The tributyltin compounds applied were 10% emulsifiable product dissolved in xylene with 5% emulsifier (Toximol 500). The triphenyltin compounds were wettable powders (10% triphenyltin compound + 85% Kaolin + 5% wetting agent Solpol-W 150) and were used as aqueous suspensions.

When the first 3 leaves had developed, the rice plants were treated with the fungicide and then inoculated with the fungus. Six days after inoculation, the infected spots were counted. The results are shown in Table XCVII.

Table XCVII. *Effects of tributyl- and triphenyltin compounds against rice blast (Piricularia oryzae Cav.); greenhouse experiments* (after Tamura 1965 b).

Preparation	Concentration (ppm)	Infected spots/leaf	Phytotoxicity
— (control)	—	7.28	—
Tributyltin acetate	50	5.42	+
Tributyltin chloride	50	5.81	+
Tributyltin oxide	50	4.76	+
Triphenyltin acetate	100	5.74	—
Triphenyltin chloride	100	6.07	—
Tributyltin acetate + triphenyltin acetate	25 50	4.48	—
Tributyltin acetate + triphenyltin acetate	50 100	4.10	+
Tributyltin chloride + triphenyltin acetate	25 50	5.03	—
Tributyltin chloride + triphenyltin acetate	50 100	4.16	+
Tributyltin oxide + triphenyltin acetate	25 50	4.35	—
Tributyltin oxide + triphenyltin acetate	50 100	3.37	+
Tributyltin oxide + triphenyltin chloride	25 50	4.46	—
Tributyltin oxide + triphenyltin acetate	50 100	3.55	+

The ratio 1:2 mixtures of tributyltin oxide and triphenyltin acetate produced the best results. Minor lesions of the leaves in this application were tolerable. The experiments were conducted with aqueous suspensions of the active ingredients; corresponding dust formulations (0.25

active ingredient in calcium hydroxide) were much less effective, however, they did not cause any lesions on the leaves.

Field trials confirmed the results described above. Two sprayings at intervals of 14 days produced the results shown in Table XCVIII. For comparison, data obtained with Hg- and As- containing fungicides are also listed.

Table XCVIII. *Control of rice blast (Piricularia oryzae Cav.) with various fungicides* (excerpt from TAMURA 1965 b).

Fungicide	Infection	Yield (t/ha)		
		Unhusked	Unpolished	Straw
TBTO (50 ppm) + TPTA (100 ppm)	21.4	6.74	4.73	5.80
TBTO (100 ppm) + TPTA (200 ppm)	17.1	7.10	5.18	5.69
HBP[a]	27.0	6.64	4.66	5.80
Azodin M[b]	17.5	6.60	4.62	5.61
Azodin[b]	40.7	6.29	4.31	5.66
Ceresankalk[b]	21.7	6.68	4.67	5.69
Control	53.8	5.89	3.91	5.83

[a] Dust preparation with 0.25% TBTO + 0.25 TPTA.
[b] Hg- or As-containing preparations.

The yields obtained after applying tributyltin oxide-triphenyltin acetate mixtures were much higher than for the untreated controls. They were also above those obtained after using other fungicides. The effects were maintained even when the tin preparation was applied during rain and when the inoculation took place 7 days after the fungicide treatment.

Tributyltin oxide and the mixture of oxide and triphenyltin acetate were also suitable against sheath blight (*Pellicularia sasaki* Shirai), where triphenyltin acetate alone proved to be less effective. The combined preparation was used in overall concentrations of 150 to 190 ppm with 2 to 3 sprayings (with 1,000 to 1,500 L/ha each). The effects of the latter in greenhouse and field tests were comparable with those from the commonly used arsenic and mercury preparations (Table XCIX). Other field tests showed similar results.

Some effects to control *Piricularia oryzae* with triphenyltin hydroxide were found by AWODERU (1974) and AWODERU and ESURUOSO (1974 and

Table XCIX. *Control of sheath blight (Pellicularia sasaki Shirai) with mixtures of (tri-n-butyltin) oxide and triphenyltin acetate* (after TAMURA 1965 b).

Compound and concentration (ppm)	% Infected bunches on		Phytotoxicity
	7/26	8/10	
TBTO (50 ppm) + TPTA (100 ppm)	6.0	14.0	++
TBTO (62.5 ppm) + TPTA (125 ppm)	4.5	9.0	++
Control	28.0	40.5	−

1975), but other fungicides were superior. Good results with triphenyltin acetate + maneb were reported by CADENA HINOJOSA and RODRIGUEZ (1975) and CADENA HINOJOSA (1975); 3 treatments with 0.5 kg/ha (60% formulation) at the shoot stage 10 days before flowering, at flowering, and 20 days later were recommended. GALVEZ and CASTAÑO (1974) used several fungicides (triphenyltin hydroxide and acetate and other organic compounds) as soil treatment to control *Piricularia oryzae*. The tin compounds were very effective. VERMA *et al.* (1976) compared the effectiveness of fungicides and antibiotics against rice blast; triphenyltin hydroxide and the other fungicides used were superior.

ASHRAFUZZAMAN (1970) was not able to control *Piricularia oryzae* Cav. sufficiently with triphenyltin hydroxide because the fungicide was applied too late.

SINGH and SHARMA (1972) conducted a field experiment to control brown spot in rice (*Helminthosporium oryzae* Hori). Triphenyltin chloride in dosages of 470 g/ha showed maximum effect after the fourth and fifth spraying. Two hundred g/ha of triphenyltin hydroxide took third place in efficiency to control the infection, but produced slightly higher yields than triphenyltin chloride. The yield after applying the hydroxide was 1,250 kg/ha, after using the chloride 1,100 kg/ha (Table C).

Table C. *Field experiment against brown spot disease (Helminthosporium oryzae) on rice with different fungicides* (excerpt from SINGH and SHARMA 1972).

Fungicide used and concentration	Infection index[a]		Yield[b] (kg/ha)
	After 3rd spray	After 4th spray	
Untreated control	68.40	71.90	850.0
Maneb (78% W.P.); 0.2%	36.40	39.10	560.0
Zineb (78% W.P.); 0.2%	44.50	46.80	850.0
Triphenyltin chloride (47% W.P.); 0.1%	32.70	34.20	1,100.0
Triphenyltin hydroxide (20% W.P.); 0.1%	43.70	46.20	1,250.0
C.D. (5%)	9.20	7.60	70.0
C.V.	14.40	10.50	4.7

[a] Mean of 4 replicates (angular transformation value).
[b] Yield is low because of late planting, death of seedlings due to heavy rain and frequent gap filling.

Triphenyltin hydroxide and chloride sprays were used as seed treatment by SINGH *et al.* (1972); the chloride was very effective. JACKSON *et al.* (1977) obtained significant reduction of stem rot of rice (*Sclerotium oryzae*) with 1 × 1.12 kg/ha of triphenyltin hydroxide (47.5% formulation); yields increased by 6 to 25%.

ββ) *Algae in rice fields:*

The algicidal effects of triphenyltin compounds were already pointed out by BAUMANN (1958). CHIAPPARINI *et al.* (1974) and CHIAPPARINI

(1965 and 1967) drew attention to the practical application of triphenyl-tin acetate to control algae in rice fields. *Spirogyra, Vaucheria, Cla-dophora, Hydrodictyon* and other kinds, more or less combined, are the main competitors against the rice plants for light and food. According to their test results, CHIAPPARINI *et al.* (1964) recommended, at a water depth of 10 cm, 0.7 to 1.0 kg of triphenyltin acetate in 200 L of liquid, which equals a concentration of about 0.7 to 1.0 ppm active ingredient. Already at 0.7 ppm all algae died within 48 hr, at higher concentrations even faster. The effect of the tin compound lasted at 0.7 ppm for approxi-mately 7 to 10 days (Fig. 19). Apparently 0.75 to 1.0 ppm tin compound

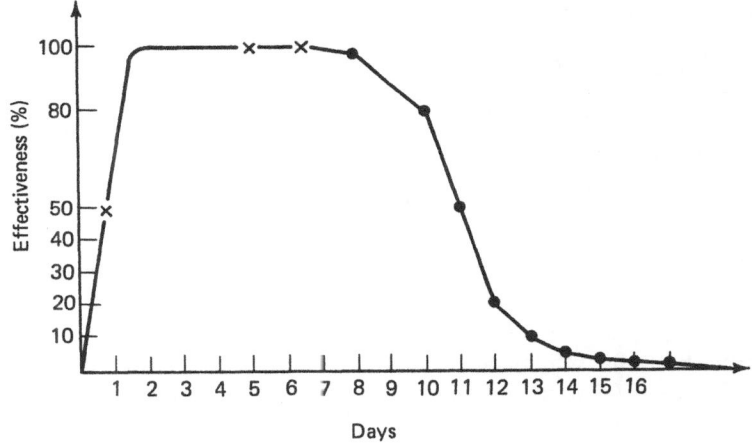

Fig. 19.   Persistence of triphenyltin acetate against algae in a rice crop (after CHIAP-PARINI *et al.* 1964).

in the water did not have phytotoxic effects on the young rice plants (1- to 4-leaf stage). Even 5.0 ppm of triphenyltin acetate in the water did not impair the development of the rice plants in the 2-leaf stage.

Based on their experiments in West Bengal, MUKHERJI and RAY (1966) and MUKHERJI (1968) came to the same conclusions; 1.0 kg/ha triphenyltin acetate, at a water level of 10 to 15 cm, proved sufficient to completely destroy *Chara* and *Nitella* strains, which in West Bengal are much more dangerous for young rice plants than other algae of the same community (Table CI).

BATALLA (1968) recommended 400 to 600 g of triphenyltin acetate or hydroxide/ha at a water level of about 5 cm to control algae.

The concentrations of triphenyltin compounds necessary to control algae had no toxic effects on fish living in the rice fields. No phytotoxic symptoms were noticed in the young rice plants (MUKHERJI and RAY 1966, MUKHERJI 1968).

Table CI. *Chemical control of algal weeds Chara and Mitella spp.*
(excerpt from MUKHERJI 1968).

Chemical and dose	Depth of water (cm)	Control of algae In 3 days	In 7 days
Zineb (78% W.P.); 3.90 kg/ha a.i.	22.9	Algae green	Algae green no control
Zineb (78% W.P.); 5.85 kg/ha a.i.	22.9	Algae turn yellow	Decomposed fully
Zineb (78% W.P.); 7.80 kg/ha a.i.	30.5	Algae turn yellow	Decomposed fully
Triphenyltin acetate (60% W.P.) 0.30 kg/ha a.i.	22.9	Algae green not affected	Algae green no control
Triphenyltin acetate (60% W.P.) 0.45 kg/ha a.i.	22.9	Algae turn yellow	Decomposed fully
Triphenyltin acetate (60% W.P.) 0.60 kg/ha a.i.	30.5	Algae turn yellow	Decomposed fully
Triphenyltin acetate (60% W.P.) 0.75 kg/ha a.i.	38.1	Algae turn yellow	Decomposed fully
Triphenyltin acetate (60% W.P.) 1.05 kg/ha a.i.	45.7	Algae turn yellow	Decomposed fully

θ) *Application of triphenyltin compounds to cacao.*—MATTA (1959) first described the effects of triphenyltin compounds against black-pod disease in cacao (caused by *Phytophthora palmivora* Butl.). In lab experiments, triphenyltin acetate was tested at different concentrations and compared with copper oxy chloride and other fungicides. The organotin compound was able to control the growth of the fungus much better than comparative fungicides. During the following field tests, 10 cacao trees and their fruit were treated with triphenyltin acetate in concentrations of 0.1, 0.08, and 0.05% (LELLIS 1959). Then, trees and fruit were inoculated with *Phytophthora palmivora*. A count of the fruit with and without disease is shown in Figure 20.

HANSEN and SILLER (1960) tested triphenyltin acetate in the field. Acetate, applied at intervals of 3 to 4 wk, reduced the infection. However, during heavy rains most of the preparation was washed off and rendered ineffective. Later, HISLOP and PARK (1962) and HISLOP (1963) conducted more extensive experiments.

In laboratory experiments with agar-cultures, HISLOP (1963) tested the toxicity of different fungicides against two strains of *Phytophthora palmivora* ("rubber group" and "cacao group"). The ED_{50} and ED_{100} values of triphenyltin acetate were lower than those of all other preparations except phenyl mercury nitrate (Table CII).

In slide germination tests, triphenyltin acetate, captan, maneb, and the copper complex of dimethyldithiocarbamate showed approximately

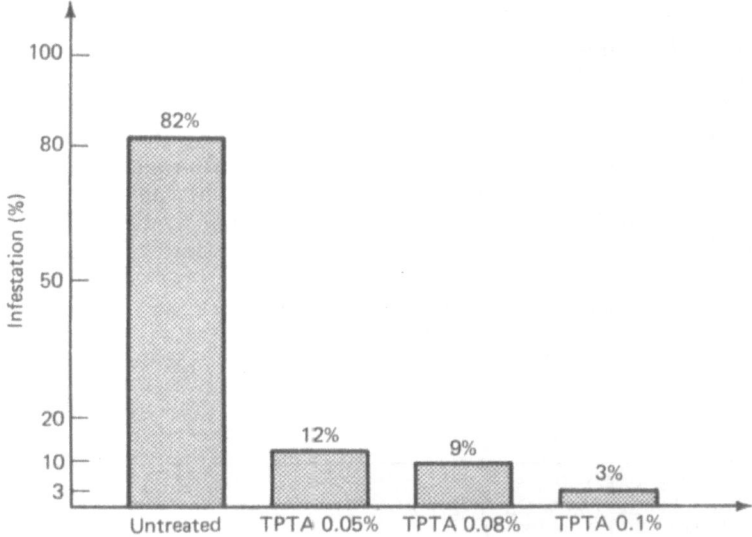

Fig. 20. Control of *Phytophthora palmivora* in cacao with triphenyltin acetate (TPTA) 20% W.P. (after LELLIS 1959): GD 0.05% = 13, GD 0.01% = 17.

Table CII. *Concentrations of fungicide active ingredients (ppm) causing 50% reduction (ED₅₀) and inhibition (ED₁₀₀) of growth of isolates of Phytophthora palmivora* (after HISLOP 1963).

Fungicide active ingredients	"Rubber group"		"Cacao group"	
	ED_{50}	ED_{100}	ED_{50}	ED_{100}
Copper (as Bordeaux mixture)	14	34–38	15–16	34–38
Copper (as cuprous oxide)	16	40–50	18–20	40–50
Copper (as oxy chloride)	15	40–50	15–18	40–50
Ziram[a]	11	30–35	13–18	65–75
Captan[b]	8	15–20	13–18	20–25
Maneb	4	10–15	9–13	10–20
Fentin acetate	0.2	1.1–1.4	0.2	1.1–2.0
Phenyl mercury nitrate	0.02	0.1–0.2	0.03–0.05	0.3–0.4
Dimethyldithiocarbamate-copper chelate	2	15–20	2–3	15–20

[a] Zinc dimethyldithiocarbamate.
[b] N-(Trichloromethanesulphenyl)cyclohex-4-ene-1,2-dicarboximide.

the same effectiveness, whereas phenylmercury nitrate was the most toxic fungicide.

After 8 applications of a triphenyltin acetate suspension (0.04 and 0.08% at intervals of 3 wk), field tests by HISLOP (1963) showed that the infection with *Phytophthora palmivora* had been reduced by about 50%.

However, the disease could not be sufficiently controlled by the tin compound or other fungicides.

MULLER et al. (1969) reported about their field tests against *Phytophthora palmivora* in cacao, that triphenyltin chloride in concentrations of 75 g/100 L of water was more effective than copper oxy chloride preparations with 500 g Cu/100 L of water (as is commonly used in Cameroun). Field experiments with triphenyltin hydroxide (MULLER and NJOMOU 1970) confirmed the superiority of triphenyltin compounds: the hydroxide was 10 times more effective than copper oxy chloride (Table CIII).

Table CIII. *Field experiments with different fungicides against Phytophthora palmivora in cacao* (after MULLER and NJOMOU 1970).

	% infection[a] with Phytophthora palmivora after days			
Fungicide and concentration	15	30	45	60
Propineb (80% W.P.); 0.24% a.i.	28.8	73.5	78.2	78.2
Copper oxy chloride (50% Cu, W.P.) 1.0% a.i.	15.4	38.3	46.1	52.2
Captafol[b] (80% W.P.); 0.2% a.i.	10.5	20.0	34.7	35.3
Triphenyltin hydroxide (20% W.P.) 0.04% a.i.	0.5	16.7	22.2	25.0

  [a] Rounded figures.
  [b] N-(1,1,2,2,-Tetrachloromethanesulphenyl)cyclohex-4-ene-1,2-dicarboximide.

ι) *Application of triphenyltin compounds to coffee.*—Several coffee diseases react to treatment with triphenyltin compounds: coffee berry disease (caused by *Colletotrichum coffeanum* Noack), coffee leaf rust (*Hemileia vastatrix* Berk. et Br.), and brown spot disease (*Cercospora coffeicola* B. et CKe).

αα) *Coffee berry disease and coffee leaf rust:*

The coffee berry disease can be controlled with 200 to 600 g of triphenyltin compound/ha (*Codex Committee* 1970); spraying intervals of 4 wk were recommended (*T.P.R.I.* Ann. Rept. 1971).

HOCKING (1965) reported that triphenyltin acetate is about as effective against coffee leaf rust as copper oxy chloride. The same holds for triphenyltin hydroxide (MULINGE 1970). However, after other fungicides had proven to be more effective against coffee berry disease than the triphenyltin compounds, it was suggested to use these only for simultaneous infections with *Colletotrichum* and *Hemileia* where the latter is the major problem (BAKER 1971 and 1973, *Coffee Research Institute Ruiru* 1972, OKIOGA and MULINGE 1974); triphenyltin compounds are further recommended for very efficient control of the giant looper (*Ascotis serenaria reciproca* Walk.) (*Coffee Research Institute Ruiru* 1972).

Under the climatic conditions of India, however, triphenyltin compounds were not quite as effective against coffee leaf rust as the copper preparations commonly used in that country (GOVINDARAJAN 1971).

In Tanzania, the coffee berry disease was successfully controlled with

Table CIV. *Effects of different fungicides on coffee berry disease (Colletotrichum coffeanum Noack) in laboratory test and field experiment; 6 treatments at intervals of 4 wk* (excerpt from T.P.R.I. 1971).

Fungicide	% Inhibition of spore germination at different fungicide concentrations						Disease control rating[a]
	1,000 ppm	500 ppm	100 ppm	10 ppm	5 ppm	1 ppm	
Untreated	—	—	—	—	—	—	10.44
Chlorthalonil[b] (75% W.P.)	100	100	100	100	99	5	16.60
Captafol (50% W.P.)	100	100	100	100	100	100	6.02
Triphenyltin acetate (60% W.P.)	100	100	100	100	100	99	2.06
Triphenyltin hydroxide (50% W.P.)	100	100	100	100	100	88	1.99
Benomyl[c] (50% W.P.)	100	100	94	89	83	80	1.27

[a] Field test; the smaller the no., the better the control.
[b] Tetrachloroisophthalonitrile.
[c] Methyl-1-(butylcarbamoyl)benzimidazole-2-ylcarbamate.

triphenyltin hydroxide and acetate (Table CIV). The organotin compounds were equal to the most efficient other fungicides.

The results of laboratory tests with triphenyltin acetate against coffee leaf rust are shown in Table CV (HOCKING 1965). They were confirmed

Table CV. *Infection of coffee leaf discs when sprayed with fungicides at 3 rates of deposit before or 1 wk after inoculation with H. vastatrix* (excerpt from HOCKING 1965).

Fungicide	Deposit (mg a.i./m²)			Infection (spots/disc)					
				Sprayed before inoculation			Sprayed after inoculation		
	A	B	C	A	B	C	A	B	C
Untreated	—	—	—	—	12.0	—	—	—	—
Dithianon[a] (75% W.P.)	20.0	2.0	0.2	0.8	3.0	8.8	2.2	5.8	12.3
Copper oxy chloride (20% W.P.)	50.0	5.0	0.5	0	0	5.0	8.5	10.0	9.8
Chlorthalonil (75% W.P.)	6.0	0.6	0.06	0	2.7	10.1	3.2	3.8	8.8
Triphenyltin acetate (60% W.P.)	20.0	2.0	0.2	0	0	6.6	2.1	1.7	7.8

[a] 2,3-Dicyano-1,4-dithia-anthraquinone.

by later studies (HOCKING 1967 b). In the field, the tin compound also proved to be about as effective as copper oxychloride (HOCKING and FREEMAN 1968).

In experiments by MULINGE (1970), triphenyltin hydroxide controlled *Colletotrichum coffeanum* and also *Hemileia vastatrix* as well as copper oxy chloride.

VINE and VINE (1971) conducted field tests to compare different fungicides against coffee berry disease and coffee leaf rust. They found that only copper oxychloride and triphenyltin hydroxide control both diseases at the same time. However, against coffee berry disease, the triphenyltin compound was inferior to other fungicides (Tables CVI and CVII).

Table CVI. *Effect of triphenyltin hydroxide on leaf rust* (after VINE and VINE 1971).

Fungicide and concentration	Leaf rust infection (%)	Clean coffee (kg/ha)
Untreated	55.7	389
Copper oxy chloride (0.35% a.i.)	11.9	714
Triphenyltin hydroxide (0.12% a.i.)	12.9	781

Experiments by BAKER (1971 and 1973) also showed that triphenyltin hydroxide is more effective against coffee leaf rust than against coffee berry disease (Table CVIII).

OKIOGA and MULINGE (1974) tried a combined treatment with Captafol and triphenyltin hydroxide against coffee berry disease, but failed to reach the effectiveness of other fungicides (Table CIX).

$\beta\beta$) *Brown spot disease:*

Triphenyltin compounds were found to be more effective against brown spot disease than the commonly used copper fungicides (GOVINDARAJAN 1971).

VENKATARAMIAH (1971) sprayed *Coffea arabica* seedlings with concentrations of 0.1 to 0.25% of triphenyltin compounds (the amount of the active ingredient is not listed). Of all tested fungicides, the tin compounds were the most effective against *Cercospora coffeicola;* the seedlings did not show any damage.

κ) *Application of triphenyltin compounds to hops.*—The first report that the primary or early infection of hop with *Pseudoperonospora humuli* Wils can be controlled with triphenyltin compounds was published by LIEBL (1968). He recommended 2 sprayings with 70 to 100 ml of liquid (0.06% triphenyltin acetate)/plant at sprouting time when the young shoots are 5 to 10 and 30 to 40 cm long (LIEBL 1968, ZATTLER and LIEBL 1970). Similar recommendations were made by KOHLMANN and LUEDERS (1973); 2 sprayings with 0.06% triphenyltin acetate (300 g of active ingredient/ha) should be applied when the sprouts are 5 and at most 30 cm high.

Table CVII. *Effect of different fungicides against coffee berry disease; summarized yield results of recommended materials* (after VINE and VINE 1971).[a]

Product	Clean coffee (kg/ha)				
	1969 Low altitude	1970 High altitude	1970 Low altitude	1970 Low altitude	1970 Low altitude
Yield potential at flowering	1,500	1,000	900	900	1,750
Captafol (80% W.P.)	2,522 (0.6)	976 (0.4)	599 (0.2)	—	1,363 (0.4)
Copper oxy chloride (50% Cu W.P.)	2,315 (1.0)	844 (1.0)	890 (0.7)	714 (0.7)	—
Triphenyltin hydroxide (50% W.P.)	2,047 (0.175)	864 (0.25)	511 (0.175)	781 (0.25)	1,520 (0.25)

[a] The values in parentheses indicate % rate product of application in a particular trial.

Table CVIII. *Effect of fungicides against coffee leaf rust* (excerpt from BAKER 1973).

Treatment and rate (% product)	Leaf rust infection (%)	Yield clean coffee (kg/ha)	Infection in pick (%)	Peak infection 8/1 (%)
Unsprayed	48.5	1,839.6	42.2	13.65
Chlorthalonil (50% W.P.); 0.4%	34.2	1,734.4	36.05	18.9
Triphenyltin hydroxide (50% W.P.); 0.25%	9.1	1,864.9	36.9	13.3
Captafol (80% W.P.); 0.4%	10.5	3,384.1	23.3	3.6
L.S.D.P. = 0.05	8.0		7.3	6.5

Table CIX. *Coffee berry disease-infection and yield in Jacaranda plot 14, Yana Estate, on plots sprayed with different fungicides* (excerpt from OKIOGA and MULINGE 1974).

Treatment and rate (% product)	Yield (kg/plot)	Infection in pick (transf.) (%)
Unsprayed	121.72	39.7
Captafol (80% W.P.) + triphenyltin hydroxide (50% W.P.); 0.3% + 0.15%	127.54	44.9
Carbendazin (50 W.P.)[a]; 0.1%	183.70	22.0
Captafol (80% W.P.); 0.4%	217.08	27.8

[a] Methyl-2-benzimidazolecarbamate.

LEIBELT (1970) recommended applying the first spraying with triphenyltin acetate (0.06%) immediately after the hop has been uncovered and pruned, using 300 ml/stock in plant-bed treatment (spraying about 50 cm wide). If single plants are treated, 100 ml/plant is sufficient. The second spraying should take place when the plants are 30 cm high, again with 300 ml in plant-bed or 150 ml in single-plant treatment.

The application of triphenyltin acetate causes minor to medium necrosis of the leaves; however, this does not interfere with the growth. The phytotoxic effects remain low if the active ingredient is applied at a concentration of no more than 0.06% and as long as the shoots have not reached more than 30 to 40 cm (ZATTLER and LIEBL 1970).

The infection with *Pseudoperonospora humuli* can also be controlled by treating the soil with a 0.05% suspension of triphenyltin compounds (*Codex Committee* 1970).

After 2 sprayings with 70 ml of a 0.06% mixture/plant (length of shoot 5 respectively 30 to 40 cm), the number of infected shoots had decreased from 120 to 9 (LIEBL 1968) (Table CX).

Table CX. *Control of primary infection of hop with Pseudoperonospora humuli Wils* (after LIEBL 1968).[a]

Fungicides and concentration	Amount/ plant	Infected shoots in 1,000	Infected leaves on[b]				
			5/11	5/15	5/23	5/29	6/2
Untreated	—	120	2	4	4	6	7
Streptomycin 0.5%	2 × 70 ml	3	0	0	0	0	0
Triphenyltin acetate (60% W.P.) 0.1%	2 × 70 ml	9	0	0	0	0	0
Test preparation 0.1%	2 × 300 ml	29	0	0	0	0	0

[a] Spraying dates 5/5 and 5/15.
[b] Infection rating: 1 = nil and 9 = 100%; this was a year with unusually severe infection of the leaves.

One-hundred ml spray volume/plant produced similar results (ZAT-TLER and LIEBL 1970; Table CXI).

Table CXI. *Effects of different fungicides against primary infection of hop* (after ZATTLER and LIEBL 1970).

Fungicides and concentration (% product)	Height of plants at each treatment (cm)			No. of infected sprouts	Infected leaves[a]
Untreated	—	—	—	104	4
Organic fungicide; 0.2%	5–10	30–40	80–100	55	1
Systemic fungicide; 0.2%	5–10	—	80–100	6	1
Triphenyltin acetate (60% W.P.); 0.1%	5–10	30–40	—	7	1

[a] Infection rating: 1 = nil and 9 = 100%.

λ) *Application of triphenyltin compounds to peanuts.*—Triphenyltin compounds have been used at various times to control leaf spot in peanuts (*Cercospora arachidicola* Hori and *Cercospora personata* [B. and C.] E. and E.). TER HORST (1961) first reported good results with the acetate in Surinam; there, the disease is a severe problem, because copper and sulphur compounds do not work satisfactorily.

The dosage of 300 to 360 g of active ingredient/ha was found to be an efficient amount (Table CXII). However, the correlation between yield and reduced infection was poor, which may be due to some phytotoxic effects of the organotin compound.

Other authors suggested lower dosages of 200 to 300 g/ha (HAERTEL 1964, *Codex Committee* 1970).

Table CXII. *Determination of optimum dosage of triphenyltin acetate to control leaf spot in peanuts* (after TER HORST 1961).

Dosage (kg a.i./ha)	Values (% of untreated)		
	Leaf infection	Wt of plant	Yield
0 (untreated)	100	100	100
0.10	84.0	118.7	129.2
0.20	71.6	118.3	118.8
0.30	69.0	94.2	116.7
0.36	59.8	109.3	150.0
0.40	62.6	128.4	133.3

Experiments by WADSWORTH *et al.* (1967) with 225 g of triphenyltin hydroxide/ha at intervals of 14 days also produced rather low yields. This again suggested phytotoxic effects which, however, were not conspicuous. Since the infection rate was generally rather low, there were no significant treatment effects compared with the control plants (see also BOYLE 1963).

Good results with triphenyltin compounds against rust (*Puccinia arachidis*) were reported by *Queensland Department of Primary Industries* (1972–73 and 1975–76) and by RAEMAKERS and PRESTON (1977).

In experiments with various fungicides (TER HORST 1961), triphenyltin acetate was rather effective against *Cercospora,* but did not produce satisfactory yields (Table CXIII).

Table CXIII. *Effect of different fungicides against Cercospora leaf spot on peanuts; 9 treatments[a] in intervals of 7 days* (excerpt from TER HORST 1961).

Fungicide and concentration	Total spots				Yield[b]
	11/23	11/30	12/3	12/4	
Untreated	179	144	102	933	100
Zineb (78% W.P.); 0.32% a.i.	40	95	55	275	95.0
Triphenyltin acetate (20% W.P.); 0.06% a.i.	144	83	45	104	104.7
Captan (50% W.P.); 0.1% a.i.	49	84	37	219	105.0
Maneb (78% W.P.); 0.1% a.i.	87	71	56	193	113.4

[a] 500 L/ha.
[b] In % of untreated.

ADDY and DASH (1966) obtained good results in controlling *Cercospora* leaf spot ("Tikka disease") by using 300 g of triphenyltin acetate + 100 g of maneb in 625 L/ha. Five treatments were applied during the vegetation period. The tin compound was able to reduce the leaf infection considerably and to increase the yield by 34.4% compared with the untreated controls (respectively, 19% and 9% compared with Bordeaux mixture and zineb treatment) (Table CXIV). However, with 1.13 or 1.56 kg/ha of triphenyltin hydroxide (20% W.P.), BOYLE (1963) obtained lower yields than from the untreated control plants.

Table CXIV. *Effect of fungicides at different intervals of spraying on Tikka disease severity in 1965–66* (after ADDY and DASH 1966).

Fungicide and amount	Leaf area diseased on the following dates[a]					Average yield (kg/plot)
	Feb. 5	Feb. 20	Mar. 7	Mar. 22	Apr. 6	
Untreated	1.62 (7.30)	6.68 (14.95)	16.33 (23.80)	24.42 (29.60)	37.36 (37.67)	3.072
Bordeau mixture; 2.25–2.25–225	0.38 (3.56)	1.57 (7.21)	3.29 (10.44)	5.38 (13.40)	8.04 (16.47)	3.476
Zineb (78% W.P.); 2.0 kg/ha ai.a.	0.20 (2.57)	0.54 (4.21)	2.00 (8.03)	3.83 (11.29)	6.05 (14.23)	3.792
Triphenyltin acetate + maneb (60% + 20% W.P.); 0.3 + 0.1 kg/ha a.i.	0.14 (2.20)	0.40 (3.61)	1.49 (7.03)	2.80 (9.63)	4.56 (12.31)	4.128
C.D. for ang. val. at 5%	0.4358	0.7860	1.0820	0.8686	1.2785	
C.D. for yield at 5%						0.3320

[a] Figures within parentheses indicate angular values.

Triphenyltin hydroxide or triphenyltin compounds combined with mancozeb and benomyl were used to control *Puccinia arachids;* yields increased by 60 to 100% (*Queensland Department of Primary Industries* 1972–73 and 1975–76, RAEMAKERS and PRESTON 1977).

WADSWORTH *et al.* (1967) tested triphenyltin acetate combined with maneb as disinfectant for seed peanuts. One-hundred kg of seed peanuts were treated with 228.8, 189.0, or 151.3 g of triphenyltin acetate and 75.6, 63.0, or 50.4 g of maneb. This improved germination, but the effects were not as good as with specific disinfectants commonly used.

μ) *Further applications of triphenyltin compounds in different areas of agriculture.*—Of the other applications of triphenyltin compounds described in the literature, only a few shall be mentioned.

SEBBEL (1975) recommended triphenyltin acetate together with a wetting and sticking agent ("Synergid") as the most effective fungicide against increasing attacks of European canker (*Nectria galligena* Bres.) in fruit trees. With the knowledge of infection conditions, experiments were conducted at 2 different times using 12 fungicides to which oil had been added. With a single treatment, triphenyltin acetate (0.06% a.i.) together with 2% Synergid produced the best effects lasting for a long period of time (Table CXV).

Table CXV. *Effects of different fungicides against European canker*
(after SEBBEL 1975).

Fungicide	Infected spots at treatment on	
	11/27/1974	11/27/1974 and 1/28/1975
Untreated	18.7	18.7
Captan + parathion oil	8.9	5.8
Preparation No. 2	11.9	14.5
Preparation No. 3	17.1	14.3
Preparation No. 4	8.3	10.6
Preparation No. 5	13.3	10.9
Preparation No. 6	17.1	23.0
Preparation No. 7	21.5	17.1
Preparation No. 8	17.5	10.5
Preparation No. 9	16.3	13.0
Preparation No. 10	13.1	17.9
Preparation No. 11	8.9	17.1
Triphenyltin acetate (60% W.P.); 0.06% a.i. + 2% sticker	4.3	6.6

The parcels treated with triphenyltin acetate showed even better results
in the control of new infections than are obvious from Table CXV after
treatment, infected spots occurred only on old wood and had been caused
by systemic fungus growth. The highest number of infections in untreated
trees occurred in one-yr-old wood and was completely eliminated with
triphenyltin acetate.

Triphenyltin acetate in aqueous suspension, with 0.1% concentration
of active ingredient, was also very effective against *Phytophthora cactorum*
(Leb. and Cohn) Schroet, the cause of collar rot in apple trees (NIEN-
HAUS 1959). The tin compound worked better than the commonly used
copper and mercury formulations and was surpassed only by copper
pastes; their use, however, requires excessive manual labor, since they
have to be brushed very carefully onto the whole trunk (Table CXVI).

KRAEMER (1970) tested the effects of triphenyltin acetate against leaf
curl in peach trees (*Taphrina deformans* [BERK.] Tul.). Three treatments
with concentrations of 0.02% were about as effective as the comparative
fungicide Dodine with 0.065% active ingredient.

DIERCKS (1957) first tried to control bark blight in osiers (*Glomerella
miyabeana* Fuk. v. Arx.) with chemicals. This disease, caused by fungi,
attacks the bark and kills the shoot tops, which lowers yield and quality
of osier rods used for basket making. Triphenyltin acetate initially was
used in concentrations of 0.17%, but this caused considerable damage.
Therefore, starting with the fifth treatment, the concentration was lowered
to 0.08%. Eleven sprayings with triphenyltin acetate produced very good
results. The fungicide effects were far better than with captan (0.2%
active ingredient); however, the yields were lower due to phytotoxic side
effects (Table CXVII).

Table CXVI. *Preventive effects of different fungicides against*
Phytophthora cactorum in apple orchards
(after NIENHAUS 1959).[a]

Fungicide and concentration (% product)	New infections on 60 trees (%)
Untreated	15
Hg-preparation I (0.2%)	13
Cu-oxy chloride A (3% and 1.5%)	7
Cu-oxy chloride B (3%)	7
Cu-preparation I (paste 3%)	11
Cu-preparation II (1.5%)	9
Cu-preparation D (paste 1:1)	0
Triphenyltin acetate (20% W.P.) (0.5%)	4

[a] Spraying dates March to May 1958; evaluation August 1958.

It was remarkable that only with triphenyltin acetate the undesirable ramification of rods diminished and the bark could be removed easily.

In a further experiment, DIERCKS (1959) pursued the question whether the phytotoxicity could be lowered or eliminated by combining triphenyltin with captan. He also wanted to find out if a spraying interval of 14 days is sufficient for control of the fungus. His results showed (Table CXVIII) that a combined preparation is less phytotoxic but its fungicidal effects are inferior to those of triphenyltin acetate alone. Since the osier grows very fast, spraying must be repeated at intervals of 7 days even with minor infections.

BACHTHALER and DAHTE (1959) were able to almost completely eliminate the branch canker of poplar (*Dothichiza populea* Sacc. and Br.) with 3 applications of triphenyltin acetate in concentrations of 0.06% (Table CXIX). One or 2 treatments of the one-yr-old trees were not sufficient. Experiments with 2-yr-old poplars were inconclusive because of infected spots in the 2-yr-old wood which could have been the result of an infection prior to treatment.

POWELL and LEBEN (1973) tested 6 fungicides against Scotch pine (*Pinus silvestris*) disease *Lophodermium pinastri* (Schrad. and Fr.). They applied 5 treatments to naturally infected 5- to 7-yr-old pines, using among others triphenyltin hydroxide (107 g/380 L of water) and maneb (720 g/380 L of water). Triphenyltin hydroxide was nearly as effective as maneb, used here as a standard comparison (Table CXX).

KUETHE (1959) used triphenyltin acetate twice (no data on dosages) against *Lophodermium pinastri* (Schrad. and Fr.) Chev. with good results, similar to those with zineb.

Experiments by JANCARIK (1969) showed that only 2 fungicides have sufficient effects on pine needle blight (*Dothistroma pini* Hulbary) in *Pinus radiata*: triphenyltin acetate and copper oxy chloride. The latter has the disadvantage of being phytotoxic to *Pinus radiata* seedlings. The

Table CXVII. Effects of different fungicides on yield and infection rate of osier caused by *Glomerella miyabeana* (excerpt from DIERCKS 1957).

Fungicide and concentration	Rods		No. of rods in %					Length of rods (cm)	Bark removal
	No.	Wt (kg)	Healthy	Mild infection	Medium infection	Severe infection			
Untreated	776	9.5	0	3.7	46.2	50.1		66	normal
Cu-oxy chloride (50% Cu W.P.); 0.5% a.i.	813	10.9	0.1	8.5	52.3	39.2		96	normal
Captan (50% W.P.); 0.2% a.i.	783	13.9	0.8	7.7	56.7	35.3		130	normal
Triphenyltin acetate (20% W.P.); 0.08% a.i.	794	12.5	8.0	17.3	50.6	24.1		130	exceptionally good

Table CXVIII. *Effects of fungicides against the osier parasites Fusicladium (All. et Tub.) Lind. and Glomerella miyabeana Fuk. v. Arx. (after* DIERCKS 1959*).*

Fungicides and concentration	Spraying interval (days)	Rods		Shoots		Rods with bark-blight (%)		
		No. (%)	Wt (%)	Without infection	Traces of infection	Mild	Medium	Severe
Untreated	—	100	100	0	1.1	43.6	42.2	13.2
Zineb (78% W.P.); 0.32% a.i.	7	102.4	142.2	6.6	58.3	21.9	9.8	3.4
	14	108.1	134.2	—	33.9	41.1	19.3	5.7
Captan (50% W.P.) + triphenyltin acetate (20% W.P.); 0.075 + 0.03% a.i.	7	111.7	130.4	25.0	49.9	19.8	4.0	1.4
	14	96.1	106.1	1.6	47.7	38.6	9.5	2.6
Triphenyltin acetate (20% W.P.); 0.06% a.i.	7	118.9	99.0	51.8	33.4	10.6	3.5	0.7
	14	115.0	120.0	—	39.4	46.9	10.8	2.9

Table CXIX. *Effects of fungicides against Dothichiza populea in poplars* (after BACHTHALER and DAHTE 1959).

Fungicide and concentration	No. of treatments	Infection (%)		
		Healthy	Mild	Severe
Untreated (average)	—	8	7.4	84.6
Copper hydroxide (35% W.P.);				
0.26% a.i.	1	9.3	12.5	79.2
	2	69.4	17.4	13.2
	3	73.8	8.7	17.5
	4	77.5	10.0	12.5
Triphenlytin acetate (20% W.P.);				
0.06% a.i.	1	12.5	4.1	83.4
	2	26.4	8.7	65.8
	3	90.8	0	9.2
	4	95.6	2.2	2.2

Table CXX. *Control of needlecast on Scotch pine with various fungicides applied monthly from June through October* (excerpt POWELL and LEBEN 1973).

Treatment and rate	Fallen or brown needles (%)	Disease[a] (%)
Check (unsprayed)	42.5 c[b]	81.5 c
Thiabendazole (43% flow);[c] 0.13% a.i.	10.0 a	31.6 b
Triphenyltin hydroxide (47.5% W.P.); 0.028% a.i.	5.9 a	8.2 a
Thiophanate-methyl[d] (70% W.P.); 0.08% a.i.	4.9 a	6.0 a
Maneb (80% W.P.); 0.19% a.i.	2.1 a	2.0 a

[a] Disease visually rated according to the Horshall-Barrett scale and converted to %.
[b] The letters indicate Duncan's multiple range groupings of treatments which do not differ significantly at the 5% level.
[c] 2-(4-Thiazolyl)benzimidazole.
[d] 4,4'-o-Phenylene-bis(3-thioallophanate).

infection with *Dothistroma* can occur immediately after germination (incubation period 4 to 5 mon). Therefore, an early treatment in monthly intervals is necessary. A concentration of 0.3% of a commercial product (60% triphenyltin acetate with 20% maneb) is sufficient (Table CXXI).

Triphenyltin acetate (810 g in 380 L of water) at spraying intervals of 7 or 14 days failed to control terminal crook disease (*Colletotrichum acutatum f. sp. pinea*) in *Pinus radiata* (GILMOUR and VANNER 1972).

Experiments by SCHMIDT (1962 a) showed that the secondary infection of phaseolus-beans with anthracnose (*Colletotrichum lindemuthianum* Bri. and Cav.) can be controlled best with triphenyltin acetate (2.5 kg/ha, 20% commercial product; 3 treatments after flowering). This favorable result was confirmed in later experiments (SCHMIDT 1962 b).

Table CXXI. *Effect of different fungicides on pine needle blight (Dothistroma pini)* (excerpt from JANCARIK 1969).

Fungicide and concentration	Spray interval (days)	Diseased trees (%)	Mean degree of burning	Persistency factor
Untreated	—	95.5	—	—
Zineb (78% W.P.); 0.16% a.i.	14	20.8	—	73
	30	47.7	—	
Maneb (78% W.P.); 0.16% a.i.	14	2.6	—	67
	30	35.8	—	
Captafol (80% W.P.); 0.16% a.i.	14	12.4	—	87
	30	25.1	—	
Triphenyltin acetate + maneb (60% W.P.); 0.18% a.i.	14	1.1	—	95
	30	6.5	—	
Copper oxy chloride (50% Cu, W.P.); 0.2% a.i.	14	0.5	1.7	96
	30	4.7	0.8	

Triphenyltin acetate was well tolerated by all varieties of beans; even a 5-fold overdose did not cause any damage (in the variety "Saxa") (Table CXXII).

ANDRÉN and OLOFSSON (1959) also obtained by far the best results against anthracnose of beans with triphenyltin acetate (in a concentration

Table CXXII. *Effects[a] of different fungicides against anthracnose in beans (variety "Brittle Wachs")* (excerpt from SCHMIDT 1972 a).

Fungicide and amount (kg/ha)	Infection (%)	
	Total	Severe infection
Untreated	80.0	37.4
Orthocid 50 (Captan[b]); 3 kg/ha	66.8	25.4
Zineb (80% W.P.); 3 kg/ha	70.6	21.8
Copper oxy chloride (50% Cu, W.P.); 5 kg/ha	64.8	20.2
Maneb (80% W.P.); 2 kg/ha	61.4	22.6
Ortho Phaltan 50 (Folpet[c]); 3 kg/ha	54.8	13.4
Triphenyltin acetate (20% W.P.); 2.5 kg/ha	24.2	2.0

[a] Median of 5 experiments.
[b] 3a,4,7,7a-Tetrahydro-N-(trichloromethanesulphenyl)phthalimide.
[c] N-(Trichloromethanesulphenyl)phthalimide.

of 0.25%; 20% commercial preparation) (Table CXXIII). With 3 spray-ings at intervals of 11 or 14 days without disinfecting the seed beans, the net yield could be increased by 145%. When the seed beans had been disinfected in addition to spraying, the net yield was 160% higher than that of the controls.

Table CXXIII. *Control of anthracnose in beans with different fungicides* (excerpt from ANDRÉN and OLOFSSON 1959).

Fungicide and concentration (% product)	Infected pods (wt %)
Untreated	63
Copper oxy chloride (50% Cu W.P.); 0.5%	43
Zineb (80% W.P.); 0.5%	15
Triphenyltin acetate (20% W.P.); 0.25%	6

Other applications of triphenyltin compounds, compiled from the literature, are shown in Table CXXIV.

c) Control of fresh water snails and sea organisms with triphenyltin compounds

1. Fresh water snails.—FLOCH and DESCHIENS (1962) recognized the molluscicidal properties of triphenyltin compounds. These compounds were tested extensively during experiments to control fresh water snails which are intermediary hosts for the pathogens of bilharziasis and helminthiasis (see STRUFE 1968, *Organisation Mondiale de la Santé* 1967).

HOPF and MULLER (1962) tested, together with several other com-pounds, triphenyltin acetate and its effects against *Australorbis glabratus* (also called *Biomphalaria glabrata;* see HOPF et al. 1967). This snail is an intermediary host for *Schistosoma mansoni*, native to South America and Africa. The scientists found that the triphenyltin compound had good molluscicidal and ovicidal effects (Table CXXV).

FLOCH and DESCHIENS (1962) and DESCHIENS and FLOCH (1963) found strong molluscicidal effects of triphenyltin compounds depending on the duration of exposure. In laboratory tests under conditions as natural as possible (sediment and vegetation according to WHO-method), 2 kinds of snails (*Australorbis glabratus* and *Bulinus contortus*) and eggs of the pond snail *Limnaea stagnalis* were exposed to aqueous solutions of tri-phenyltin acetate or chloride. With normal procedure (24 hr exposure to the active ingredient, then rinsing of the animals and 72 hr further observation), triphenyltin chloride killed the snails at a concentration of 1 ppm. With the same length of contact, but longer observation (4 to 5 days), all animals died already at a concentration of 0.5 ppm. If the exposure was extended to 3 days and the observation period to 6 to 8

Table CXIV. *Various agricultural applications of triphenyltin compounds.*

Plant	Disease	Agent	References
Snap beans	Root rot	*Rhizoctonia solani* Kuehn	ANDES & GILMORE (1963)
Banana	Sigatoka disease	*Cercospora musae* Zimm.	HAERTEL (1962), KRANZ (1965)
Pineapple	Pineapple disease	*Ceratostomella paradoxa* (Des.) Dade	ANTOINE (1957)
Apple	Scab	*Venturia inaequalis* (Looke) Wint.	BRUECKNER & HAETEL (1955)
Apple	Scab	*Venturia inaequalis* (Looke) Wint.	LUIJTEN (1960)
Apple	Scab	*Venturia inaequalis* (Looke) Wint.	PIETERS (1962)
Apple	Codling moth	*Carpocapsa palmonella* L.	MARFURT & TOSCANI (1966/67)
Apple	Oriental fruit moth	*Grapholita molesta* (Busck)	MARFURT & TOSCANI (1966/67)
Cucumber	Downey mildew	*Pseudoperonospora cubensis* (Berk. and Curt.) Rostowzew	NUGENT (1963), SITTERLY (1963 a)
Cucumber	Anthracnose	*Colletotrichum obiculare*	HORN (1963)
Cucumber	Powdery mildew	*Erysiphe cichorarearum* DC.	NUGENT (1963), SITTERLY (1963 a)
Cucumber	Gummy stem blight	*Mycosphaerella melonis*	SITTERLY (1963 a)
Cucumber	Fruit rot	*Pythium* and *Rhizoctonia* spp.	SITTERLY (1963 b)
Watermelon	Downey mildew	*Pseudoperonospora cubensis* (Berk. and Curt.) Rostowzew	SCHENK & CRALL (1963)
Tomato	Early blight	*Alternaria solani* (J.a.Gr.)	ANDRADE (1960 b)
Tomato	Late blight	*Phytophthora infestans* (de Bary)	PIETERS (1962)
Tomato	—	*Cladosporium fulvum* (Cooke)	PIETERS (1962)
Tomato	Anthracnose	*Colletotrichum* spp.	SCHROEDER (1963)
Tomato	Early blight	*Alternaria solani* (J.a.Gr.)	SCHROEDER (1963)

Table CXIV. (*continued*)

Plant	Disease	Agent	References
Orange	Scab	*Elsinoe fawcetti* (Bintacourt and Jenkins)	FISHER (1969)
Iris (bulbs)	Blue mold disease	*Penicillium* sp.	MILLER & GOULD (1963)
Tobacco	Brown spot disease	*Alternaria alternata* (Fr.) Keissl	SPURR (1972), SPURR & WELTY (1972)
Fox glove	—	*Septoria digitalis*	KRÁL et al. (1974)
Sweet maize	Corn downy mildew	*Sclerospora philippinensis* Weston	EXCONDE et al. (1975/76)
Cypress wheat	Common root rot	*Cochliobolus sativus*	CHINN (1977)
Poplar	—	*Cytospora chrysosperma* (*Valsa sordida*)	PRASAD & MOODY (1975)
Rose	Leaf blight	*Alternaria alternata* (Fr.) Keissler	SAHNI (1973)
Sunflower	Rust disease	*Puccinia helianthi*	RAMASAMY & MATHAR (1973)
Strawberry	Red core	*Phytophthora fragariae*	MONTGOMERIE & KENNEDY (1975)
Barley	Spot blotch	*Bipolaris sorokiniana*	COUTURE & SUTTON (1978)
Soybean[a]	—	—	HEPPERLY & SINCLAIR (1977)
Soybean	Frog eye leaf spot	*Cercospora sojina* Hara	HORN et al. (1975)
Soybean	Target spot	*Corynespora cassiicola* (Burk. et Curt.) Wei	HORN et al. (1975)
Soybean	Pod and stem blight	*Diaporthe phaseolorum* (Cke. et Ell.) Sacc.	HORN et al. (1975)
Grass[b]	Blind seed disease	*Gloeotinia temulenta*	HARDISON (1972)
Grapevine	Downey mildew	*Peronospora viticola*	HAETEL (1963 c)
Wheat	Node canker	*Septoria nodorum* (Berk.)	OBST & KEES (1972)
Wheat	Leaf rust	*Puccinia recondita* (Rob. a.Desm.)	BOŠCOVIĆ (1962)
Fingermillet	Blast	*Piricularia* sp.	KESHI & MOHANTY (1967)

Beets	Downey mildew	*Peronospora farinosa* (Fr.)Fr.	Byford (1965)
Chrysanthemum	Leaf spot	*Septoria chrysanthemella*	Van Eylen (1963)
Horseradish	White blister	*Albugo candida* (Pers.O.Ktze.)	Kalch-Schmidt & Krause (1976)
Corn	Smut	*Ustilago maydis* (DC Corda)	Koenig (1969)
Corn	—	*Helminthosporium sp.* (*turcicum* Nisik u.Mivake)	Anonymous (1971)
Clover	Dodder	*Cuscuta campestris*	Muromtsev & Agnistokova (1969)
Clover	Dodder	*Cuscuta monogyna*	Muromtsev & Agnistokova (1969)
Clover	Sclerotinia wilt	*Sclerotinia trifoliorum* Eriks.	Malmus (1959)
Safflower	Alternaria blight	*Alternaria carthami* (Chowdbury)	Chauhan (1970)
Digitalis lanata	Septoria blight	*Septoria digitalis* (Pass.)	Baumann (1958)
Hevea	Tappong panel disease	*Phytophthora palmivora* (Butl.)Butl.	Peries *et al.* (1962)
Citrus	Citrus bud mite	*Asceria sheldoni* (Erwins)	Sternlicht (1966)
Pinus radiata	Crook disease	*Colletotrichum acutatum* f. sp. pinea	Gilmour & Vanner (1972)

[a] Seed treatment.
[b] *Lolium perenne.*

Table CXXV. *Screening test results with selected compounds*
(after HOPF and MULLER 1962).

| Compound | Young snails | | Adult snails | | Eggs |
	LC$_{50}$ (ppm)	LC$_{100}$ (ppm)	LC$_{50}$ (ppm)	LC$_{100}$ (ppm)	LC$_{100}$ (ppm)
Sodium pentachlorophenolate	0.15	0.35	1.0	2.5	1.0
Pentachlorophenol	0.1	0.3	1.0	2.5	1.0
Copper sulfate	0.2	0.3	2.0	4.0	*ca.* 10
Tri-*n*-propyltin oxide	0.02	0.05	0.05	0.1	—
Triphenyltin acetate	—	—	0.05	0.09	0.1–1.0
Tri-*n*-butyltin acetate	0.025	0.05	0.05	0.1	—

days, a concentration of 0.05 ppm was sufficient. The eggs of *Limnaea stagnalis* were killed after 1 day of exposure to a triphenyltin chloride concentration of 1 ppm.

The effects of the triphenyltin compounds lasted about 8 mon at an initial concentration of 1 ppm, 100 days at 0.5 ppm.

Field tests by CROSSLAND *et al.* (1962) in running water showed that triphenyltin acetate (concentration about 0.5 ppm) has good effects against *Biomphalaria pfeifferi* and *Bulinus tropicus* (adult animals) and very good effects against the eggs of these snails (Table CXXVI). Further, the *Lanista ovum* snail in rice fields could be controlled with approximately 1 kg of triphenyltin acetate/ha (CROSSLAND 1964).

Table CXXVI. *Estimation of ovicidal efficacy in field experiments*
(after CROSSLAND *et al.* 1962).

Treatment	No. of eggs counted	Observed mortality (%)	Corrected mortality (%)
I.C.I. 24223 at 0.5 ppm for 8 hr	1.450	55	40
Bayer 73 at 0.5 ppm for 8 hr	1,400	84	79
Triphenyltin acetate at 0.5 ppm for 8 hr	2,200	99.9	99.9
Control			
(a) Before treatment	2,200	25	—
(b) After treatment	2,200	23	—

For *Bulinus tropicus*, DE VILLERS and MACKENZIE (1963) established an LC$_{50}$ of triphenyltin acetate and hydroxide of 0.075 and 0.005 ppm, and a LC$_{100}$ of 1 and 0.1 ppm.

CAMEY and PAULINI (1964) observed that the sensitivity of *Australorbis glabratus* to triphenyltin acetate depends on their age and—to a certain extent—also on the distribution of active ingredient in the water. An alcoholic emulsion of triphenyltin acetate was more than twice as effective as a suspension. For snails in the embryo stage, the LC$_{50}$ was 0.022 to 0.05 ppm, the LC$_{90}$ 0.04 to 0.075 ppm. For mature animals, the values were 0.26 (0.05 ppm after HOPF and MULLER 1962) and 0.58 ppm.

FRICK and JIMENEZ (1964) also reported that the toxicity of triphenyl-tin acetate against *Australorbis glabratus* varies very much with the age of the eggs and the snails. The LC_{50}, LC_{90}, and LC_{100} for eggs were, depending on age, 0.12 to 1.6, 0.22 to 3.5, and 0.5 to 4.0 ppm; for snails, the corresponding values were 0.07 to 0.66, 0.155 to 1.5, and 0.2 to 4.0 ppm. Alkyltin compounds showed distinctly better effects. However, the test results of these authors are opposed to the findings of HOPF and MULLER (1962), FLOCH and DESCHIENS (1962), and FLOCH *et al.* (1964) who achieved 100% mortality of young and adult snails with much lower concentrations.

RITCHIE *et al.* (1964) determined the effects of triphenyltin hydroxide and acetate against *Australorbis glabratus* at different concentrations and periods of contact. Whereas during 24 hr exposure several tenths ppm of active ingredient were sufficient to kill all the snails, the triphenyltin compounds had little effect in exposure of 1 to 6 hr. Tripropyl- and tributyltin compounds were much more toxic.

In experiments to control snails in rice fields, triphenyltin compounds with different anions (acetate, fluoride, pentachlorophenolate) showed about the same toxicities towards *Australorbis glabratus* (HOPF *et al.* 1967). However, because of the relatively high phytotoxicity of the tin compounds the experiments were discontinued.

SEIFFER and SCHOOF (1967) tested the effects of triphenyltin acetate and chloride against *Australorbis glabratus* in the laboratory, in ponds, and in running water. At a contact period of 6 hr, the acetate had an LC_{50} of 0.4 ppm and an LC_{95} of 1.4 ppm. In ponds, all snails were killed within 48 hr at concentrations of 1 to 2 ppm. In running water after 2 hr exposure, 93% of the naturally occurring snails were killed 300 m below the point of application when the concentration was 5 ppm. Animals were still killed when placed in a cage at a distance of over 1,500 m below the point of application.

VAN DER MAAS *et al.* (1972) described experiments to determine the concentration of triphenyltin hydroxide which does not have any detect-able effects on the pond snail *Limnaea stagnalis*. They found that as little as 0.002 ppm still caused some damage. All animals had died at 0.50 ppm after 18 hr, at 0.25 ppm after 24 hr, at 0.125 ppm after 48 hr, and at 0.01 ppm after 9 days. The triphenyltin compound was accumulated in the body of the snails (mainly in head, foot, and intestines) without any detectable degradation.

According to BALLESTEROS *et al.* (1969), the control of snails and other pests is also possible in brackish fish ponds. Two methods were established and applied with success. The spot or canal method was recommended at an infestation level of more than 50 snails/m². For this purpose, the fish pond has to be drained very gradually, so that the snails congregate in the lowest spots covered by no more than 5 to 10 cm of water. Only these places are then sprayed with triphenyltin acetate (200 to 600 g) in 19 L of water/500 to 1,000 m². The pond may not be refilled with water for 7 days after treatment. For fish ponds that are heavily infested with snails (100 snails/m²), the blanket method was recommended. Here, the water is drained to a level of 5 to 10 cm for a period of 7 days and sprayed with 400 to 600 g of triphenyltin acetate (in 19 L of water/500 to 1,000 m²). On the 8th day, the remaining water is drained and the pond has to stay dry for 7 days. After the bottom of the pond has been fertilized, it can be refilled. Triphenyltin acetate and chloride were used successfully as antifeedants against the snail *Opeas gracile* (Hutton) by ASARI and DALE (1974).

2. **Marine organisms.**—Organotin compounds (among them triphenyltin linoleate) as components in paint for ships were first recommended by TISDALE (1943). These additives prevent the growth of various organisms such as barnacles (*Balanides, Lepatides*), tubeworms (*Serpulides*), conchs (*Conchoderma*), sea moss (*Hydroids*), green algae (*Entermorpha*), brown algae (*Ectocarpus*), seaweed, and many others on ship hulls.

Later experiments confirmed the usefulness of mainly triphenyltin chloride which protects the paint on ships for a long period of time if applied in high concentrations (6 to 18%, preferably 12 to 18%) (SPARMANN 1957 and 1958, *Schering AG* 1958). Hexaphenyldistannoxane is also effective, and in concentrations of 4% (based on dry paint) it prevents the growth of pests completely (BRUECKNER *et al.* 1959).

Contrary to copper-containing paints, these compounds do not cause galvanic corrosion, which is a positive side effect for ship hulls made of aluminum (EVANS 1970). Presently, however, trialkyltin compounds may be the most commonly used (EVANS 1970, PLUM 1972).

Triphenyltin chloride or acetate can provide good protection for wooden constructions in sea water against shipworms (*Teredo diegensis, Martesia*). The wood is impregnated with a 2% solution of the compound in xylene (VIND and HOCHMANN 1962 and 1963). For simultaneous protection against *Limnoria tripunctata*, a combination of the organotin compound with a chlorinated hydrocarbon is recommended.

d) Effects of triphenyltin compounds on insects

1. **General.**—Of all tested fungicides, triphenyltin acetate has the lowest risk index for bees (BERAN 1958, STEINER 1960/1961) and was recognized by the *Biologische Bundesanstalt, Braunschweig* (1956) as "not harmful to bees." However, it does show some effects on several other kinds of insects. In this respect, 3 effects have been observed: toxicity, sterilization, and anti-feeding.

2. **Toxicity.**—Just as in warm-blooded animals and in micro-organisms, triphenyltin compounds are much more toxic to insects than the tetra-, di-, and monophenyl compounds, and the influence of the anion is also relatively low.

Experiments were mainly conducted with the common housefly (*Musca domestica* L.). According to data of several authors, the LD_{50} of triphenyltin compounds for adult flies is approximately 0.1 to 1 μg/fly in external application; the LC_{50} in food is several 100 ppm for adult animals, for larvae it is reduced by the factor 10.

Other insects for which the toxicity of triphenyltin compounds was tested are: *Spodoptera littoralis* Boisd., *Culex pipiens berbericus,* the larvae of dragonflies, *Tineola bisselliella* Humm., *Sitophilus oryzae* L., *Cellosobruchus chinensis* L., *Leptinotarsa decemlineata* Say., *Heliothis, Trichoplusia,* and others.

By dripping 1 μl of an acetone solution of triphenyltin chloride onto common houseflies (*Musca domestica* L.), BLUM and PRATT (1960) obtained an LD_{50} of approximately 0.15 μg/animal. Monophenyltin trichloride, diphenyltin dichloride, and tetraphenyltin were less toxic by several orders of magnitude.

KOCHKIN *et al.* (1964) used a different technique by evaporating organotin compound solutions on the bottom of the insect container and determining the amount of triphenyltin compound/cm^2 which kills 100% of the insects. Flies were kept in the container for 5 min, other insects for 15 min, then transferred to a clean container and observed for 24 to 96 hr. Triphenyltin acetate proved to be rather toxic at 50 μg/cm^2; it was lethal to flies, bed bugs, and cockroaches.

PIEPER and CASIDA (1965) determined the toxicity of phenyltin chlorides for common houseflies by dripping 1 μl of an acetone solution onto their back and observing them for 48 hr or by injecting 1 μl of the solution into the thorax. The triphenyltin compound showed by far the highest toxicity (LD_{50} *ca.* 1 μg/fly), whereby the mode of application had no significant influence. The toxicity trend is paralleled by the inhibiting effect on phosphoric acid cleavage from ATP through ATPase and Mg^{2+}.

KENAGA (1965 a and b) determined the toxicity of triphenyltin compounds with different anions for common houseflies by feed experiments. Table CXXVII shows the concentrations in the food which caused 95% mortality within 8 days.

Table CXXVII. *LC_{95}-Values of triphenyltin compounds for the common housefly (Musca domestica); duration of experiment 8 days* (excerpt from KENAGA 1965 b).

Triphenyltin compound	LC_{95} (ppm in the food)
$(C_6H_5)_3SnF$	1,000
$(C_6H_5)_3SnCl$	250
$(C_6H_5)_3SnJ$	250
$(C_6H_5)_3SnOH$	1,000
$(C_6H_5)_3SnOCOCH_3$	250
$(C_6H_5)_3Sn$-stearate	1,000
$(C_6H_5)_3SnSSn(C_6H_5)_3$	1,000
$[(C_6H_5)_3Sn]_3BO_3$	250

Feeding experiments with triphenyltin chloride and acetate (37.5 and 42.5 μg/ml in milk/H_2O, ratio 1:1) showed after 10 days mortalities of 73.3 and 66.6% in male, and 52.0 and 45.6% in female houseflies (KISSAM and HAYS 1966).

High mortality in houseflies was observed when the food contained 1% triphenyltin hydroxide (FYE et al. 1966) or 2.5% triphenyltin acetate or chloride (HAYS 1968).

According to ASCHER et al. (1968), almost all houseflies died after feeding on milk with > 500 ppm of triphenyltin acetate for 3 days or with 500 ppm for 5 days. The acetate and the hydroxide were found to be toxic to larvae of flies at concentrations of 50 ppm in the food; at 200 to 300 ppm, almost all larvae died (ASCHER and MOSCOWITZ 1968). Females of *Musca domestica* L. were placed in glass jars containing triphenyltin acetate in concentrations of 0.01 to 5%. Paradoxically, the mortality rate was 0% at 0.02%, *ca.* 100% at 0.15 to 0.5%, and 0% at 2.5% concentration (ASCHER and NEMNY 1976).

PAUSCH (1969) fed triphenyltin hydroxide to the little housefly (*Fannia canicularis* L.). At concentrations of 0.5%, the mortality was 0% after 3 days, 50% after 7 days. At 1% of active ingredient, 20% of the insects died within 3 days, 80% within 7 days.

Acetone solutions of triphenyltin hydroxide were injected into the stomach of cotton leaf worm larvae (*Spodoptera littoralis* Boisd.); in other experiments, the tin compound was applied externally to the third thorax segment (ABO ELGHAR and RADWAN 1971). In both application methods, the number of insects killed increased with higher doses; injection into the stomach was more harmful (Table CXXVIII).

Table CXXVIII. *Daily recorded mortality of cotton leaf worm larvae from contact and stomach action at different concentrations of triphenyltin hydroxide; active ingredient in acetone* (after ABO-ELGHAR and RADWAN 1971).

Dose/larva (ppm)	Mortality (%)					
	Stomach action after			Contact action after		
	24 hr	48 hr	72 hr	24 hr	48 hr	72 hr
100	10	20	52	0	0	8
200	15	25	48	0	4	14
400	25	50	80	4	12	26
600	68	70	88	8	18	56
800	75	100	—	6	24	64
1,000	82	100	—	8	36	72
Check	0	0	0	0	0	0
Acetone	0	0	0	0	0	0

EL-SEBAE and AHMED (1972/73) observed in field tests that triphenyltin hydroxide is toxic for 4th instar larvae of the cotton leaf worm. When pupae of *Spodoptera littoralis* Boisd. are immersed in 0.5% suspensions of triphenyltin acetate or hydroxide, a considerable number of them is killed (FINDLAY 1968).

ELBADRY et al. (1972) observed histological changes in some tissues of Spodoptera following treatment with triphenyltin hydroxide (0.125% suspension). Midgut tissues and muscles were destroyed, and there was some damage to the fat body and the Malpighian tubules.

For Culex pipiens berbericus, the LC_{50} values for triphenyltin acetate and bis(triphenyltin) sulfide were found to be 0.25 and 0.28 ppm (CASTEL et al. 1963).

Dragonfly larvae are killed within 24 hr with aqueous suspensions of 0.5 ppm triphenyltin chloride (FLOCH and DESCHIENS 1962).

One % of triphenyltin compounds in wool proved to be of high feeding toxicity to larvae of the moth (Tineola bisselliella Humm.). Depending on the tin compound used, a lower or higher percentage of the larvae had died after 14 days (mortality at impregnation of wool with triphenyltin thiophenoxide 76%, with chloride 67%, with the acetate 61%, with the bromide 53%, and with the hydroxide 50%). The bis(triphenyltin) sulfide was almost ineffective (GARDINER and POLLER 1964).

Filter paper discs, impregnated with a 1% solution of triphenyltin compounds in benzene and then dried, did not show any toxicity against Sitophilus oryzae; however, the experiments may have been impeded by adsorption of the tin compounds on the paper (GARDINER and POLLER 1964).

Thiophosphoric acid triphenyltin compounds were toxic to the Azuki bean weevil (Cellosobruchus chinensis L.). The acetone or chloroform solutions of the tin compounds were evaporated in Petri dishes of 9 cm diameter where the insects were then kept for 24 hr. The LD_{50} values for different organotin compounds were 55 to 160 μg/dish (KUBO 1965).

Colorado beetles (Leptinotarsa decemlineata Say.) showed high mortality (ca. 50%) in the larvae as well as in the beetle stage after contact with potato plants treated with triphenyltin hydroxide (BYRDY et al. 1965).

According to ASCHER and NISSIM (1964), triphenyltin acetate and hydroxide are toxic to the cutworm (Agrotis segetum Schiff.) of beets. JENKINS (mentioned by ASCHER and NISSIM 1964) found that the tobacco hornworm (Protoparce sexta) was killed by triphenyltin hydroxide.

WOLFENBARGER et al. (1968) determined the toxicity of triphenyltin compounds to larvae of the bollworm (Heliothis zea Boddie) and the tobacco budworm (Heliothis virescens F.). The compounds were either applied externally in acetone solution (1 μl) or sprayed onto the cotton plants hosting the larvae. In the first experiment, mainly chloride and hydroxide were found to be toxic; in the feeding tests, chloride, acetate, and hydroxide showed about the same effects (Table CXXIX).

For larvae of the cabbage looper (Trichoplusia ni Huebner), LD_{50} of triphenyltin hydroxide was 0.33 mg/g, of the beet army worm (Spodoptera exigua Huebner) 0.35 mg/g. A dose of 0.43 mg/g was not toxic for

Table CXXIX. *Mortality of bollworm and tobacco budworm larvae in laboratory tests with various triphenyltin compounds* (excerpt from WOLFENBARGER *et al.* 1968).

Triphenyltin compound	Insect	Rate of application (mg/g)	% killed after topical application after		% killed by spray on plant foliage after	
			24 hr	48 hr	24 hr	48 hr
Acetate	bollworm	1.12	0	1	9	18
	tobacco budworm	1.07	0	0	2	2
Chloride	bollworm	0.01	22	34	9	25
	tobacco budworm	0.17	16	22	0	0
Hydroxide	bollworm	0.18	34	39	1	23
	tobacco budworm	0.32	27	32	0	0

Heliothis virescens F. The LD_{50} of triphenyltin chloride was 1.09 mg/g for *Heliothis virescens* F., and 0.019 mg/g for *Heliothis zea* Boddie. The latter was more sensitive to organotin compounds than *Heliothis virescens* F.

Field tests with high dosages of triphenyltin hydroxide (5.6 kg/ha) on cotton showed a marked reduction in the number of *Heliothis* larvae, and the damage to the plants was decreased by 10 to 15%.

After external application of 10 μg triphenyltin acetate or hydroxide, GRAVES *et al.* (1965) found the mortality of the tobacco budworm (*Heliothis virescens* F.) to be 30 or 35%. However, attempts to reproduce the results of this insect and also for *Heliothis zea* Boddie were not successful.

Tribolium confusum Jacquelin du Val (confused flour beetle) dies at dosages of 1,000 ppm of triphenyltin acetate in the food (KENAGA 1965 b).

Triphenyltin hydroxide in a concentration of 0.05 to 0.1% in the food is toxic for the third instar nymphs of the pea aphid (*Acyrtosiphon pisum* Harris) (BHALLA and ROBINSON 1968).

Thirty % or more of adult boll weevils (*Anthonomus grandis* Boheman) died having received food with 0.05 to 0.2% triphenyltin acetate for 3 days. The same result was found with 0.001 to 0.1% triphenyltin hydroxide (KLASSEN *et al.* 1968). The data listed in Table CXXX were obtained by immersing the same insect for 5 sec in solutions containing triphenyltin compounds or feeding the active ingredient in a 10% sugar solution for 48 hr (HAYNES *et al.* 1971).

Only 0.0005% triphenyltin acetate, chloride, or bis(triphenyltin) sulfide in flour caused 50% of the larvae of the rust-red flour beetle (*Tribolium castaneum*) to die (McINTOSH 1966).

LARGE (1965) has published data about the control of aphids on pecan trees with triphenyltin hydroxide.

The mortality of the Indian mealworm (*Corcyra cephalonica* S.) was 100, 80, and 20% when fed with flour containing 0.02 to 0.04, 0.01 respectively 0.0025% triphenyltin acetate (DALE and SARADAMMA 1974).

Table CXXX. *Toxicity of triphenyltin compounds to Anthonomus grandis Boheman* (excerpt from HAYNES *et al.* 1971).

Triphenyltin compound	Concentration and solvent	Treatment of the insects	Mortality after 7 to 14 days (%)
Acetate	0.1% in ethanol	dipping	100
	0.5% in sucrose solution	feeding	100
Chloride	0.5% in ethanol	dipping	100
	0.5% in sucrose solution	feeding	98
Hydroxide	1% in ethanol	dipping	100
	0.5% in sucrose solution	feeding	100
bis-Triphenyltinsulphide	10% in DMF	dipping	96
	0.5% in sucrose solution	feeding	4

Some mortality was observed after dipping convergent lady beetles (*Hippodamia convergens* Geuren-Meneville) into 0.3% suspensions of triphenyltin hydroxide or when keeping the insects in Petri dishes containing > 0.47 µg/mm² of the triphenyltin compound. Soya leaves treated with suspensions of $C_6H_5)_3SnOH$ (conc. $\geq 0.3\%$) were toxic for this animal (LIVINGSTON *et al.* 1978 a).

The toxicity of *Bacillus thuringiensis* for larvae of *Plutella xylostella* (L.) was increased when combined with triphenyltin hydroxide; there was synergism with the factor 3.7 (HAMILTON and ATTIA 1977).

LIVINGSTON *et al.* (1978 b) found insecticidal properties of triphenyltin hydroxide on 3 lepidopterous pests of soybean: *Heliothis zea* Boddie, *Pseudoplusia includens* (Walker), and *Trichoplusia ni* (Huebner).

Suspensions of triphenyltin hydroxide in water (1.0 to 9.5 mg/ml) sprayed on pine needles were toxic for larvae of the swaine jack sawfly (*Neodiprion swainei* Middleton) (ALL and BENJAMIN 1976).

3. Sterilization.—To control insects by releasing sterilized specimens, great numbers of insects are required that cannot reproduce yet are not damaged seriously in any other way. Sterilization can be achieved with chemicals if there is a considerable difference between the concentration that causes sterilization and the concentration leading to general damage.

MURBACH and CORBAZ (1963, and MURBACH 1967) first noticed sterilizing effects of triphenyltin compounds when the number of eggs of the Colorado beetle was reduced considerably during experiments to control phytophthora in potato fields. However, the effect was attributed to reduced eating of the females due to the fungicide's anti-feeding effect (see section 4). KENAGA (1965 a) first reported explicitly about sterilization of insects with the use of this compound group: Various kinds of insects showed diminished or altogether no reproduction after feeding on triphenyltin compounds. External application of the active ingredient proved to be ineffective.

KENAGA (1965 b) and other authors used mainly the common house-fly (*Musca domestica* L.) for their experiments. At concentrations of about 100 to 400 ppm of triphenyltin acetate or hydroxide in the food, the hatching rate of the larvae is lowered considerably provided the active ingredient is given to both sexes. After sterilization of the common house-fly proved successful, similar experiments were conducted with many other kinds of insects (see ASCHER 1969 b).

So far, however, triphenyltin compounds have not found practical application as chemosterilizing substances.

KENAGA (1965 a and b) fed common houseflies (both sexes) continuously with triphenyltin compounds of various anions. He then determined the concentration for each compound causing a 95% reduction in the hatching rate of the eggs. The data thus obtained ranged between 62 and 250 ppm (Table CXXXI).

Table CXXXI. *Lowered reproduction rate of the common housefly (Musca domestica L.) by use of triphenyltin compounds* (excerpt from KENAGA 1965 b).

Compound	Concentration required for a 95% reduction (ppm in the food)
$(C_6H_5)_3SnF$	62
$(C_6H_5)_3SnCl$	62
$(C_6H_5)_3SnJ$	125
$(C_6H_5)_3SnOH$	62
$(C_6H_5)_3SnOCOCH_3$	125
$(C_6H_5)_3Sn$-stearate	250
$[(C_6H_5)_3Sn]_2S$	62
$[(C_6H_5)_3Sn]_3BO_3$	62

A comparison of Tables CXXVII and CXXXI shows a wide difference between toxic and sterilizing concentrations.

BYRDY et al. (1965 and 1966 b) fed common houseflies of both sexes for 3 days with a 1:1 mixture of sugar and milk, containing 100 ppm of triphenyltin compound. Thereafter, they placed "larval medium" in the cage and, after 6 days, determined the number of larvae. A reduction in the number of larvae of 87 to 100% (compared with controls) was obtained with several triphenyltin compounds (thiocyanate, hydroxide, methoxide, ethoxide, acetate, benzoate, hydrogenphthalate, and phthalate).

FYE et al. (1966) observed that the choice of food is of great importance. For 3 days, flies were treated with a triphenyltin compound either added to a mixture of sugar, milkpowder, and egg yolk ("fly food") or to pure sugar. Later, untreated fly food was used. The experiments were evaluated as to whether egg laying, hatching of larvae from the eggs, or pupation of larvae had been prevented (Table CXXXII). The effective

Table CXXXII. *Decrease in fertility of common housefly (Musca domestica L.) by triphenyltin compounds* (excerpt from FYE *et al.* 1966).

Compound	Minimum concentration[a] required to prevent					
	Oviposition		Hatching		Pupation	
	Fly food	Sugar	Fly food	Sugar	Fly food	Sugar
$(C_6H_5)_3SnOCOCH_3$	25	2,500	1	500	1	250
$(C_6H_5)_3SnCl$	500	2,500	1	100	1	100
$(C_6H_5)_3SnOH$	100	5,000	50	100	50	100
$[(C_6H_5)_3Sn]_2S$	10,000	> 10,000	10	2,500	10	2,500

[a] ppm.

concentrations were found to be much higher with sugar alone than with the fly food.

KISSAM and HAYS (1966) experimented with both sexes of flies, feeding them for 10 days with triphenyltin acetate (425 ppm in condensed milk/H_2O at a ratio of 1:1) or with triphenyltin chloride (375 ppm in the same mixture). They noticed a considerable decrease in the number of larvae (1.02 respectively 2.15 larvae/day and female fly) compared with controls (44.07 larvae/day and female insect).

HAYS (1968) and ASCHER *et al.* (1968 and 1971) conducted extensive studies on the influence of the sex on the fertility rate of common houseflies treated with triphenyltin compounds. HAYS (1968) fed a mixture of sugar, milk powder, and egg powder with 0.025% triphenyltin chloride or acetate for 48 hr to 20 flies of one sex. Then the same number of untreated flies of the opposite sex was placed in the cage and fed with normal food. The number of eggs, the hatching rate, and the mortality were then determined (Table CXXXIII). The effects of triphenyltin acetate varied considerably, depending on the treated sex.

Table CXXXIII. *Effect of triphenyltin acetate and chloride on each sex of the housefly* (after HAYS 1968).

Treatment	Treated sex	No. of eggs	Hatch (%)	No. dead in 17 days
Acetate	F	4,200	21.5	2.5
	M	2,600	70.9	2.5
Chloride	F	1,600	28.6	4.5
	M	1,300	50.4	0.5
None	F	4,640	99.4	1.5
	M	5,920	95.3	3.5

Triphenyltin chloride (0.025%) fed to both sexes in combination with another sterilizing compound (Tepa), caused the whole fly population to be extinct within 35 days.

According to Ascher *et al.* (1971), feeding of triphenyltin acetate or hydroxide (3 days 0.05% acetate or 0.045% hydroxide in milk, then untreated milk) to male animals results in increasing damage to the sperm within the spermatheca of untreated female insects after copulation. As a consequence, the hatching rate of the eggs is slowly reduced.

By contrast, the fertility is reduced quickly when the tin compound is fed only to females. Later, however, the insects recover gradually, and the hatching rate of the eggs slowly increases. The difference in these reactions explains that feeding both sexes with a triphenyltin compound produces the best results.

Bořkovec (1966), however, believes that only female flies are sterilized by triphenyltin compounds and that sterilization is only temporary, necessitating continuous feeding of active ingredient.

Abo-Elghar and Radwan (1971) conducted extensive experiments about the sterilizing effects of triphenyltin hydroxide on the cotton leaf worm (*Spodoptera littoralis* Boisd.). Third-instar larvae of the insect were fed for 48 hr with leaves that had been immersed in triphenyltin hydroxide suspensions of various concentrations. Then they were allowed to develop and reproduce. Various combinations of treated and untreated male and female insects showed that the male offspring of treated larvae were affected more than the respective females (Table CXXXIV). Concentrations of 0.1% triphenyltin acetate and more led to high mortality.

Table CXXXIV. *The effect of triphenyltin hydroxide (50%) on Spodoptera littoralis Boisd. when used at larval feeding period on fecundity of adults obtained* (after Abo-Elghar and Radwan 1971).

Concentration (% used for larvae)	Average no. of eggs laid/pair[a]						Eggs/ pair (mean)
	F & TM		TF & M		TF & TM		
	Laid	reduction (%)	Laid	reduction (%)	Laid	reduction (%)	
0.0500	314	76.1	309	76.5	238	81.9	287
0.0100	372	71.7	402	64.4	341	74.1	272
0.0050	469	64.7	566	57.0	345	73.8	458
0.0010	856	35.0	1,003	23.8	769	41.6	876
0.0005	994	24.5	1,182	10.2	1,187	9.8	1,121
Check	—	—	—	—	—	—	1,317

[a] T = adults from treated larvae, F = female, and M = male.

Similar experiments where adult insects instead of the larvae were fed for 24 hr with triphenyltin hydroxide (in sugar solution) also produced more severe damage in male animals. Sterilization was most efficiently achieved (up to 100%) when both sexes were fed with the tin compound.

Elbadry *et al.* (1971 b) obtained similar results in feeding experiments with 3rd instar larvae of *Spodoptera littoralis* Boisd. At concentrations of 0.005 to 0.5% triphenyltin hydroxide (50% preparation), partial sterilization was found, combined with rather high mortality. Male animals were more affected than females.

MITRI and KAMEL (1973 and 1974 a and b) supplied *Spodoptera* larvae with food containing 0.5% triphenyltin hydroxide, triphenyltin acetate + maneb or decyltriphenylphosphoniumchlorobromotriphenylstannate; they found almost complete sterility of the insects.

When pupae of *Spodoptera littoralis* are immersed for 30 sec in suspensions of triphenyltin hydroxide or acetate, the insects developing show a rate of sterilization that depends on the concentration of the tin compound. Concentrations of 0.5 to 1% were highly toxic. Feeding experiments with adult animals (0.01 to 0.5% triphenyltin acetate or hydroxide in the food) also produced a considerable reduction in the number of eggs.

The hatching rate of larvae from eggs treated for 30 seconds with suspensions of organotin compounds was also reduced. However, rather high concentrations of active ingredient (up to 2.5%) had to be used to obtain considerable effects (FINDLAY 1970).

Table CXXXV shows results of sterilization experiments with numerous other insects. It also lists observations about reduced numbers of eggs and lowered hatching rates.

Triphenyltin compounds proved to be less or not at all suitable to sterilize boll weevil (*Anthonomus grandis* Boheman) (KLASSEN *et al.* 1968, HAYNES *et al.* 1971), red bollworm (*Diparopsis castanea* Hmp.) (CAMPION and OUTRAM 1968), and eye gnat (*Hippelates collusor* Townsend) (MULLA 1968), *Stethorus loxtoni* (Britton and Lee) (WALTERS 1976), *Stethorus nigripes* Kapur (WALTERS 1976), and *Stethorus vagans* (Blkb.) (WALTERS 1976).

4. Anti-feeding and repellent effects.—MURBACH and CORBAZ (1963) found that triphenyltin acetate reduces consumption in Colorado beetle (*Leptinotarsa decemlineata* Say.), and SOLEL (1964) noticed the same effects on cotton leaf worm (*Spodoptera littoralis* Boisd.[4]). Many kinds of insects eat less or entirely reject food containing these compounds, which leads to reduced infestation and decreased damage in the plants by the remaining insects. If reduced consumption is caused by nonsystemic compounds (like triphenyltin compounds), only the leaves covered with active ingredient are protected. The young leaves developing after spraying can be attacked by the insects.

The behavior of the insects is the same as that of warm-blooded animals, which also show reduced consumption and decreased weight gain when triphenyltin compounds are added to the food.

The reason for reduced consumption and rejection by the insects

---

[4] This insect is also called *Prodenia litura* F. (ASCHER and RONES 1964; SOLEL 1964; ASCHER and NISSIM *et al.* 1964).

Table CXXXV. *Fertility reduction in different insects with triphenyltin compounds.*

Insect	Trivial name	Triphenyltin compound and concentration, respectively, amount	Tested	References
Tribolium confusum Jacquelin du Val	Confused flour beetle	Hydroxide (0.025%)	hatching rate	KENAGA (1965 b)
Blatella germanica L.	German cockroach	Hydroxide (0.1%)	eggs	KENAGA (1965 a)
Leptinotarsa decemlineata Say	Colorado beetle	Hydroxide (0.02%); oxide (0.02%)	eggs	BYRDY *et al.* (1965)
Cellosobruchus chinensis L.	Azuki bean weevil	Hydroxide (0.3 µg/insect)	eggs; hatching rate	NAGASAWA *et al.* (1965 and 1967)
Ceratitis capitata Wied.	Mediterranean fruit fly	Hydroxide (0.5%)	eggs; hatching rate	ORPHANIDIS & PATSAKOS (1969 and 1970)
Dacus oleae. Gmel.	Olive fly	Hydroxide (0.5%)	eggs	ORPHANIDIS & PATSAKOS (1969 and 1970)
Dacus oleae. Gmel.	Olive fly	Acetate (0.4%)	eggs	ORPHANIDIS (1965)
Plutella maculipennis Curt.	Diamond back moth	Hydroxide (0.025%–0.050%)	eggs; hatching rate	BONNEMAISON (1966)
Popillia japonica Newman	Japanese beetle	Acetate, hydroxide, chloride (6 µg/insect)	eggs; hatching rate	LADD (1968)
Tenebrio molitor L.	Yellow meal worm	Acetate, hydroxide		FINDLAY (1968)
Heliothis virescens F.	Tobacco budworm	Hydroxide (3.8 µg/g); acetate (0.5% µg/g); chloride (42 µg/g)	hatching rate	WOLFENBARGER *et al.* (1968)
Heliothis zea Boddie	Bollworm	Hydroxide (3.5 µg/g); acetate (3.3 µg/g); chloride (25 µg/g)	eggs; hatching rate	WOLFENBARGER *et al.* (1968)

Species	Common name	Compound (concentration)	Effect	Reference
Acyrthosiphon pisum Harris	Pea aphid	Hydroxide (0.01–0.05%)	eggs	BHALLA & ROBINSON (1968)
Fania canicularis L.	Little housefly	Hydroxide (0.5–1%)	hatching rate	PAUSCH (1969)
Anthonomus grandis Boheman	Boll weevil	Hydroxide (1%)	hatching rate	HAYNES et al. (1971)
Agrotis ipsilon Rott	Greasy cutworm	Hydroxide (0.005–0.02%)	eggs	SHAABAN et al. (1975)
Spodoptera litura (Fabricius)	Tobacco caterpillar	Acetate (0.075%)	hatching	JOSHI et al. (1973)
Macrosiphum euphorbiae (Thomas)	Potato aphid	Hydroxide (0.06–3.0 µg/aphid) Fluoride (0.06–1.8 µg/aphid)	eggs	CHAWLA et al. (1974)
Corcyra cephalonica S.	Indian mealworm	Acetate (0.0025–0.02%)	eggs; hatching	DALE & SARADAMMA (1974 b)
Corcyra cephalonica S.	Indian mealworm	Hydroxide; acetate; chloride (0.05%; 0.1%)	hatching	ABDUL-KAREEM et al. (1977 a)
Earias insulana (Boisd.)	Spiny bollworm	Hydroxide (0.03%–0.25%)	eggs	SALEM et al. (1976)
Menochilus sexmaculatus (F.)	Ladybird beetle	Acetate; hydroxide; chloride (0.025%; 0.05%; 0.10%)	predation	ABDUL-KAREEM et al (1977 b)

seems to be the bad taste of the treated leaves (ALEXANDRESCU *et al.*
1973). ASCHER and ISHAAYA (1973) also pointed out that triphenyltin
acetate reduces the protease and amylase activity in larvae of *Spodoptera
littoralis* Boisd. This effect, however, was not caused by a direct influence
on ferment activity, but by reduced ferment production.

Anti-feeding effects were tested mainly for Colorado beetle (*Leptinotarsa decemlineata* Say.), for cotton leaf worm (*Spodoptera littoralis*
Boisd.), and for giant looper (*Ascotis serenaria reciprocaria* Walk.), to a
lesser extent for several other insects. Contrary to the sterilizing effects
of triphenyltin compounds, which are sufficiently strong only at rather
high and therefore often toxic concentrations, the anti-feeding properties
have gained practical importance for several agricultural crops.

In potato fields, feeding damage from Colorado beetle and its larvae
can be prevented almost completely by application of triphenyltin compounds in amounts normally used to control early blight or late blight.
There is no need for separate treatment with an insecticide against Colorado beetle (MURBACH and CORBAZ 1963, MURBACH 1967).

When triphenyltin compounds are used in sugarbeet crops to control
Cercospora leaf spot, they also prevent damage from owlet moths *Noctuidae*, which are—because of their particular behavior pattern—hard to
control with common insecticides. The amounts of 240 to 360 g/ha normally used against leaf spot are also sufficient against these insects (SOLEL
1964, PIVAR *et al.* 1965).

In East Africa, triphenyltin compounds (acetate and hydroxide) were
recommended to simultaneously control coffee rust (*Hemileia vastatrix*
Berk. and Br.) and giant looper (*Coffee Research Institute Ruiru* 1972).
The dosage of 1.25 kg of triphenyltin compound/ha used against the
coffee rust is also sufficient against the fourth and fifth larval stage of the
parasite. Other commonly used insecticides are inefficient against this
advanced stage. In earlier developmental stages of the giant looper (first
to third larval stage), merely 500 to 750 g of triphenyltin compound/ha
is sufficient. Apart from the giant looper, larvae of the leaf skeletonizer
(*Leucoplema dohertyi*) in coffee plantations are also eliminated. Useful
parasites of leaf miners and other pests—contrary to commonly used
insecticides—will not be affected even at the high dosage of 1.25 kg
triphenyltin derivative/ha (ANONYMOUS 1972, ABASA and MULINGE 1973,
BARDNER and MATHENGE 1974 a).

Field trials by MURBACH and CORBAZ (1963) and by MURBACH (1967)
with 400 to 500 g of triphenyltin acetate/ha (2 to 3 applications) resulted
in a significant decrease of Colorado beetle larvae, probably attributable
to the beetle's reduced food intake and egg production. The larvae ate
very little of the treated foliage. These results were confirmed by laboratory experiments (MURBACH 1975).

BYRDY *et al.* (1966 a) sprayed potato plants with 0.2% suspensions of
triphenyltin compounds until the leaves were completely soaked. After
drying, 10 larvae of the Colorado beetle were placed on each plant. The

larvae were weighed before the experiment, after 24 hr and after 48 hr. Three of the triphenyltin compounds tested caused considerable growth reduction or even weight loss of the larvae (Table CXXXVI).

In a succeeding field trial, 600 L of liquid with 0.2% active ingredient/ ha were sprayed. After the liquid had dried, Colorado beetles and their larvae were placed on the plants. The number of beetles and larvae found in the parcels on the fourth day after spraying is also listed in Table CXXXVI.

Table CXXXVI. *Effects of triphenyltin compounds on larvae of Colorado beetle (Leptinotarsa decemlineata Say.) (after* BYRDY *et al. 1966 a).*

Compound	Wt of larvae[a]			Wt change after 48 hr (%)	Field test[b]	
	Before test	After 24 hr	After 48 hr		No. of beetles	No. of larvae
$(C_6H_5)_3SnOCH_3$	62.41	60.60	48.30	− 22	2	5
$(C_6H_5)_3SnOCOCH_3$	65.36	81.20	75.00	+ 15	0	6
$(C_6H_5)_3SnOCOC_6H_5$	69.00	83.10	65.10	− 5.6	4	2
Control	68.20	117.20	141.00	+107	14	10

[a] Average weight of 1 larva in mg (of 10).
[b] Counted 4 days after spraying.

KOULA and RAJCHARTOVÁ (1971) tested the anti-feeding effects of 28 organotin compounds in adult Colorado beetles. Triphenyltin hydroxide was more effective than acetate; however, some trialkyltin compounds proved to be the most efficient (e.g., tri-*n*-propyltin acetate).

Rather unexpected were the findings of MEISNER and ASCHER (1965) that triphenyltin acetate (0.05%) and hydroxide prevent the potato tuber moth (*Gnorimoschema operculella* Zell.) from getting into the leaves of potato plants, but do not protect the tubers against this pest.

ALEXANDRESCU and BAICU (1973) used suspensions of 0.3% triphenyltin hydroxide, 0.3% triphenyltin acetate + maneb, or 1% copper oxychloride + 0.3% triphenyltin acetate to protect potatoes from the attack of Colorado beetles.

When controlling *Cercospora* leaf spot of sugar beets with triphenyltin acetate (preparation of 60% acetate + 20% maneb; 600 g/ha), SOLEL (1964) observed that the treated parcels—contrary to the controls—were not damaged by noctuids (*Spodoptera littoralis* Boisd.).

ASCHER and RONES (1964) immersed sugar beet leaves in triphenyltin acetate suspensions of different concentrations. 24 hr later, they placed *Spodoptera* larvae on the leaves. After another 48 hr, the larval weight was checked and the leaf surface consumed by the single animal determined. Concentrations of 0.035 to 0.1% active ingredient almost completely prevented the larvae from feeding; at 0.1% their wt loss was

Table CXXXVII. *Effect of triphenyltin acetate on leaf surface consumption and larval weight of Spodoptera littoralis Boisd.* (excerpt from Ascher and Rones 1964).[a]

Triphenyltin acetate (%)	Mean larval wt		Mean % of area consumed
	Initial (mg)	After 48 hr (mg)	
0	192	526 (+334)	75
0.01	185	394 (+209)	44
0.02	187	247 (+ 60)	15
0.035	184	166 (− 18)	6
0.05	175	158 (− 17)	2
0.1	177	137 (− 40)	2
Starved	178	124 (− 54)	—

[a] 40 replications/concentration and/control.

almost as high as for larvae that had not received any food at all (Table CXXXVII).

In comparing the anti-feeding effects of triphenyltin acetate and hydroxide, the area of sugarbeet leaves consumed by *Spodoptera* larvae was also measured; acetate proved to be more effective than hydroxide (Ascher and Nissim 1965).

Mitri *et al.* (1970) compared the anti-feeding effects of triphenyltin hydroxide and acetate by dipping one batch of cotton leaves in 0.031 to 0.54% suspensions of active ingredient and spraying another with 0.1 to 0.5% suspensions. *Spodoptera* larvae consumed much less of the immersed leaves than of those sprayed which apparently contained less active ingredient. The protective effects of hydroxide were slightly higher than those of acetate; regardless of the nature of triphenyltin compound the larvae lost 83 to 87% wt (compared with controls). The kind of foliage (cotton or castor bean) did not seem to have any influence. These results were confirmed by further laboratory and field tests (Mitri and Kamel 1972 a and b, and 1974, Kamel *et al.* 1974, Radwan and Shaaban 1973). Rizk and Radwan (1975) used *ca.* 600 g of triphenyltin hydroxide/ha.

Abo Elghar *et al.* (1971 a) obtained a reduction in leaf surface consumed by *Spodoptera* larvae within 3 days from 10.90 cm² (untreated controls) to 1.29 respectively 0.01 cm² when the larval stomachs had been injected with 100 respectively 1,000 ppm of triphenyltin hydroxide. In feeding experiments with leaves of various plants (tomato, corn, castor bean, cotton, potato, grape, snap beans, egg plant, cabbage, and sweet potato), the effects of hydroxide did not depend on the plant family (Abo Elghar *et al.* 1971 b). Similar results were obtained using sprays of triphenyltin acetate on various plants by Dale and Chandrika (1972).

Tobacco leaves treated with triphenyltin acetate (0.7 L of 0.075% suspension per m²) suffered minimal attack from *Spodoptera* larvae.

However, the anti-feeding effect was reduced with increasing time between application of active ingredient and introduction of the larvae; the effect disappeared altogether after an average of 12.7 days (Joshi et al. 1967 and 1971).

Elbadry et al. (1971 a) fed Spodoptera larvae with leaves of the castor bean which had been sprayed with a 0.125% suspension of triphenyltin hydroxide. After 5 days, the consumption was 0.36 cm² of leaf/larva (control: 29.5 cm²) and all larvae had died. At what stage the larvae received the active ingredient was of hardly any consequence.

Regupathy (1973 a) obtained 90% control of Spodoptera larvae on castor leaves after treatment with 0.2% suspensions of triphenyltin acetate. Ammal and Dale (1974) reported similar results with triphenyltin acetate and chloride; the persistence was 16 to 18 days (50% protection).

Castor bean plants sprayed with 0.125% triphenyltin hydroxide suspension were protected against feeding by Spodoptera littoralis larvae to a great extent provided the whole plant had been treated with active ingredient. Otherwise, the larvae moved on to untreated parts (Elbadry et al. 1971 c).

Verma and Jain (1972) observed effects against Spodoptera larvae already with 0.01% suspensions of triphenyltin acetate; concentrations of 0.06% led to wt loss.

El-Sebae and Ahmed (1972 and 1973) conducted feeding experiments with Spodoptera larvae on castor bean leaves which had been treated with triphenyltin acetate and hydroxide. The results do not show significant differences in the anti-feeding effects of both compounds, although the hydroxide may have been slightly more effective.

Pivar et al. (1965), after treatment of sugarbeets with triphenyltin acetate, obtained a significant decrease in attack by various noctuids (Noctuidae spp.). The caterpillars had appeared in great numbers, and timely application of the active ingredient prevented them from feeding. The effects of the triphenyltin compound were supported by laboratory tests.

Marfurt and Toscani (1966 and 1967) treated apple trees with triphenyltin acetate to control codling moth (Carpocapsa pomonella L.) and oriental fruit moth (Grapholita molesta Busk). Seven applications of a 0.25% suspension (60% preparation) reduced the number of affected "Delicious" apples from 55.9% (untreated control) to 3.5%. In "Granny Smith" apples, a 0.25% suspension caused a damage reduction from 56.7% to 1.24%; a 0.20% suspension led to a reduction to 3.5%. The results can be attributed to reduced food intake.

To prevent or reduce the feeding damage by Hylobius pales Herbst, Thomas (1969) conducted laboratory experiments with triphenyltin acetate and hydroxide in concentrations of 0.05 to 0.0125%. He used artificial food wafers prepared from the marrow of elder (Sambucus candensis L.) containing various pine extracts. At 0.05% triphenyltin compound, consumption was reduced very effectively; 0.025% also produced satisfactory results. However, the lowest concentration of 0.0125% was not enough to reduce consumption.

VINE and VINE (1971) observed anti-feeding effects on the giant looper, a lepidoptera larva (*Ascotis serenaria reciprocaria* Walk.), while controlling coffee rust (*Hemileia vastatrix* Berk. and Br.) and coffee berry disease (*Colletotrichum coffeanum* Noack) with triphenyltin compounds at the recommended dosages of 1.25 kg active ingredient/ha. Experimental results about anti-feeding effects of triphenyltin hydroxide on giant looper in coffee plants were published by ABASA and MULINGE (1972 and 1973). Three-yr-old coffee plants that had been treated with aqueous suspensions of 0.125% triphenyltin hydroxide were not attacked by larvae of the 3rd and 4th stages. The larvae died within 120 hr (Fig. 21). In

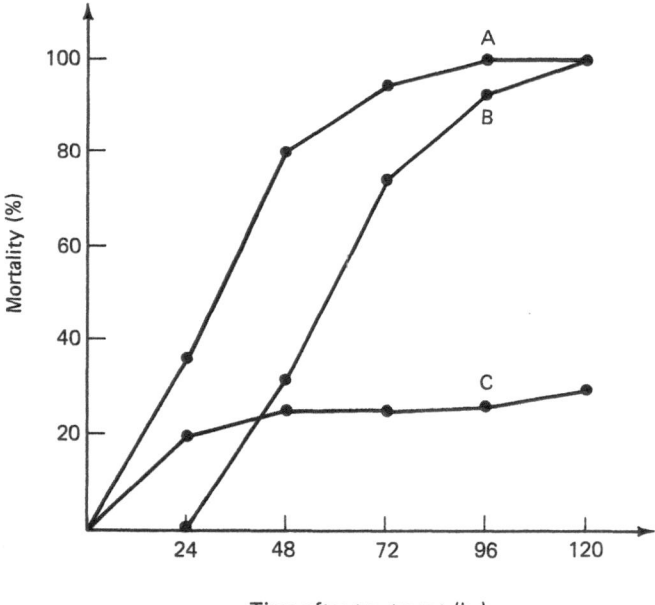

Time after treatment (hr)

Fig. 21. Mortality of giant looper fed on seedlings sprayed with triphenyltin hydroxide at 0.25%: A = seedlings sprayed after infestation, B = seedlings infested 24 hr after spraying, and C = unsprayed (after ABASA and MULINGE 1972).

another experiment, leaves treated with 0.125% triphenyltin hydroxide were offered to larvae of the fifth stage. Fig. 22 shows the anti-feeding effects of triphenyltin hydroxide. All larvae of the fifth stage were dead within 5 days (Table CXXXVIII).

ABASA (1972) was able to show that far lower amounts (0.06% tri-

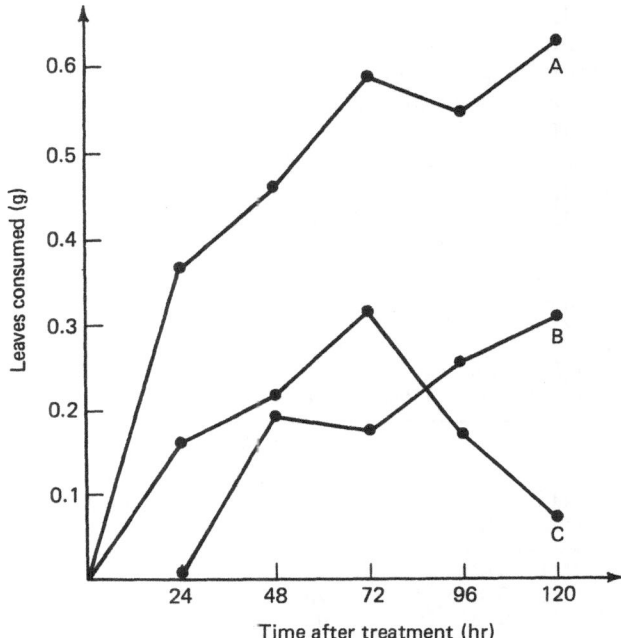

Fig. 22. Mean daily consumption of leaves by 5th instar giant loopers: A = unsprayed leaves, B = sprayed leaves presented 24 hr after spraying, and
C = sprayed leaves on same day as sprayed (after ABASA and MULINGE
1972).

Table CXXXVIII. *Mortality of 5th instar Ascotis larvae fed on triphenyltin hydroxide
sprayed leaves starting at different times after spray application.*
(after ABASA and MULINGE 1973).[a]

Treatment	Larvae observed	Mortality after					
		24 hr	48 hr	72 hr	96 hr	120 hr	144 hr
Fed starting imme-diately after spray	40	2	7	20	35	40	—
Fed starting 24 hr after spray	40	—	—	25	37	40	—
Fed on untreated leaves	40	—	—	—	1	3	11
Starved	20	—	—	5	8	10	20

[a] Figures are cumulative.

phenyltin acetate or hydroxide) are sufficient to eliminate giant looper larvae of the fourth and fifth stage from coffee plantations. At this concentration, 94.5 to 96.5% of the leaf surface is spared from feed damage. Later, 0.02 to 0.5% suspensions of triphenyltin hydroxide or acetate were used (ABASA 1975).

BARDNER and MATHENGE (1974 a) confirmed these findings. Based on their experiments they believe that the amount of 1.25 kg/ha of triphenyltin acetate or hydroxide, recommended against coffee rust and coffee berry disease, is justified for the fifth instar stage of giant looper larvae only. Lower larval stages die of starvation already at 500 to 750 g/ha of triphenyltin compound (Table CXXXIX).

Table CXXXIX. *Effects of leaves treated with triphenyltin hydroxide or acetate on 3rd to 4th instar giant looper larvae* (after BARDNER and MATHENGE 1974 a).

Treatment (triphenyltin compound)	Dead larvae		Dry wt of excrement (mg)
	Untransformed (%)	Angular transformation (%)	
Hydroxide (50% W.P.) 0.01%	1	7	192
Hydroxide (50% W.P.) 0.05%	75	60	40
Hydroxide (50% W.P.) 0.25%	95	77	21
Acetate (60% W.P.) 0.01%	0	0	445
Acetate (60% W.P.) 0.05%	51	43	56
Acetate (60% W.P.) 0.25%	71	57	23
Standard error of differences		±5.4	±68

Concentrations of 1.25 kg/ha triphenyltin compound also kill most of the larvae of leaf skeletonizer (*Leucoplema dohertyi*) which cause considerable damage in coffee plantations, as the following experiment shows. Since this insect feeds mainly on the underside of the leaves, the upper side alone was treated with the triphenyltin compound. One-half of the larvae was then placed on the upper, the other half on the underside of the leaves. Within 3 days, 73% of the first and 68% of the last group died; the difference between the two results is not significant.

BARDNER and MATHENGE (1974 a) also conducted experiments with 1.25 kg/ha of triphenyltin hydroxide against the larvae of the stinging caterpillar (*Parasa vivida*). The wt of the excrements of these larvae after 6 days amounted to only 0.02 g on treated leaves (compared with 1.81 g for larvae on untreated leaves). Furthermore, BARDNER and MATHENGE (1974 b) prevented feeding of *Phytometra orichalcea* (F.) on coffee plants with triphenyltin hydroxide. BHAT and CHACKO (1976) obtained protection of coffee leaves against larvae of *Eupterote canaraica* Moore with 0.2 to 0.4% suspensions of triphenyltin hydroxide.

Strong anti-feeding effects of triphenyltin compounds were also found on larvae of the moth (*Tineola bisselliella* Humm.) (GARDINER and

POLLER 1964). Chloride, acetate, and 10 other triphenyltin compounds with different anions were effective. Damage on wool treated with 1% tin compound was between 0.9 and 4.5% (untreated = 100). Strangely, only hexaphenyldistannoxane was ineffective.

HUECK and BRJIN (1972) confirmed the anti-feeding effects of triphenyltin acetate on moth larvae and found that this compound also affects larvae of carpet beetle (*Anthrenus vorax* Waterh.). The anti-feeding properties of wool treated with 1.2% triphenyltin acetate was not affected considerably by repeated laundering and dry cleaning. In these experiments, too, hexaphenyldistannoxane proved to be ineffective.

Further experiments with several other insects are compiled in Table CXL.

e) Control of mites with triphenyltin compounds

The citrus rust mite (*Phyllocoptruta oleivora* Ashmead) can be controlled with triphenyltin hydroxide (FISHER 1963 and BULLOCK 1964). Spraying with a 0.12% hydroxide suspension in the fall provides protection for 21 wk; a 0.036% suspension protects for 16 wk. Application of the active ingredient in spring provides protection for approximately 9 wk (0.024 to 0.048% suspension).

REED et al. (1967) described experiments to control mites in citrus plantations (*Aculus pelekassi* Keifer and *Phyllocoptruta oleivora* Ashmead). Both kinds are sensitive to triphenyltin hydroxide and chloride ($LD_{50} = 2$ ppm); *Aculus pelekassi* Keifer is less sensitive to acetate ($LD_{50} = 20$ ppm).

According to TOSCANI and MARFURT (1969) infestation with *Panonychus ulmi* mite can be prevented with triphenyltin acetate. LARGE (1965) noted without quantification that triphenyltin hydroxide is effective against mite in pecan plantations.

Experiments by STERNLICHT (1966) showed that triphenyltin acetate in concentrations of 0.06% (60% spraying powder) was not sufficiently effective against the citrus bud mite (*Aceria sheldoni* Ewing).

f) Effects of triphenyltin compounds against rodents

Of the more than 8,000 tested compounds, triphenyltin chloride and tributyltin oxide proved to be the most efficient against rodents feeding on textiles, paper, and synthetic containers (ZEDLER 1961). According to TIGNER and BESSER (1962), however, triphenyltin acetate is rather ineffective against mice (required concentration to prevent half of the mice in a test group from chewing through a burlap bag is 6.9 mg/cm² compared with 0.56 mg/cm² of tributyltin chloride).

g) Effects of triphenyltin compounds on plants

1. Phytotoxicity.—

α) *General.*—Triphenyltin compounds can be phytotoxic. If more than a certain maximum concentration—varying with the kind of plant—

Table CXL. *Triphenyltin compounds causing reduced food consumption of different insects*

Insect	Trivial name	Plant	Tin compound and amount (concentration)	References
Agrotis ypsilon Rott.	Greasy cutworm	Sugarbeet	Acetate (0.03 to 0.1%)	ASCHER & RONES (1964)
Agrotis ypsilon Rott.	Greasy cutworm	Castor	Hydroxide, acetate, chloride (0.2 to 10 mg/ml)	ABDEL-MEGEED et al. (1974)
Gnorimoschema operculella Zell.	Potato tuber moth	Egg plant	Hydroxide, acetate, oxide, chloride (0.05%)	ASCHER & MEISNER (1969)
Chilo agamemnon Bles.	Striped maize borer	Corn	Hydroxide, acetate (0.03 to 0.05%)	MEISNER & ASCHER (1965)
Musca domestica vicina	Housefly	—	Hydroxide, acetate (6.7 to 37 ppm)	ASCHER et al. (1967) ASCHER & MOSCOWITZ (1968)
Heliothis zea Boddie	American bollworm	Cotton	Hydroxide, acetate, chloride (1.23 kg/ha)	WOLFENBARGER et al. (1968)
Heliothis virescens F.	Tobacco budworm	Cotton	Hydroxide, acetate, chloride (1.23 kg/ha)	WOLFENBARGER et al. (1968)
Heliothis armigera Huebner	—	Wheat	Acetate, chloride (0.05 to 0.1%)	CHARI & PATEL (1973)
Trichoplusia ni Huebner	Cabbage looper	Cotton	Hydroxide (3.36 kg/ha)	WOLFENBARGER et al. (1968)
Athalia rosae L.	—	Rape	Hydroxide, acetate (1%)	JERMY & MATOLCSY (1967)
Phytodecta formicata Brueggm.	—	Alfalfa	Hydroxide, acetate (1%)	JERMY & MATOLCSY (1967)
Hyphantria cunea Drury	—	Mulberry	Hydroxide, acetate (1%)	JERMY & MATOLCSY (1967)
Ceutorrhynchus macula-alba Hrb.	—	Poppy	Hydroxide, acetate (1%)	JERMY & MATOLCSY (1967)
Antigastra catalaunis Dup.	—	Sesame	Acetate (0.0125 to 0.4%)	MATHUR & SAXENA (1972)
Pseudaletia separata Wlk.	—	Corn	Acetate (0.0125 to 0.4%)	MATHUR & SAXENA (1972)
Hylobius pales Herbst.	Pales weevil	Pine	Acetate, hydroxide	THOMAS (1969)
Atherigona soccata Rond.	—	Sorghum	Acetate, hydroxide	ABDUL-KAREEM et al. (1974)

Table CXL. (*Continued*)

Insect	Trivial name	Plant	Tin compound and amount (concentration)	References
Neodiprion swainei (Middleton)	Saw fly	Jack pine	Hydroxide (1.0 to 9.5 mg/ml)	ALL & BENJAMIN (1976)
Scolytus mediterraneus Eggers	Fruit bark beetle	Peach leaves	Hydroxide, acetate (0.005 to 0.2%)	ASCHER et al. (1975)
Euproctis fraterna Moore	—	Castor	Acetate (0.0125 to 0.2%)	DALE & CHANDRIKA (1973)
Pericallia ricini (F.)	—	Castor	Acetate (0.0125 to 0.2%)	DALE & CHANDRIKA (1973)
Pericallia ricini (F.)	—	Castor	Acetate (0.3 to 0.4%) Hydroxide (0.15 to 0.2%)	REGUPATHY (1973 a)
Pericallia ricini (F.)	—	Cacao	Hydroxide, acetate (0.0125 to 0.2%)	KUMAR (1974)
Euproctis fraterna, Moore	—	Castor	Acetate (0.0125 to 0.2%)	DALE & CHANDRIKA (1973)
Pericallia ricini (F.)	—	Bitter gourd	Acetate (0.0125 to 0.2%)	DALE & CHANDRIKA (1973)
Epilachna vigintioctopunctata (F.)	—	Bitter gourd	Hydroxide, acetate	DALE & SARADAMMA (1973)
Porthetria dispar L.	Gypsy moth	Oak	Acetate (0.03%)	MEISNER & SKATULLA (1975)
Callosobruchus chinensis L.	Pulse beetle	Cowpea seeds	Acetate, chloride (0.01 to 0.04%)	REGUPATHY (1973 b)
Achoea janata Linn.	—	Castor	Acetate, chloride	AMMAL & DALE (1975)
Pectinophora gossypiella (Saund.)	Bollworm	Cotton	Hydroxide (600 g/ha)	RIZK & RADWAN (1975)
Papilio demolens L.	Citrus defoliator	Citrus	Chloride (0.046 to 0.065%)	REGUPATHY (1974)
Spodoptera litura Fabricius	Tobacco caterpillar	Tobacco	Hydroxide, acetate (0.075%)	JOSHI et al. (1978)
Spodoptera mauritia B.	—	Paddy	Acetate, chloride (0.01 to 0.2%)	AMMAL & DALE (1974)
Pericallia ricini F.	—	Castor	Acetate, chloride (0.01 to 0.2%)	AMMAL & DALE (1974)

is used there may be changes in the plant appearance and/or burns on the leaves. However, by using suitable formulations and the right dosage, damage to several kinds of plants can be avoided without losing the fungitoxic effects.

The phytotoxicity of a preparation does not only depend on the chemical formula of the active ingredient but also on several other factors.

Evaluation of comparative experiments to determine the phytotoxicity of different fungicides is usually impeded by the absence of formulation data. Little or no information is available on the particle size especially of unformulated, insoluble substances suspended in the spray liquid. Further, series of experiments conducted at different times and/or different locations usually cannot be compared because of weather influence (see below). Finally, some of the experimental results described in the literature have not been sufficiently evaluated towards their statistical significance. Such evaluation might help our understanding why, for example, of 2 tested triphenyltin compounds one causes more damage in the plants and at the same time higher yields.

In summary, all comparative experiments aimed at establishing the phytotoxicity of different triphenyltin compounds are subject to uncertainties of varying degree.

BAUMANN (1958) conducted the first systematic studies on the phytotoxicity of organotin compounds. In greenhouse experiments, at first distinct differences appeared in the effects of various organotin compounds on tomatoes (Table CXLI). Tripropyltin acetate and several tributyltin compounds were highly phytotoxic, while triphenyltin compounds, mainly acetate and hydroxide, were tolerated much better.

Table CXLI. *Phytotoxicity of various organotin compounds in experiments with tomatoes; exposure 8 days* (after BAUMANN 1958).

	Phytotoxicity[a] at various concentrations		
Compound	1.0%	0.5%	0.1%
Tripropyltin acetate	5	4–5	4
Tributyltin acetate	5	4–5	4
Tributyltin-o-nitrophenolate	5	5	3
Tributyltin-p-nitrobenzoate	5	5	3–4
Triphenyltin chloride	4	3	2
Triphenyltin acetate	3	2	1
Triphenyltin hydroxide	3	1	1

[a] Rating: 0 = no burns and 5 = total damage.

Another series of experiments with triphenyltin acetate in different concentrations showed that, among the tested plants, potatoes, beets, and celery were the least sensitive to the active ingredient (Table CXLII).

Table CXLII. *Phytotoxicity of triphenyltin acetate at different concentrations on cultivated plants; exposure 8 days* (after BAUMANN 1958).

Plant	Phytotoxicity[a] at various concentrations			
	1.0%	0.5%	0.25%	0.1%
Grape vine seedling	4–5	4	3	2
Hops	4–5	4	3	2
Apple	4	3	2	1–2
Kohlrabi	4	3	2	1
Dwarf bean	3	2	2	1
Tomato	3	2	1	1
Potato	2	1	0–1	0–1
Beet	1	0–1	0	0
Celery	0–1	0	0	0

[a] Rating: 0 = no burns and 5 = total damage.

Other experiments should be mentioned in which the influence of different factors and test conditions on the phytotoxicity of triphenyltin compounds was determined.

$\alpha\alpha$) *Effects of the anion:*

The effects of X in triphenyltin compounds of the type $(C_6H_5)_3SnX$ have been studied extensively. Of various inorganic acid residues, chloride and sulfate proved to be strongly phytotoxic. Where X was an organic carbonic acid, distinct differences could be noticed between the less phytotoxic dicarbonic acids and the more phytotoxic monocarbonic acids. The phytotoxicity of triphenyltin sulfides and disulfides was very low (Table CXLIII), but these compounds did not have strong fungitoxic effects (VAN DER KERK 1961).

According to HAERTEL (1962), the phytotoxicity of triphenyltin compounds towards tomato plants can be reduced considerably by suitable choice of the group X; however, experiments about the fungitoxicity of these compounds towards *Pernospora* in grape vines, conducted at the same time, showed that reduced fungitoxicity goes along with a decrease in phytotoxicity. Thus, variation of the group X is of no practical advantage (Table CXLIV). This evaluation may be vulnerable to the extent that results are compared which were obtained from different plants and with varying concentrations of active ingredient. Several contrasting results appear to show that the above conclusions cannot be generalized.

BYRDY *et al.* (1966 a) tested the phytotoxicity of different triphenyltin compounds towards mustard plants (*Sinapis alba*); the preparations were applied as 30% spraying powder in 3 different concentrations. The mean degree of damage in the leaves and the average wt loss of the plants were determined. In addition, fungitoxicity toward *Venturia inaequalis* (Cooke) Aderh. and *Alternaria tenuis* Nees. was determined with the drip method

Table CXLIII. *Phytotoxicity of triphenyltin acetate, sulfide, and disulfide* (after VAN DER KERK 1961).

Triphenyltin compound	Concen-tration (%)	Phytotoxicity[a]				
		Bean	Chick-weed	Potato	Beet	Cucumber
Acetate	0.1	0	1	0	0	3
	0.3	2	1	0	0	3
	1	3	2	1	0	4
	3	5	2	2	1	6
Sulfide	0.1	0	0	0	0	0
	0.3	0	0	0	0	1
	1	0	0	0	0	2
	3	0	0	0	0	3
Disulfide	0.1	0	0	0	0	2
	0.3	0	0	0	0	2
	1	0	0	0	0	3
	3	0	0	0	0	3

[a] Arbitrary scale, 10 days after spraying: 0 = no damage and 6 = completely burned.

Table CXLIV. *Comparison of phyto- and fungitoxicity of various triphenyltin compounds* (excerpt from HAERTEL 1962).

Triphenyltin compound	Damage in tomato plants at 50 ppm active ingredient		*Pernospora* infection of grape vines at 1.25 ppm active ingredient (%)
	Degree[a]	Leaf wt[b] (%)	
Hydroxide	3	70	18
Chloride	3	78	2.1
Acetate	2	83	14
Sorbinate	2	84	12
Perchlorate	0–1	106	53

[a] Rating: 0 = no damage and 5 = 100%.
[b] Compared with untreated plants.

by Blumer-Kundert and towards *Phytophthora infestans* de Bary on potato plants in the greenhouse (Table CXLV). Differences in phyto- and fungitoxicity were found which are difficult to explain and contrary to other results mentioned in literature (BAUMANN 1958, HAERTEL 1962 and 1963 c, PIETERS 1961, SCHICKE *et al.* 1968).

Of special interest is the fact that triphenyltin hydroxide shows the strongest phytotoxicity of all the listed compounds, while its fungitoxicity is rather low. The same authors emphasized, on the other hand, that they chose triphenyltin hydroxide for their chemosterilization experiments because of its low phytotoxicity (BYRDY *et al.* 1966 a).

Table CXLV. *Phytotoxicity of triphenyltin compounds $(C_6H_5)_3SnX$ in mustard plants and fungitoxicity towards different fungi* (after BYRDY *et al.* 1966 a).

X	Phytotoxicity		Order of fungitoxicity[a]		
	Degree of damage	Wt loss	V. inae-qualis	A. tenuis	Ph. infestans
OCOC₆H₅	0.00	0.50	—	—	3
OCH₃	0.00	0.54	1	3	2
OCOCH₃	0.00	0.80	3	1	1
OC₂H₅	0.21	4.53	4	6	—
SCN	0.50	8.13	2	2	—
OSn(C₆H₅)₃	0.55	10.56	6	4	—
OH	0.57	10.68	6	4	—

[a] Scale: 1 = highest and 6 = lowest toxicity.

SCHICKE *et al.* (1968) studied the fungitoxicity of triphenyltin acetate, hydroxide, and chloride and of the complex compounds of chloride with dimethylsulfoxide, quinoline-*N*-oxide, and decyltriphenylphosphonium bromide (micro organisms: *Septoria apii* in celery; *Phytophthora infestans* in tomatoes and potatoes; *Cercospora beticola* in sugarbeets and feed beets; *Alternaria solani* in tomatoes). The same compounds were tested for their phytotoxicity towards dwarf beans and bush tomatoes. There was no correlation in efficacy against fungi and tolerance by the plants.

Replacement of one acyl group in triphenyltin compounds by the respective acetylamino group (for example benzoate by 4-acetylamino-benzoate) decreases the phytotoxicity without an adverse change in the fungitoxicity (KOOPMANS 1957 and 1958).

ββ) Effects of concentration; synergism:

Of course, the phytotoxicity of triphenyltin compounds can be lowered by reducing the concentration in the spray liquid and of the dosage (see Table CXLII). In order to obtain sufficient fungitoxic effects, a second fungicide can be added to the preparation; a combination of triphenyltin acetate and manganous ethylenebisdithiocarbamate is particularly recommended (BRUECKNER *et al.* 1960 and 1961, HARDON *et al.* 1962, HAERTEL 1963 c and 1964 a and b).

Surprisingly, a synergistic intensifying of the fungitoxic effects could be noticed in these mixtures; also, the manganous compound reduced the phytotoxicity of triphenyltin acetate even at high concentrations which, if the acetate were used alone, would be rather toxic (Table CXLVI).

Similar results are reported by PAQUET and WILKIN (1970). By mixing 1 part triphenyltin compound with 3 parts propineb, the phytotoxicity of the triphenyltin compound could be reduced significantly and the effects against fungi improved.

The opposite effect—increase of phytotoxicity by a second compound —was observed by SPURR (1972). When tobacco plants were treated

Table CXLVI. *Reducing phytotoxicity of triphenyltin acetate by adding manganous ethylenebisdithiocarbamate; test plant: potato* (excerpt from BRUECKNER *et al.* 1961).

Triphenyltin compound (600 mg/L)	Degree of damage[a] when mixed with manganous ethylenebisdithiocarbamate			
	0	25 mg/L	50 mg/L	100 mg/L
Acetate	3	1	0–1	0
Chloride	4	2	0–1	0
Hydroxide	3–4	2	1	0
Oxide	1–2	1	0	0
8-Hydroxyquinolate	2	1	0–1	0
Sorbinate	3	1–2	0–1	0

[a] Rating: 0 = no damage and 5 = 100% damage.

with triphenyltin hydroxide and at the same time (or shortly before) with 1-decanol, increased leaf damage occurred. Presumably, the toxicity of the tin compound was intensified when dissolved in decanol droplets.

TER HORST (1961) pointed out that peanut plants showed severe damage after a wetting agent had been added to the triphenyltin acetate mixture with which they were sprayed. Without the wetting agent, the triphenyltin compound caused no visible leaf damage. The phytotoxicity of triphenyltin compounds can also be intensified by combining them with oil-containing insecticide pastes or emulsifiable products (for example DDT-lindane pastes). This was shown in a field trial using triphenyltin acetate combined with DDT-lindane paste, as Table CXLVII shows (HAERTEL 1963 a).

Table CXLVII. *Phytotoxic field tests on tomatoes with triphenyltin acetate 0.12 and 0.06%, combined with DDT-lindane paste (0.2%[a] product) spraying mixture 600 L/ha* (after HAERTEL 1963 a).

Preparation	Degree of damage[b] at concentration of % triphenyltin acetate		Plant growth in % of untreated at concentration of % triphenyltin acetate		Plant wt in % of untreated at concentration of % triphenyltin acetate	
	0.12	0.06	0.12	0.06	0.12	0.06
Triphenyltin acetate	2.8	2.3	80	98	64	85
Triphenyltin acetate + DDT-lindane paste 0.2%[a]	10	10	0	0	26	26
DDT-lindane paste 0.2%[a]	0	—	88	—	110	—
Untreated	0	0	100	100	100	100

[a] The components and their concentrations in the DDT-lindane paste are unknown. The commercial product was used at the concentration recommended, *i.e.*, 0.2%.

[b] Rating: 0 = no damage and 10 = total damage.

Adhesive agents (stickers), which are often used with fungicides, are able to increase or decrease the phytotoxicity of triphenyltin compounds depending on their structure. This was shown in a greenhouse experiment using 2 adhesives of unknown composition (HAERTEL 1963 b) (Table CXLVIII). Further, JARVIS *et al.* (1967) noticed that the phytotoxicity of triphenyltin hydroxide to potatoes is increased in the presence of aphids.

Table CXLVIII. *Effects of wetting and adhesive agents (stickers) on phytotoxicity of triphenyltin acetate; field tests on tomatoes* (after HAERTEL 1963 b).

Product	Degree of damage[a] at concentrations of % triphenyltin acetate			Plant wt in % of untreated at concentrations of % triphenyltin acetate			Plant growth in % of untreated at concentrations of % triphenyltin acetate		
	0.4	0.2	0.1	0.4	0.2	0.1	0.4	0.2	0.1
Triphenyltin acetate	4.1	3.5	1.3	58	66	82	74	102	89
Triphenyltin acetate + AT 87[b] 1.0%	2.7	2.1	1.0	72	73	84	87	100	83
Triphenyltin acetate + Lovo[b] 190 1.0%	9.5	8.4	3.4	23	46	68	0	21	74
Triphenyltin acetate + Lovo[b] 192 1.0%	8.5	4.8	2.5	32	61	83	24	72	85
Untreated	0	0	0	100	100	100	100	100	100

[a] Index of phytotoxicity: 0 = no damage and 10 = 100% damage.
[b] Components of the product and their concentration unknown.

γγ) *Influence of pH-value:*
According to DUYFJES and DE LANGE (1961), the phytotoxicity of triphenyltin compounds can be reduced by raising the pH-value of the spray liquid to 8 to 12. Alkali metal carbonates, hydrogencarbonates, borax, calcium hydroxide, or organic bases, for example triethanolamine, are recommended for this purpose.

δδ) *Influence of weather:*
Experiments by BAUMANN (1958) showed that the phytotoxicity of triphenyltin acetate for tomatoes increases with rising humidity and temperature (Table CXLIX). This result was confirmed by SENEVIRATNE (1970); in field trials with potatoes; triphenyltin acetate and hydroxide caused the most severe leaf damage at very humid weather conditions.

εε) *Influence of solubility:*
The rather low phytotoxicities of bis-(triphenyltin) sulfide and disulfide (see Table CXLIII) were explained by VAN DER KERK (1961) with extremely low solubilities of both compounds. Similar results were

Table CXLIX. *Phytotoxic effects of triphenyltin acetate on tomatoes, depending on humidity and temperature; duration of exposure 8 days* (after BAUMANN 1958).

	Leaf damage at different concentrations of active ingredient[a]			
Conditions	2.0%	1.0%	0.5%	0.3%
30°C; up to 100% rel. humidity	4–5	4	3–4	3
18–30°C; up to 100% rel. humidity	4	3–4	3	2–3
18–30°C; 60–70% rel. humidity	3–4	3	1–2	1–2

[a] Rating: 0 = no damage and 5 = 100% damage.

also described for triphenyltin phosphoric acid ester compounds (KUBO 1965).

ζζ) Influence of formulation and particle size:

Undoubtedly, the phytotoxicity is influenced by the formulation of the active ingredient. Yet, this fact is only rarely mentioned in literature. Under various conditions BAUMANN (1958) noticed distinct differences between formulations when using 2 different carriers; in all experiments (Table CL), the addition of T2 was more favorable than T1 (components not listed). Further, adding $CaCO_3$ to the active ingredient produced favorable results. However, exact data are missing, and nothing is mentioned about the influence of particle size.

Table CL. *Phytoxicity of triphenyltin compounds with two different carriers towards tomatoes; concentration 0.2%; duration of exposure 8 days* (after BAUMANN 1958).

Triphenyltin compound	Carrier	Phytotoxicity[a] at different conditions			
		a	b	c	d
Acetate	T 1	4–5	4	4	3–4
	T 2	3	2–3	2	1–2
Hydroxide	T 1	4	3–4	3–4	3
	T 2	3	1–2	2	1
Chloride	T 1	5	5	4–5	4
	T 2	4	4	3	3

[a] Scale: 0 = no burns and 5 = total damage; a = 24 hr greenhouse, then 6 days climate room; b = 24 hr greenhouse, then 6 days humidity room; c = 24 hr climate room, then 6 days greenhouse; and d = 7 days greenhouse.

β) Phytotoxicity of triphenyltin compounds in potatoes.—After 3 sprayings with triphenyltin acetate mixture (600 g of active ingredient/ha) leaf burns were noticed. However, at 360 g/ha (3 times) these appeared only occasionally. The crop yields in both experiments were equal within acceptable tolerances (BAUMANN 1957).

PIETERS (1962) noticed severe leaf burns after 7 sprayings with 360, 400, 440, and 4 × 480 g of triphenyltin acetate/ha, whereas the hydroxide showed very little phytotoxicity. The yields, however, were slightly better with acetate. A second series of experiments (similar amounts) showed slightly increased leaf burns with acetate in comparison with the hydroxide. Amounts of 240 to 300 g/ha of acetate caused light leaf damage; hydroxide, none at all. Finally, 4 sprayings (300, 320, 340, and 360 g of active ingredient/ha) caused the same (low) damage with acetate and hydroxide, whereas the chloride was more phytotoxic.

In experiments by CETAS (1962 a and b), however, triphenyltin hydroxide was phytotoxic. Initially, 4 applications with 1,600 to 2,400 g/ha (probably 20% preparation) were conducted, during the following experiments 5 applications with 800 to 1,200 g/ha or 1,100 to 1,500 g/ha. With similar dosages, MANZER and MERRIAM (1963) noticed moderate phytotoxicity of triphenyltin hydroxide. BYRDY et al. (1966 a) obtained discoloration and minor burns with emulsifiable preparations.

After 7 applications of triphenyltin acetate, CALLBECK (1963) noticed leaf damage during a dry, hot summer, which did not appear during the cooler, more humid weather of another year. Experiments by SENEVIRATNE (1970) showed severe damage with triphenyltin acetate and light burns with hydroxide where the active ingredients had been applied very early to the young potato plants (7 applications with 110 g of acetate respectively 90 to 110 g of hydroxide/380 L of liquid; time interval 7 days). The damage caused by the acetate was much lower with 14-day intervals between sprayings. In this case, however, and given the extremely unfavorable conditions, the control effect was unsatisfactory.

In a year without blight, VENTURA and HERVÉ (1962) compared the yields of untreated plots with those of fields treated with triphenyltin acetate. After 4 applications of 60 g of active ingredient at intervals of 10 days, distinctly lower yields were found in several potato varieties, in others the decrease was too low to be significant.

Leaf damage in potatoes can be prevented by using the above mentioned combined preparation of triphenyltin acetate and manganous ethylenebisdithiocarbamate (see Tables LX and CXLVI).

The complex of triphenyltin chloride with decyltriphenylphosphonium bromide is not toxic for potatoes in dosages of 250 to 375 g/ha (CELA).

γ) *Phytotoxicity of triphenyltin compounds in sugarbeets.*—Sugarbeets are especially insensitive to triphenyltin compounds (PICCO 1957, HAERTEL 1962). In 2 series of experiments by BAUMANN (1957) with 4 applications (360 or 600 g of acetate/ha), no leaf damage could be found. Three applications of 400 g of acetate in 400 L of mixture also caused no toxic symptoms (PECHINEY-PROGIL 1960).

Of several kinds of plants tested, beets were the least sensitive to triphenyltin derivatives of phosphoric acid and thiophosphoric acid esters (KUBO 1965). The complex of triphenyltin chloride with decyltriphenylphosphonium bromide is not toxic for beets at dosages of 250 to 375 g/ha (CELA).

δ) *Phytotoxicity of triphenyltin compounds in celery.*—After 6 sprayings with 600 g of triphenyltin acetate/ha, changes of the leaves (deformation of feathered leaves, leaf growth similar to ginkgo tree) of celery were noticed sporadically, but no burns. The leaf deformities did not appear at lower dosages (360 g/ha) (BAUMANN 1957).

After 7 sprayings with 0.1% triphenyltin acetate suspensions, SCHUPP (1957) noticed only sporadic leaf deformities in celery. The leaves normalized during the growth period. VON HOESSLIN (1960), however, found damage when the concentration of the active ingredient exceeded 0.04%. PICCO (1957) noticed light damage after 3 applications with 0.4% of a 20% formulation.

HARDON et al. (1962) used suspensions with 0.04% triphenyltin acetate, since 0.06% had still caused considerable damage. The dosages of active ingredient were 500 to 800 g/ha and 680 to 1,200 g/ha, respectively.

BYRDY et al. (1966 a) found discolorations and light burns in celery leaves after applying triphenyltin hydroxide.

The complex of triphenyltin chloride with decyltriphenylphosphonium bromide is not toxic for celery in concentrations of 0.04 to 0.06% (CELA).

ε) *Phytotoxicity of triphenyltin compounds in rice.*—CHIAPPARINI et al. (1964) and CHIAPPARINI (1965) did not notice any damage in rice plants when controlling algae in rice fields with triphenyltin acetate (0.67 to 1 ppm in water). Plants in the second leaf stage showed no damage even at 5 ppm.

When rice seedlings are planted in fields that have been treated with triphenyltin acetate against snails, the growth is already slowed down at concentrations of 0.25 ppm of active ingredient. The damage can be reduced or eliminated by planting the seedlings several days after applying the tin compound; during this waiting period most of the active ingredient is absorbed by mud particles and thus eliminated from the water (CROSSLAND et al. 1962, HOCKING and WHITE 1967). HOPF et al. (1967) also reported damage in rice seedlings by using triphenyltin acetate.

Experiments by TAMURA (1965 b) with rice plants produced no or only light (acetate) respectively distinct (chloride) leaf damage after triphenyltin acetate and chloride suspensions with 100 ppm of active ingredient had been applied. Dust preparations with 0.25 to 0.50% acetate or 0.25% chloride caused no damage, mixtures with 0.50% chloride caused distinct damage. Mixtures or triphenyltin acetate + tributyltin oxide (50 + 25 ppm) did not burn the leaves, 100 + 50 ppm caused none to distinct damage, and 200 + 100 ppm more severe leaf damage.

Triphenyltin compounds with different phosphoric acid esters usually showed low phytotoxicity when applied in amounts of 25 to 250 μg/cm^2 leaf surface (KUBO 1965).

ζ) *Phytotoxicity of triphenyltin compounds in other plants.*—Table CLI lists data compiled from literature about the phytotoxicity of triphenyltin compounds in numerous other plants.

2. **Systemic properties.**—There is no evidence so far that triphenyltin compounds are systemic agents. In experiments with radioactively labeled as well as unlabeled organotin compounds, no detectable or only extremely low concentrations were found in the tested plants. Isolated results to the contrary, obtained with qualitative biological methods (see below), are hard to explain. There should be no doubt as to the reliability of the chemical and radiochemical methods employed.

α) *Sugarbeets.*—For several weeks, HEROK and GOETTE (1963) conducted extensive studies with labeled triphenyltin acetate of high specific activity (1.3 mC ^{113}Sn/mg Sn). In some of their experiments, leaves were treated and the neighboring leaves tested; in others, roots were treated and the leaves tested. The applied amounts of active ingredient were 60 to 80 μg/plant during leaf treatment. In the untreated leaves of the same plant, 0.3 ppb of ^{113}Sn was found, in the roots 0.8 to 1.1 ppb. For the root treatment, 150 μg active ingredient was applied to each plant within 1 wk in 3 equal proportions; after that, the sprouts contained 0.4 to 1.4 ppb of ^{113}Sn.

No attempts were made to identify the chemical nature of the ^{113}Sn derivatives. Thus, the above values represent maximum concentrations of active ingredient, calculated from the overall ^{113}Sn concentrations, disregarding the possible presence of lower molecular weight degradation products.

FREITAG (1972) and FREITAG and BOCK (1974 b) obtained similar results in greenhouse experiments with triphenyltin chloride which was applied to isolated sugarbeet leaves in amounts of 288 or 576 μg with activities of 155,000 respectively 310,000 cpm. All the tin remained on the leaves. The activities of test samples, obtained from stem or beets 28 to 42 days after application, ranged in the area of background radiation, *i.e.*, approximately 40 to 50 cpm.

Experimental results by SOLEL (1970 b and 1971 b) contrast with these findings. When sugarbeet leaves were treated with triphenyltin acetate on one side only, the other was also partly protected against *Cercospora* infection. Also, when the roots were immersed into a suspension of active ingredient for 3 days, the whole plant was protected to a certain degree.

MUKHOPADHYAY and THAKUR (1972) described similar observations with triphenyltin chloride. Sugarbeet seedlings were immersed to ½ of the root length in suspensions of active ingredient (750 and 1,000 μg/ml). Juice pressed from different parts of the plants after 24 or 48 hrs and applied to *Alternaria tenuis* resulted in slightly reduced growth.

Finally, SOLEL (1970 a) inoculated sugarbeet leaves with *Cercospora beticola*. At different times after infection, he treated the plants with triphenyltin acetate (+20% manganous ethylenebisdithiocarbamate) in concentration of 360 ppm. By applying the active ingredient 3 days after inoculation, the disease could be controlled efficiently. However, after an interval of 5 days, the infection had already spread to 70%, and after 6 days to 97% of the untreated controls. The development of *Cercospora*

Table CLI. *Phytotoxicity of triphenyltin compounds for various plants.*

Plant	Triphenyltin compound	Concentration or amount	Phytotoxicity	References
Tomato	Acetate	0.1–1%	Slightly to highly toxic	BAUMANN (1958)
Tomato	Acetate	0.08%	Toxic	PICCO (1957, 1958, and 1965)
Tomato	Hydroxide, acetate, chloride et al.	50–100 ppm	Toxic	HAERTEL (1962)
Tomato	Hydroxide	0.2–0.3 kg/ha	Toxic	SCHROEDER (1963)
Tomato	Acetate, hydroxide, chloride, chloride complexes	250–1,000 ppm	Toxic	SCHICKE et al. (1968)
Tomato	Chloride	25–250 μg/cm^2	Very toxic	KUBO (1965)
Tomato	Phosphoric acid ester	25–250 μg/cm^2	Very toxic	KUBO (1965)
Tomato	Acetate, hydroxide et al. + Mn^a	0.03%	Not toxic	BRUECKNER et al. (1961)
Egg plant	Acetate, hydroxide, chloride, oxide, sulfide, disulfide	0.05%	Not toxic	ASCHER & MEISNER (1969)
Watermelon	Hydroxide	0.3 kg/ha	Toxic	SCHENK & CRALL (1963)
Cucumber	Hydroxide	0.38 kg/ha	Toxic	SITTERLY (1963 a and b)
Cucumber	Phosphoric acid ester	25–250 μg/cm^2	Very toxic	KUBO (1965)
Radish	Phosphoric acid ester	25–250 μg/cm^2	Very toxic	KUBO (1965)
Soybean	Phosphoric acid ester	25–250 ppm	Slightly toxic	KUBO (1965)
Bean	Acetate, hydroxide, chloride, chloride complexes	250–1,000 pm	Slightly toxic	SCHICKE et al. (1968)
Bush bean	Acetate	0.1–1%	Slightly to highly toxic	BAUMANN (1958)
Bush bean	Acetate	0.05%	Not toxic	ANDRÉN & OLOFSSON (1959)

Bush bean	Acetate	0.05–0.25%	Not toxic	Schmidt (1962 a and b)
Kohlrabi	Acetate	0.1–1%	Slightly to highly toxic	Baumann (1958)
Carrot	Acetate	0.04–0.06%	Not toxic	Schmidt (1965)
Carrot	Acetate	—	Not toxic	Ascher & Nissim (1964)
Carrot	Acetate	0.04–0.06%	Not toxic	Franz (1972)
Grape vine	Acetate	0.1–1%	Distinctly to highly toxic	Baumann (1958)
Grape vine	Acetate	—	Toxic	Haertel (1958 b and 1962)
Grape vine	Acetate + Mn[a]	—	Not toxic	Haertel (1964 a)
Hops	Acetate	0.1–1%	Distinctly to highly toxic	Baumann (1958)
Hops	Acetate + Mn[a]	—	Not toxic	Haertel (1964 a)
Hops	Acetate + Mn[a]	0.06%	Not toxic	Liebl (1968)
Hops	Acetate + Mn[a]	0.06%	Slightly toxic	Zattler & Liebl (1970)
Fruit	Acetate	—	Toxic	Haertel (1958)
Fruit	Acetate + Mn[a]	—	Not toxic	Haertel (1964 a)
Apple	Acetate	0.1–1%	Distinctly to highly toxic	Baumann (1958)
Apple	Acetate, hydroxide	0.01–0.02%	Slightly toxic	Pieters (1962)
Apple	Oxide	—	Not toxic	Luijten (1960)
Apple	Acetate	0.15%	Not toxic	Marfurt & Toscani (1966/67)
Apple	Acetate + Mn[a]	1.5%	Slightly toxic	Hardon et al. (1962)
Pecan nut	Hydroxide	0.02%	Not toxic	Cole (1965)
Pecan nut	Hydroxide	0.02–0.03%	Not toxic	Diener & Garrett (1967)
Pecan nut	Acetate	0.03%	Not toxic	Diener & Garrett (1967)
Tobacco	Acetate	63–500 ppm	Slightly to moderately toxic	Kubo (1965)

Table CLI. (*continued*)

Plant	Triphenyltin compound	Concentration or amount	Phytotoxicity	References
Tobacco	Phosphoric acid ester	63–500 ppm	Slightly to moderately toxic	Kubo (1965)
Tobacco	Acetate	0.005–0.08%	Toxic	Kroeber & Massfeller (1961)
Tobacco	Acetate	0.008%	Moderately toxic	Baets (1961)
Different water plants	Acetate	0.01–1 ppm	Toxic	Floch & Deschiens (1962), Deschiens & Floch (1963), Strufe (1968)
Different water plants	Acetate	1–5 ppm	Not toxic	Seiffer & Schoof (1967)
Flax	Acetate	30–120 g/ha	Highly toxic	Estienne & Hennebert (1959)
Cacao	Acetate	0.15%	Not toxic	Hislop (1963)
Cacao	Chloride	0.075%	Not toxic	Muller et al. (1969)
Cacao	Hydroxide	0.04%	Not toxic	Muller & Njomou (1970)
Coffee	Hydroxide	0.125%	Not toxic	Vine & Vine (1971), Baker (1973)
Coffee	Acetate	0.125%	Not toxic	Abasa & Mulinge (1972 and 1973)
Coffee	Acetate; hydroxide	0.05–0.125%	Not toxic	Bardner & Mathenge (1974)
Barley	Acetate	25–250 μg/cm³	Slightly to highly toxic	Kubo (1965)
Barley	Phosphoric acid ester	25–250 μg/cm³	Slightly to highly toxic	Kubo (1965)
Alfalfa	Acetate	—	Slightly toxic	Ascher & Rones (1964)

Willow	Acetate	0.17%	Toxic	DIERCKS (1957)
Willow	Acetate	0.06–0.08%	Slightly toxic	DIERCKS (1957 and 1959)
Poplar	Acetate	0.06%	Not toxic	BACHTHALER & DAHTE (1959)
Citrus	Acetate	0.06%	Not toxic	STERNLICHT (1966)
Peanuts	Acetate	0.3 kg/ha	Not toxic	ADDY & DASH (1966)
Peanuts	Acetate	0.1–0.4 kg/ha	Slightly toxic	TER HORST (1961)
Peanuts	Acetate	0.18–0.25 kg/ha	Toxic	WADSWORTH et al. 1967)
Papaya	Acetate		Highly toxic	BOLKAN et al. (1976)

[a] Combined preparation with manganous ethylenebisdithiocarbamate.

beticola spores seems to be prevented via the vapor phase of triphenyltin acetate; the effect disappears within 11 days.

β) *Potatoes.*—THOMAS and TANN (1971) measured the content of triphenyltin compounds in potato tubers harvested from fields that had been treated with triphenyltin acetate or hydroxide. By analyzing large quantities (250 g), a detection limit of about 1 ppb was obtained. Of 44 samples, 35 (= 80%) contained residues of ≦ 1 ppb (counted as triphenyltin hydroxide); in 6 samples (= 13%), the content was between 2 and 5 ppb, in 3 samples between 6 and 8 ppb. The mean values were 1.6 ppb for the acetate and 1.5 ppb for the hydroxide.

These values were not corrected for losses through concentrating the triphenyltin compounds (yields 70 to 92%). Further, the tubers had been stored before analysis for 3 mon at 2° to 10°C and the initial tin content may have been slightly higher. On the other hand, the tubers were not peeled but only washed, and residual amounts of triphenyltin compounds may have remained on the outside of the potatoes, thus increasing the analytical findings.

After treating potato plants with triphenyltin acetate, HOLMES and STOREY (1962) found no tin in any of the tested samples (detection limit 0.1 ppm). JOHNSON (1962) was also unable to detect tin in potato tubers. KUBO (1965) conducted diffusion experiments on potato petioles with triphenyltin acetate and several triphenyltin compounds of phosphoric acid and thiophosphoric acid esters. None of the active ingredients were found to penetrate into the plant tissue.

HEROK (1965) used radioactively labeled triphenyltin acetate and chloride ([113]Sn, specific activity 40 and 43 mCi [113]Sn/g Sn) to determine whether these active ingredients are absorbed by the potato plants, transferred systemically, and stored in potato tubers. Under field conditions, 360 g of these substances, formulated as wettable powder, were applied to the plants grown in plastic tubs. The plants were sprayed 3 times during a growing period with a total precipitation of 74.2 mm. No evidence was found that triphenyltin acetate and chloride are absorbed and transferred by potato plants (Table CLII). Also, no absorption has been observed on potato tubers grown in soil containing these substances. The detection limit was below 0.0001 ppm for acetate/chloride and below 0.00003 ppm for tin (Table CLIII).

γ) *Celery.*—Using the technique described for potatoes, HEROK and GOETTE (1963) found only 0.4 to 0.7 ppb of [113]Sn in celery tubers after the leaves had been treated with 60 to 80 μg of triphenyltin acetate/plant. When 150 μg of labeled triphenyltin acetate had been applied to the roots, 0.3 to 1.7 ppb of [113]Sn were detected in the leaves.

After treating celery plants with triphenyltin acetate, KROELLER (1960 and 1963) could not find any triphenyltin compounds in the tubers (detection limit 0.05 ppm Sn = 0.15 ppm triphenyltin acetate); the overall tin content within the tubers had not increased after application of the active substance.

Table CLII. *Result of experiments with radioactivity labeled triphenyltin acetate and chloride (^{113}Sn) in potatoes* (after HEROK 1965).

Test series	Remainder/plant (mg of compound)	Compound[a] (ppm)	
		Potato peelings	Peeled potatoes
Triphenyltin chloride, soil not covered	2.0–3.0	0.0009 (0.0007–0.0015)	not detectable
Triphenyltin chloride, soil covered	1.7–3.3	0.002 (0.0006–0.005)	not detectable
Triphenyltin acetate, soil not covered	1.6–2.3	0.004 (0.002–0.009)	not detectable (<0.00006)
Triphenyltin acetate, soil covered	1.7–2.5	0.009 (0.004–0.018)	not detectable (<0.0001)

[a] Mean values; limiting values stated in parentheses.

BAUMANN (1958) obtained contradicting results. When parts of celery plants were treated with triphenyltin acetate, reduced *Septoria* infection occurred in the untreated parts of the same plant; in hydro cultures, the active ingredient appeared to enter via the roots, and juices pressed from plants treated with triphenyltin acetate reduced the germination of *Botrytis cinerea* spores. Based on this, systemic properties have been attributed to the active ingredient.

δ) *Beans.*—After applying 60 to 80 μg of triphenyltin acetate to the leaves of bean plants, HEROK and GOETTE (1963) found 0.5 ppb of ^{113}Sn

Table CLIII. *Detection limits for (^{113}Sn)-triphenyltin acetate and chloride in potatoes* (after HEROK 1965).

Compound	Counts/min of 1 μg compound	Background radiation (counts/min)	Detection limits		Notes
			μg compound/test sample	ppm[a]	
Triphenyltin chloride	20,200	265	0.013 ±5%	0.0005 ±5%	Effect = background radiation
			0.0013 ±33%	0.00005 ±33%	Effect = 1/10 of background radiation
Triphenyltin acetate	17,500	265	0.015 ±5%	0.0006 ±5%	Effect = background radiation
			0.0015 ±33%	0.00006 ±33%	Effect = 1/10 of background radiation

[a] Since 25 g of potato tuber or peelings were used for each test sample.

in the roots; application of 150 μg of acetate to the roots resulted in 0.7 to 2.4 ppb of [113]Sn in stems and leaves.

ε) *Rice.*—CHIAPPARINI *et al.* (1964) did not find tin in rice plants after applying 780 g of triphenyltin acetate/ha. SINGH and SHARMA (1973) observed that triphenyltin chloride is transported from the roots of rice seedlings to the leaves, but again these results were indicated only by the effects on a fungus (*Cochliobolus miyabeanus*), not by chemical or radiochemical analysis.

ζ) *Cacao.*—Triphenyltin chloride is not systemic in cacao plants and shows no evidence of penetrating into the cacao pods (MASSAUX 1971).

3. Stimulation of growth.—Because of the development of distinctly dark green leaves and the sometimes unexpectedly high improvement of yields—mainly in sugarbeets—it was occasionally assumed that triphenyltin acetate stimulates growth besides its fungitoxicity (SCHUPP 1957, PICCO 1957, SCHLOESSER *et al.* 1957, ESTIENNE and HENNEBERT 1958, BRUECKNER *et al.* 1960). This assumption is supported to some extent by observations of BAUMANN (1958) that in some plants respiration is increased by the active ingredient.

KAARS SIJPESTEIJN (1959) pointed out that dinitrophenol, similar to triphenyltin acetate, stimulates the respiration of plant tissue, causes the foliage to turn dark green and increases the yields. Dinitrophenols are also strong inhibitors of oxidative phosphorylation, just as triethyltin sulfate (ALDRIDGE and CREMER 1955) and possibly triphenyltin acetate. Therefore, the yield increase caused by triphenyltin acetate may be due to similar biochemical properties and plant physiological effects.

BAUMANN (1958), on the other hand, did not find increased yields in disease-free cultures of potatoes, sugarbeets, and celery after applying triphenyltin acetate. No growth stimulation of soybeans by triphenyltin hydroxide was found by HORN *et al.* (1978).

BRUINSMA (1961) investigated whether the conspicuous development of dark green leaves, observed after application of triphenyltin acetate and other fungicides, is due to an increase in chlorophyll content. He concluded, however, that fungicides do not affect the chlorophyll content.

All the above seems to leave the question unresolved whether there are truly stimulating effects of triphenyltin compounds.

VI. Other applications of phenyltin compounds

a) General

The following section will deal with the biological behavior of various phenyltin compounds and with proposed applications that have not gained significant importance. Applications other than biological will not be mentioned (for example, stabilizing of polyvinylchloride, protection of polychlorinated aromatics against corrosion, catalysts in ethylene-polymerization, and others).

b) Tetraphenyltin

Tetraphenyltin in 0.1% suspension has no fungicidal effects on *Fusarium culmorum* Sacc., *Alternaria tenuis* Nees, and *Rhizoctonia solani* Kuehn (CZERWIŃSKA *et al.* 1967).

The insecticidal effects of tetraphenyltin were studied several times. The LD_{50} for houseflies (*Musca domestica* L.) is $> 290 \times 10^{-10}$ mole/fly (BLUM and PRATT 1960) respectively $> 100 \times 10^{-10}$ mole/fly (PIEPER and CASIDA 1965). This compound is therefore much less toxic than other organotin compounds (e.g., LD_{50} of triphenyltin chloride 26×10^{-10} mole/fly); also, restricting effects were noticed towards ATPase, but they are weaker than those of trialkyl or triaryltin compounds. Feeding experiments with 1% $(C_6H_5)_4Sn$ in the food showed almost no toxicity for common houseflies ($LD_{95} > 1,000$ ppm), and no sterilizing effects could be found (KENAGA 1965 b).

HARTMANN *et al.* (1925) proposed to use tetraphenyltin to protect textiles against moths; however, according to HUECK and LUIJTEN (1958), this is ineffective. It is possible that the preparations used by HARTMANN contained triphenyltin compounds as impurities.

GARDINER and POLLER (1964) who fed wool containing 1% tetraphenyltin to larvae of the textile moth (*Tineola bisselliella* Humm.) did not find mortality or reduced wt compared with the control within 14 days. The compound was also ineffective as contact insecticide against *Sitophilus oryzae* L.

Tetraphenyltin is practically ineffective against the snails *Australorbis glabratus* and *Bulinus contortus* (FLOCH *et al.* 1964).

c) Triphenyltin compounds

Triphenyltin chloride is suitable to protect canvas and burlap against rotting (CHALMERS 1967). From 0.25 to 0.50% of a triphenyltin compound (chloride, borate, etc.), added to dispersed dyes (polyvinyl acetate basis) or to varnish (alkyd basis) protects against different fungi (*Speira heptaspora, Fusarium culmorum* Sacc., *Penicillium funiculosum*); however, tributyltin compounds are more effective (GIESEN 1966).

In several experiments, triphenyltin compounds were tested as wood preservatives. FAHLSTROM (1958) impregnated wood with different organotin compounds; triphenyltin chloride, though effective against the tested fungi *Lentinus lepideus* (Bull.) Fr., *Lenzites trabea*, and *Poria monticula* Murr., was inferior to tributyltin oxide. RICHARDSON (1964 a and b) obtained similar results with the same fungi and also with *Coniophora cerebella* (Pers.) Schroet. (RICHARDSON 1968). NISHIMOTO and FUSE (1966) used the agar test to determine the minimum toxic concentration of triphenyltin acetate against *Coriolellus palustris, Merculius lacrymans* (Wulf.), *Coniophora cerebella* (Pers.) Schroet., *Polystictus versicolor* (L.), *Irpex consors,* and *Polystictus sanguineus;* the values ranged between 0.02 and 0.05%, but trialkyltin compounds were much more toxic. Triphenyltin acetate showed little effect against *Coniophora olivacea* (Fr.) Karst., *Poria monticula* Murr., and *Fomes lividus* (Calch.) Sacc. in experiments by DA COSTA and OSBORNE (1971 and 1972).

HOLMES and STOREY (1962) observed that *Senecio senecionis* (groundsel) was totally absent in a potato field which had been treated with triphenyltin acetate. Untreated neighboring fields were covered with this weed.

The application of triphenyltin compounds as seed dressing and preservative for wood, leather, cardboard, and textiles has been patented (BRUECKNER and HAERTEL 1956).

KERR and WALDE (1956) studied the effects of triphenyltin hydroxide and chloride on chickens infected with worms (*Ascaridia galli* and *Raillietina cesticillus*). The tin compounds were administered partly with food, partly as a single dose in gelatin capsules. The feeding experiments resulted in 60 to 80% elimination of the parasites at 1 to 3 mg of active ingredient/kg. In capsule application, 1 mg/kg was sufficient to eliminate *Ascaridia galli*, whereas 50 to 100 mg/kg were necessary for *Raillietina cesticillus*. Triphenyltin laurate, however, was completely ineffective.

d) Diphenyltin compounds

Diphenyltin dichloride in concentrations of 10 to 20 ppm prohibits the growth of *Botrytis allii* Munn., *Penicillium italicum* Wehm., *Aspergillus niger* v. Tigh., and *Rhizopus nigricans* Ehrenb. ex Fr.; the compound is also effective against several gram-positive bacteria such as *Bacillus subtilis*, *Mycobacterium phlei*, and *Streptococcus lactis* (KAARS SIJPESTEIJN et al. 1969).

Diphenyltin dichloride was tested for potato blight control but its efficacy was only 12% compared to triphenyltin acetate (McINTOSH 1971). BRUECKNER and HAERTEL (1956) reported considerable yield increase in sugarbeet fields infected with *Cercospora* after treatment with diphenyltin-*p*-aminobenzoate.

KERR and WALDE (1956) tested diphenyltin dichloride and oxide for its effectiveness against worm infection in chickens (*Ascaridia galli* and *Raillietina cesticillus*). Both compounds were not very effective (necessary dose about 50 to 200 mg/kg). Diphenyltin diacetate and dilaurate were completely ineffective. These results were confirmed and expanded by GRABER and GRAS (1964, 1965, and 1966): whereas the 2 tin compounds mentioned above were still effective against several other worm parasites, they failed completely against infections with *Strongyloides* sp., *Subulura brumpti*, and *Acuaria spiralis*. Against *Ascaridia styphlocerca*, only the oxide was totally ineffective. The toxicity of these compounds to chickens, however, was too high, and aliphatic organotin compounds were more effective. CASTEL et al. (1960) tried to control worm infections in various warm-blooded animals with diphenyltin oxide, but the compound was too toxic.

According to GARDINIER and POLLER (1964), wool samples impregnated with 1% diphenyltin dichloride reduced consumption by moth larvae (*Tineola bisselliella* Humm.); even if they reached the chrysalis stage the moths never hatched. Diphenyltin dichloride was also proposed for impregnation of wood to protect against shipworms (VIND and HOCHMANN 1962 and 1963).

e) Monophenyltin compounds

FOELDESI and STRÁNER (1965) discovered that phenyltin trioxinate inhibits the growth of various fungi on agar for an extended period, whereas the corresponding diphenyltin compound is completely ineffective. This effect might be attributed to the 8-hydroxyquinoline groups.

Phenyltin trichloride and phenylstannonic acid were ineffective against worm infections (*Ascaridia galli* and *Raillietina cesticillus* in chickens (KERR and WALDE 1956).

VII. Residues of triphenyltin compounds; behavior in food processing; tolerances and waiting periods

a) Residues of triphenyltin compounds on and in plants

When used for pest control in agriculture, the concentration of triphenyltin compounds on the plants decreases rapidly because of losses in wind and rain and because of decomposition by atmospheric influence and sunlight. Field experiments have shown that the content is usually reduced to ½ within 3 to 4 days (KROELLER 1960 and 1963, HAERTEL 1963 c and 1964 a, WIT and VAN LIER 1960).

As can be expected, the "half-life periods" in greenhouse experiments are longer; they can be 7 to 8 days (DORMAL and NANGNIOT 1961, HAERTEL 1964 b and c), or even 2 to 3 wk (FREITAG and BOCK 1974 b).

The numerous studies on residues, described in the literature, can only partly be mentioned. Quite often the results cannot be compared because of great variations in dosage of active ingredient and in the time intervals between sprayings.

1. Residues in sugarbeets.—Decomposition and formation of degradation products of triphenyltin chloride on sugarbeet leaves in greenhouse experiments have already been mentioned (see above "metabolism"). A field trial with 3 applications of 400 g of triphenyltin acetate/ha (in 400 L of liquid) at intervals of 4 respectively 2 wk resulted in the residue values listed in Table CLIV.

Findings from numerous other field trials are compiled in Tables CLV and CLXIII. They confirm, in general, the above observations on "half-

Table CLIV. *Decrease of triphenyltin acetate concentration on sugarbeet leaves depending on time* (after PECHINEY-PROGIL 1960).

Days after last spraying	Triphenyltin acetate concentration (ppm)
0	41.3
3	22.3
5	6.5
7	5.1
14	2.2
21	1.8
28	0.96
35	0.63
42	0.28

life periods." The reports do not mention whether the triphenyltin contents of sugarbeets were found in the beet or on its surface.

2. Residues in potatoes.—COUSSEMENT (1972) determined the decrease of triphenyltin acetate concentration on potato leaves as a function of time; within 10 days, the content decreased to about ⅓ of the initial value (Fig. 23).

LLOYD *et al.* (1962) reported similar results after 3 treatments of potatoes with triphenyltin acetate in a 0.06% suspension (Table CLVI).

The concentration of triphenyltin compounds in potato tubers after treatment of the plants has been determined in many other experiments.

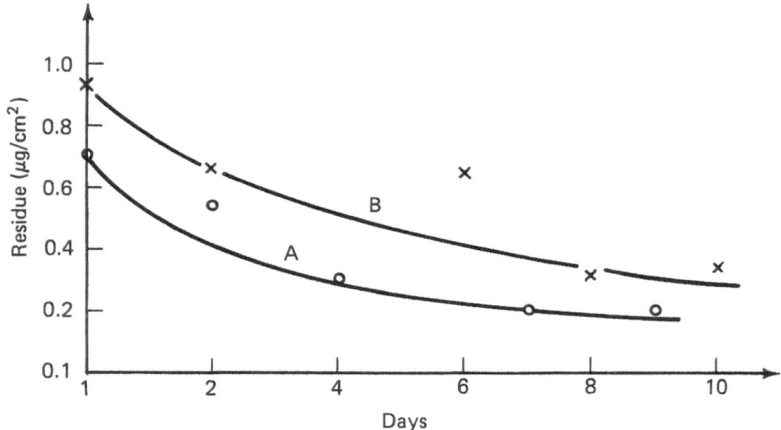

Fig. 23. Reduction of triphenyltin acetate residues on potato leaves depending on time (COUSSEMENT 1972): A = after single spraying with 0.34 kg/ha, and B = after 1 × 0.34 and 6 × 0.36 kg/ha.

Table CLV. *Decrease of triphenyltin acetate concentration in sugarbeets and on sugarbeet leaves depending on time* (excerpt from *Codex Committee* 1970).

Triphenyltin compound and dosage (kg/ha)	Days after application	Concentration (ppm)	
		Beets	Tops and leaves
Acetate 1 × 0.324	0	0.3	4.5
	7	0.2	1.8
	14	0.2	0.6
	21	0.04	0.3
Acetate 2 × 0.336	0	—	1.8
	7	—	0.8
	14	—	0.4
	21	—	0.3
Acetate 3 × 0.216–0.336	3	—	4.0
	5	—	4.0
	10	—	1.0
	14	—	2.0
	20	—	0.7
Acetate 4 × 0.336	0	—	6.0
	7	—	0.8
	14	—	0.3
	21	—	0.2
	28	—	0.2
Hydroxide 5 × 0.30	3	0.11; 0.06	5.43; 5.07
	15	0.22; 0.34	1.07; 2.22
	29	0.06; 0.07	2.00; 1.91
	43	0.05; 0.16	0.87; 1.36
	57	0.18; 0.07	1.17; 0.59
Hydroxide 5 × 0.336	14	<0.03–0.06	—
Hydroxide 6 × 0.224–0.448	26	<0.03	—
Hydroxide 5 × 0.224	43	<0.03–0.04	—
Hydroxide 5 × 0.336	43	0.06–0.09	—
Hydroxide 2 × 0.3–0.4	44	—	1.31–2.45
Chloride 1 × 0.27	0	0.3	2.0
	7	0.2	1.3
	14	0.06	0.5
	21	—	0.5

Since most reports fail to mention whether the potatoes had been peeled before analysis, it is possible that the findings are due to residues on the surface.

Experiments by HEROK (1965) and THOMAS and TANN (1971) which showed no or only negligibly low amounts of triphenyltin acetate, hydroxide, and chloride have already been discussed [see above, Chapter V. g) 2]. Tables CLVII and CLXIII summarize the results from other

Table CLVI. *Triphenyltin acetate residues on potato leaves treated 3 times (7/21; 8/8; 8/25/1960) with triphenyltin acetate (0.06%)* (after LLOYD et al. 1962).

Plot	Leaves	Triphenyltin acetate (ppm)[a] on				
		8/26/60	8/27/60	8/30/60	9/2/60	9/9/60
A	Top	19	27	16	15	12
	Bottom	30	19	18	13	15
B	Top	48	29	18	13	8
	Bottom	36	36	19	10	13
C	Top	29	23	17	18	7
	Bottom	34	30	23	15	11
D	Top	26	28	15	10	12
	Bottom	42	33	18	13	9
Average from sprayed plots A, B, C and D	Top	31	27	17	14	10
	Bottom	36	30	20	13	12

[a] Top and bottom leaves from unsprayed plots gave a value of less than 2 ppm (limit of detection).

Table CLVII. *Residues of triphenyltin compounds in potato tubers* (excerpt from *Codex Committee* 1970).

Triphenyltin compound	Residue (ppm)	No. of analyses
Hydroxide	<0.01	23
	<0.03	30
	<0.06	12
	0.01–0.06	7
	0.11	1
Acetate	<0.03	15
	<0.04	8
	0.02–0.07	6
Chloride	<0.03	24
	0.04–0.05	2

residue determinations which have been obtained with different dosages of active ingredient and after waiting periods of 2 to 169 days.

3. Residues in celery.—After one treatment with 360 g of triphenyltin acetate/ha and after 2 sprayings with the same amount at intervals of 1 mon, KROELLER (1960) observed a decrease in concentration of the active ingredient within 45 days below the detection limit of 0.01 ppm. Both experiments resulted in almost identical persistence curves (Fig. 24). Increasing amounts of diphenyltin compounds were found from the 11th to the 25th day after starting the experiment; however, after 45 days, their concentration had also fallen below the detection limit. Later experiments (KROELLER 1963) showed residual diphenyltin compounds of 0.06 ppm

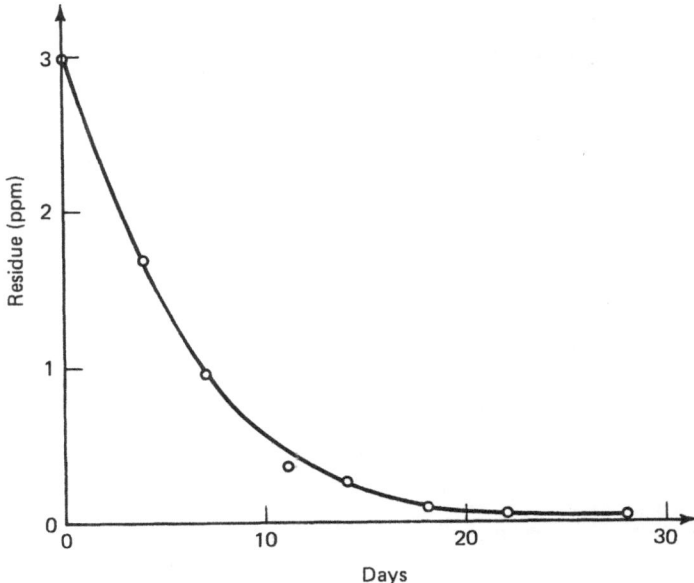

Fig. 24.   Decrease of triphenyltin acetate residues on celery leaves depending on
time (after KROELLER 1960).

after 14 days and 0.30 ppm after 21 days. The "half-life period" for the
decrease of triphenyltin compounds was approximately 4 days. No tin
content beyond the values established for untreated plants could be found
in the tubers.

By statistical evaluation of 70 analyses conducted 14 days after the
last spraying, WIT and VAN LIER (1960) established "half-life periods"
of 3.9 to 4.3 days for residues of triphenyltin acetate on celery leaves.
The values decreased from 10.3 ppm immediately after the last spraying
to <1 ppm within 14.5 days and to <0.1 ppm within 29 days. The respective times were 9.4 days (<1 ppm) and 22 days (<0.1 ppm) at
lower initial concentrations (5.3 ppm).

DORMAL and NANGNIOT (1961) studied the decrease of triphenyltin
acetate on stems and leaves of celery. Within 18 days, the residual
$(C_6H_5)_3Sn^+$ on the leaves was reduced from 24.5 to 2.6 ppm (Table
CLVIII).

Experiments by HARDON et al. (1962) with different amounts of active
ingredient showed a high reduction of triphenyltin acetate residues within
the first week after application; after that the decrease was less distinct
(Table CLIX).

In numerous analyses, RYAN et al. (1972) were unable to find triphenyltin compound (acetate, hydroxide, chloride) in celery 4 wks after

Table CLVIII. *Reduction of residual triphenyl-tin on celery leaves and stems, depending on time* (after DORMAL and NANGNIOT 1961).

Waiting time (days)	$(C_6H_5)_3Sn^+$ (ppm)	
	Leaves	Stems
0	24.5	1.1
1	24.1	1.7
3	17.2	2.7
7	13.5	3.9
10	6.6	3.5
18	2.6	1.7

Table CLIX. *Reduction of residual triphenyltin acetate on celery leaves depending on time* (after HARDON et al. 1962).

Active ingredient dosage (kg/ha)	Triphenyltin acetate concentration (ppm) (weeks after last spraying)				
	0	1	2	3	4
1 × 0.75	10.2	3.0	3.0		
1 × 0.75	29.3	3.3	2.7		
8 × 0.75	29.4	3.5	0.9	0.4	1.0
8 × 0.50	13.5	7.5	1.4	0.9	0.3
8 × 0.094	3.6	0.3	0.8	0.3	0
8 × 0.11	3.1	0.7	1.6	0.7	
9 × 1.2	37.3	17.4	4.3	1.9	
9 × 0.8	24.8	1.9	4.2	1.2	
9 × 0.15	6.8	1.0	1.5	0.9	
9 × 0.18	6.0	1.8	1.7	1.4	
1 × 0.24	5.4	2.7	1.1	1.1	0.9
1 × 0.24	8.2	1.2	1.5	1.0	0.9
6 × 0.68	40.7	1.6	0.4	0.3	0.4
6 × 0.085	3.0	0.2	0	0	
6 × 0.104	4.6	0.9	0.7	0.3	0.2
6 × 0.30	13.6	1.1	0.7	0.6	0.1
6 × 0.30	8.2	1.6	0.9	0.9	0.4

the last spraying; only 2 samples contained 0.3 respectively 0.4 ppm acetate (detection limit not mentioned).

Tables CLX and CLXIII show the results of further experiments.

4. **Residues in carrots.**—After treatment of the fields with triphenyltin compounds, the carrots were usually found to be free of residues; occasionally, values of several hundredths ppm were determined, but rarely more. It is usually not mentioned whether the residues were found inside or on the surface of the carrots. Tables CLXI and CLXIII list several results obtained with varying amounts of active ingredient and waiting periods.

Table CLX. *Reduction of residual triphenyltin on celeriac leaves and tubers depending on time* (excerpt from *Codex Committee* 1970).

Triphenyltin compound and dosage (kg/ha)	Days after last application	Concentration (ppm)	
		Tuber	Leaves
Acetate (1 × 0.324)	0	—	10
	3	0.2	3
	7	0.1	2
	14	—	1
	21	<0.1	3
	28	—	2
	35	—	0.6
Chloride (1 × 0.27)	0	0.1	7
	3	—	4
	7	<0.1	2
	14	—	3
	21	<0.1	1
	28	0.1	1
	35	—	1

Table CLXI. *Triphenyltin residues in carrots* (after *Codex Committee* 1970).

Triphenyltin compound	Residue (ppm)	No. of analyses
Acetate	<0.02	4
	<0.03	5
	0.02–0.06	5
	0.15	1
Hydroxide	<0.05	5
	0.07	1
Chloride	<0.02	1
	0.02–0.05	2
	0.08	1
	0.3	1

5. Residues in pecans.—In 17 analyses of pecan nuts from trees treated with triphenyltin hydroxide, residues could not be found (detection limit 0.03 ppm) (*Codex Committee* 1970, FAO/WHO Evaluations 1971) (Table CLXIII).

6. Residues in rice.—Triphenyltin acetate and hydroxide concentrations of 0.4 to 1.2 ppm in the irrigation water of rice fields did not result in any detectable residues on the rice 90 days after application. Eight samples were examined, the detection limit being 0.08 ppm (*Codex Committee* 1970, FAO/WHO Evaluations 1971) (Table CLXIII).

7. Residues in peanuts.—Treatment of peanuts with triphenyltin

Table CLXII. *Triphenyltin hydroxide residues in peanuts* (excerpt from *Codex Committee* 1970).

Dosage (kg/ha)	Vines	Residue (ppm)	
		Hulls	Nuts
4 × 0.224	—	0.18	<0.01
4 × 0.224	—	0.13	<0.01
4 × 0.34	—	0.07	<0.01
4 × 0.34	—	<0.01	<0.01
4 × 0.34	—	0.12	<0.01
4 × 0.34	—	0.11	<0.01
6 × 0.34	—	0.18	<0.01
6 × 0.34	—	0.29	<0.01
4 × 1.0	—	0.95	0.02
4 × 1.0	—	1.1	<0.01
3 × 0.34	>2.5	0.38	0.03
3 × 0.34	>2.5	0.28	0.02
3 × 0.34	—	0.36	<0.01
2 × 1.0	—	0.86	<0.01
4 × 0.3	—	—	<0.01

hydroxide caused light residues on stems and shells; however, no or only extremely low amounts of active ingredient were found inside the nuts (Tables CLXII and CLXIII).

8. Residues in cacao.—In 8 analyses of cacao after treatment with triphenyltin acetate, 7 samples were found to contain residual amounts of less than 0.1 ppm; one had 0.1 ppm (*Codex Committee* 1970, FAO/WHO EVALUATIONS 1971) (Table CLXIII).

9. Residues in coffee.—After the bushes had been treated with triphenyltin acetate, a residue of 0.06 ppm active ingredient was found in the coffee beans (*Codex Committee* 1970, FAO/WHO EVALUATIONS 1971) (Table CLXIII).

10. Residues in tomatoes.—YONEDA *et al.* (1966) found triphenyltin acetate in tomatoes by qualitative analysis 1 to 3 days after spraying.

11. Residues in apples.—Apples treated with low dosages (0.1 kg/ha; combined formulation with manganous ethylenebisdithiocarbamate) contained 0.30 ppm of active ingredient immediately after the last spraying, 0.19 ppm after 1 wk, 0.21 ppm after 2 wk, 0.22 ppm after 3 wk, and 0.07 ppm after 4 wk (HARDON *et al.* 1962).

12. Summary of residues.—Table CLXIII contains a list of residue values in various commodities and countries (after FAO/WHO EVALUATIONS 1971).

b) Behavior of triphenyltin compounds in food processing

The residues of triphenyltin compounds remaining on the plants can be consumed by human beings directly or get into human food through animals fed with these plants.

Table CLXIII. *Summary of triphenyltin compound residues in supervised trials* (after FAO/WHO EVALUATIONS 1971).

Crop	Country	Rate of application (kg a.i./ha)	No. of treatments	Pre-harvest interval (days)	Plant part analyzed	Detection limit (ppm)	No. of analyses	No. of residues under detection limit	Highest residues found (ppm)
				Residues of triphenyltin hydroxide					
Potatoes	Holland	0.20–0.46	4–7	2–98	tubers	0.01	23	17	0.04
Potatoes	Holland	0.08–0.38	6–7	98–110	tubers	0.06	14	12	0.11
Potatoes	England	0.22–0.56	5	21–43	tubers	0.01	5	5	
Potatoes	England	0.28	1	30–35	tubers	0.03	6	6	
Potatoes	Canada	0.15–0.25	?–12	?–169	tubers	0.03	7	7	
Potatoes	U.S.A.	0.17–0.34	5–11	13–15	tubers	0.03	17	17	
Potatoes	Belgium	0.30–0.36	5	22	tubers	0.01	1	1	
Sugarbeets	Holland	0.30	5	43–57	beets	0.03	4		0.18
Sugarbeets	Holland	0.30	5	43–57	leaves	0.06	4		1.36
Sugarbeets	Germany	0.30	1–4	46–64	beets	0.03	5	2	0.18
Sugarbeets	Germany	0.30	1–4	46–64	leaves	0.06	5	2	0.81
Sugarbeets	Italy	0.30–0.40	?	31	beets	0.03	10	1	0.20
Sugarbeets	Switzerland	0.30–0.40	2	44	leaves	0.02	4		2.45
Sugarbeets	U.S.A.	0.22–0.34	3–6	14–45	beets	0.03	17	8	0.09
Carrots	Israel	0.25–0.30	5–7	12–28	carrot	0.05	4	3	0.07
Pecans	U.S.A.	0.22–0.54	2–8	46–256	nuts	0.03	17	17	

Table CLXIII. (*continued*)

Crop	Country	Rate of application (kg a.i./ha)	No. of treatments	Pre-harvest interval (days)	Plant part analyzed	Detection limit (ppm)	No. of analyses	No. of residues under detection limit	Highest residues found (ppm)
				Residues of triphenyltin hydroxide					
Rice	Italy	0.4–1.2 ppm in irrigation water	1	90	hulled rice	0.08	5	5	
Groundnuts	U.S.A.	0.22–0.34	3–7	9–72	nuts	0.01	14	10	0.03
Groundnuts	U.S.A.	0.22–0.34	3–7	9–72	hulls	0.01	12	1	0.38
Groundnuts	U.S.A.	0.34	3–7	9	vines	0.01	3		2.5
Groundnuts	Israel	0.30	4	5–23	nuts	0.01	4	4	
Groundnuts	Israel	0.30	4	5–23	vines	0.16	4		2.6
Coffee	Kenya	0.22–0.60	?	35	roasted beans	0.03	3	3	
				Residues of triphenyltin acetate					
Potatoes	Germany	0.216–0.324	1–5	7–21	tubers	0.03–0.04	12	8	0.07
Potatoes	Ireland	0.3	3	not stated	tubers	0.03	2	2	
Sugar beets	Germany	0.216–0.324	6	38	beets	0.1	6	6	
Celeriac	Germany	0.3	1–2	21–45	tubers	not stated	3	3	
Celeriac	Germany	0.324	1	21–35	tubers	0.1	3	3	

Residues of triphenyltin acetate

Crop	Country								
Celeriac	Belgium	not stated	3	21–28	stems	not stated	6	6	0.19
Celeriac	Germany	0.3	1–2	21–45	leaves	not stated	13	2	3
Celeriac	Germany	0.324	1	21–35	leaves	0.1	3		1.0
Celeriac	Holland	0.24–0.30	1–6	21–28	leaves	0.1	2		0.30
Celeriac	Holland	0.27	6	21–28	leaves	0.01	2		0.05
Celeriac	Belgium	not stated	3	21–28	leaves	not stated	6	5	0.15
Carrots	Germany	0.216–0.324	1–3	7–28	[a]	0.03	20	17	0.06
Carrots	Germany	0.216–0.324	1–3	7–28	[b]	0.02	6	5	0.02
Carrots	Germany	0.216–0.324	1–3	7–28	[c]	0.02	6	5	0.02
Carrots	Germany	0.216–0.324	1–3	7–28	[d]	0.02	6	5	0.02
Rice	Italy	0.4–1.2 ppm in irrigation water	1	appr. go	hulled rice	0.08	3	3	

Residues of triphenyltin chloride

Crop	Country								
Cocoa	Nigeria	not stated	not stated	not stated	not stated	0.1	8	7	0.1
Potatoes	Germany	0.22–0.27	1–3	7–14	tubers	0.04	8	6	0.05
Sugar beets	Germany	0.18–0.27	6	38	beets	0.10	4	4	
Celeriac	Germany	0.27	1	21–35	tubers	0.1	3	2	0.1
Carrots	Germany	0.20	3	10–31	washed carrots	0.02	4	2	0.08

[a] Washed unpeeled carrots.
[b] Peeled/scraped carrots.
[c] Peels, calculated on total wt carrot.
[d] Scrapings, calculated on total wt carrot.

There is little danger of directly consuming residues in harmful amounts since the parts of the plants used for human food do not normally contain detectable concentrations of triphenyltin compounds; this holds mainly true for potatoes, celery tubers, carrots, and rice.

Further, any residues that might be present are usually completely or partly decomposed when the food is processed.

1. **Residues in human food.**—If celery (after it has been washed) is sterilized for 37 min at 118.3°C, 92 to 98% of the triphenyltin acetate residues present are destroyed (DORMAL and NANGNIOT 1961).

During normal cooking of food (20 to 30 min), about 10 to 20% of the triphenyltin acetate residues are decomposed, as experiments by SCHMITT-STRECKER (1976) showed [see Section **III.**_b_].

In sugar production from beets triphenyltin compounds are completely decomposed to inorganic tin (IV) during the evaporation process (*Codex Committee* 1970, FAO/WHO EVALUATIONS 1971) (Table CLXIII).

Coffee beans, after roasting, were found to contain no triphenyltin compounds (detection limit 0.03 ppm) (*Codex Committee* 1970, FAO/ WHO EVALUATIONS 1971) (Table CLXIII).

In manual husking of cacao beans part of the residues (an average of about 0.18 ppm, calculated from the overall tin content as triphenyltin chloride) are transferred onto the beans. However, the percentage of organotin compound decreases from an initial 78% in the fresh beans to 35, 2.7 and 2.3% in the fermented, dried, and roasted beans. The concentration of organotin compounds in commercial beans is, therefore, approximately 0.005 ppm (MASSAUX 1971); this concentration is likely to be reduced even further during continued processing.

Several kinds of food (potatoes, carrots, rice, peanuts, pecans) are washed and peeled or scraped before consumption, eliminating most of the residual triphenyltin compounds.

2. **Residues in animal food.**—Indirect intake of triphenyltin compounds by humans is possible if sugarbeet leaves have been fed to cattle or cacao bean pods to pigs, cattle, and sheep.

The fate of triphenyltin acetate on sugarbeets consumed by cows has been examined by BRUEGGEMANN *et al.* (1964 a). The milk was found to contain phenyltin compounds (triphenyltin compounds and breakdown products) in concentrations close to the detection limit (approximately 4 ppb). In meat, no tin content higher than the blank value of untreated animals could be determined. Only in kidneys and liver the overall tin level rose significantly to 0.5 to 2.4 respectively 0.8 to 1.5 ppm.

However, the animals can be fed with fresh leaves only for a short period of time; after being siloed for only 5 wk, sugarbeet leaves do not contain any detectable amount of organotin compounds. Besides, these are quickly eliminated from the animal organism. Human consumption of organotin compounds in meat is therefore possible only through animals that have been slaughtered after consumption of fresh sugarbeet leaves.

Even then the amounts transferred to the human body are extremely low (BRUEGGEMANN *et al.* 1964 b). It can further be assumed that any phenyltin compound residues still present will be partly or completely destroyed when the meat is cooked or fried.

Cacao bean pods are used in limited amounts (*ca.* 6 to 20 g/day and kg body wt) as supplement to pig food. Under most unfavorable conditions, the animals would consume about 80 μg of triphenyltin compound/day and kg live wt.

Cacao bean hulls which on average contain no more than 0.02 ppm of triphenyltin compounds are used as supplement to animal food with 1 to 3 kg/day for cattle, 0.30 to 0.60 kg/day for pigs, and 0.30 to 0.45 kg/day for sheep. The amounts consumed by animals are, therefore, only 60 μg/day, 12 μg/day or 9 μg/day, respectively, which means less than 0.3 μg/day and kg live wt (MASSAUX 1971).

c) Tolerances and waiting periods

Various countries have regulated the maximum residual concentration of active ingredient for their crops at harvest time and also the waiting periods after the last spraying. The limits are based on toxicity, recommended dosages, and the decrease of residues depending on time.

The values applicable for different countries at this time (1977) are compiled in Table CLXIV, recommendations by FAO/WHO in Table CLXV.

VIII. Analytical methods

a) Analysis of technical products

Triphenyltin compounds which are the main component of a formulation can be measured quantitatively with gravimetric or titrimetric methods.

For gravimetric analysis, the sample is dissolved in a suitable solvent, for example benzene. Then the di- and monophenyltin compounds are separated by extraction with weak alkaline tartrate or citrate EDTA-solution. Finally, the hexaphenyldistannoxane, present in the organic solution, is transferred to an aqueous solution by repeated extraction with $0.2N$ H_2SO_4. From this, a mixture of $(C_6H_5)_3SnOH$ and $(C_6H_5)_3SnOHSn-(C_6H_5)_3$ is precipitated with alkali hydroxide, which is weighed as oxide after drying at 100°. Tetraphenyltin can be determined in the remaining benzene phase (BOCK *et al.* 1962).

Titrimetric analysis of triphenyltin hydroxide and hexaphenyldistannoxane is conducted with hydrochloric acid in *i*-propanol/ethyleneglycol solution with potentiometric indication using glass/calomel electrodes (FRIEBE and KELKER 1963). This process is the basis for the Hoechst AG analytical procedure for determination of triphenyltin acetate in technical

Table CLXIV. *Permissible maximum concentration of triphenyltin compounds and waiting periods for various countries* (after a tabulation by *Hoechst AG* 1975).

Country	Triphenyltin compound	Commodity	Tolerance (ppm)	Waiting period
Belgium	$(C_6H_5)_3Sn^+$	Fruit, vegetable	0	—
Belgium	$(C_6H_5)_3Sn^+$	Celery	1.0	—
Federal Republic of Germany	Acetate, chloride, hydroxide (all figured as hydroxide)	Celery tubers	1.0	21 days
		Celery leaves	1.0	21 days
		Sugarbeet	0.2	49 days
		Carrot	0.1	35 days
		Potato	0.1	7 days
		Oil seed	0.1	—
		Raw coffee	0.1	—
England	Acetate, hydroxide	Hops	0.1	—
		Potato	0.1	—
Yugoslavia	Acetate	Hops	?	42 days
	Hydroxide	Potato, celery, sugarbeet	?	42 days
Netherlands	$(C_6H_5)_3Sn^+$	Potato	0.1	4 wk
		Celery	1.0	4 wk
Austria	Acetate	Potato	—	5 wk
		Celery	—	5 wk
		Sugarbeet	—	5 wk
Switzerland	Acetate	Potato	—	6 wk
		Celery	—	6 wk
South Africa	Acetate	Potato	—	42 days
		Onion	—	42 days
U.S.A.	Hydroxide	Peanut	0.05	14 days
		Peanut shell	0.4	
		Potato	0.05	7 days
		Carrot	0.1	14 days
		Pecan	0.05	[a]
		Sugarbeet	0.1	14 days

[a] Not to be applied after green shell has opened.

and formulated products (GORBACH and HOMMEL 1963): the weighed sample (4 g in products with about 20% active ingredient, 1 g at about 60% active ingredient) is slurried with CCl_4 and then transferred to a sinter-glass filter with 4×25 ml CCl_4. Allowing slow filtration, the CCl_4 extracts are collected and then extracted by shaking with 70 ml of a weak alkaline sodium tartrate solution. In this manner the anions and the di- and monophenyltin compounds are removed. The remaining solution is brought to 250 ml with additional CCl_4. An aliquote of 100 ml is mixed

Table CLXV. *FAO recommendations for triphenyltin residues* (FAO/WHO report of the 1972 joint meeting, annex 1, p. 34).

Maximum acceptable daily intake (mg/kg body wt)	Commodity	Tolerances (mg/kg)
0.0005	Celery	1[a]
	Sugarbeet, carrot	0.2[a]
	Potatoes, celeriac	0.1[a]
	Cocoa beans, coffee (raw beans), rice (in husk)	0.1[b]
	Peanuts, pecans (shelled)	0.05[b]

[a] Referred to as triphenyltin in FAO/WHO 1965 b. Tolerances on root crops expressed on "soil free" basis. Tolerances refer to the total amount of triphenyltin compounds present, expressed as triphenyltin hydroxide. (Inorganic tin is not included in these tolerances.)

[b] Level at or about the limit of determination.

with 100 ml of *i*-propanol/ethyleneglycol (1:1) and then slowly potentiometrically titrated with 0.1N HCl in *i*-propanol/ethyleneglycol. Any tetraphenyltin that may be present is not determined with this method.

Further, triphenyltin chloride can be titrated potentiometrically with tetraphenylarsonium chloride in acetonitrile solution; however, with this method di- and monophenyltin compounds are also determined (TAGLIAVINI and ZANELLA 1968).

Indirect analytical methods determine the anionic part of the triphenyltin compound. For example, crude triphenyltin chloride is dissolved in benzene or carbon tetrachloride. Then the chloride and the di- and monophenyltin compounds are separated by extraction with weak alkaline aqueous tartrate solution. The hexaphenyldistannoxane remaining in the organic phase is reconverted to the chloride with diluted HCl. Then the chloride-ion is extracted from the organic phase with aqueous alkali and analyzed in the aqueous solution (BOCK et al. 1962).

In a similar procedure for analyzing the anionic fraction of triphenyltin compounds the alkaline solution is passed over a cation exchanger (H+ form) and the resulting free acid determined by titration (GORBACH and HOMMEL 1963).

The chloride content of phenyltin chlorides can also be determined directly by potentiometric titration in acetone solution (VERDONCK and VAN DER KELEN 1965).

1. Purity tests of triphenyltin compounds.—Impurities of di- and monophenyltin compounds in triphenyltin compounds can be determined by titration in methanol solution with aqueous EDTA. Triphenyltin compounds and tetraphenyltin do not react (BUERGER 1961 a).

Further, thin-layer chromatography has been used to determine impurities in triphenyltin chloride (FREITAG and BOCK 1974 a; see below).

2. Analysis of mixtures of triphenyltin hydroxide and hexaphenyldistannoxane.—Analysis of mixtures of triphenyltin hydroxide and hexaphenyldistannoxane poses a special problem. The hydroxide can be determined using the Karl Fischer method (GILMAN and MILLER 1951), whereby 1 mole of iodine is consumed by 1 mole of tin:

$$(C_6H_5)_3SnOH + I_2 + SO_2 + CH_3OH \rightarrow (C_6H_5)_3SnI + HI + HSO_4CH_3 \tag{33}$$

Hexaphenyldistannoxane also reacts with Karl Fischer-reagent, however, here 1 mole of iodine is consumed by 2 moles of tin:

$$(C_6H_5)_3SnOSn(C_6H_5)_3 + I_2 + SO_2 + CH_3OH \rightarrow 2(C_6H_5)_3SnI + HSO_4CH_3 \tag{34}$$

This difference in behavior can be utilized for analysis of mixtures of triphenyltin hydroxide and hexaphenyldistannoxane (KUSHLEFSKY and ROSS 1962, KUSHLEFSKY et al. 1963).

Solid hydroxide/oxide mixtures can be analyzed by IR spectroscopy. The hydroxide can best be found by the O-H stretching vibration at 3,620 cm^{-1} and the O-H bending vibration at 896 cm^{-1} (according to other sources 898 cm^{-1}) (KRIEGSMANN and GEISSLER 1963). Quantitative analysis is best conducted in Nujol mull; the extinction at 896 cm^{-1} is determined using the base-line technique and then evaluated by comparing with the oxide band at 1,076 cm^{-1}, which is used as internal standard (FRIEBE and KELKER 1963).

3. Analysis of tin in triphenyltin compounds.—The tin in triphenyltin compounds is analyzed with well-known methods using oxidation decomposition of the organic substance. Various oxidation procedures are in use; satisfactory results have been obtained by decomposition with Na_2O_2 (FRITZ and SCHEER 1964), $HCl + HClO_3$ (GEYER and SEIDLITZ 1964), H_2SO_4 (GILMAN and ROSENBERG 1953), $H_2SO_4 + NH_4NO_3$ (KOHAMA 1963), $H_2SO_4 + HNO_3$ (KOCHESHKOV 1928, LUIJTEN and VAN DER KERK 1955, MARR 1975), $H_2SO_4 + H_2O_2$ (ASMUS et al. 1971, MARR 1975), or $H_2SO_4 + HNO_3 + HClO_4$ (BOWEN 1972, HEIMES and BRAUN 1971). DOSIOS and PIERRI (1930) decomposed tetraphenyltin dissolved in CCl_4 with acidified bromine water.

Treatment of tetraphenyltin with bromine, followed by decomposition with $H_2SO_4 + HNO_3$ (GILMAN and KING 1929) appears to yield incomplete results (HEIMES and BRAUN 1971). Oxidation in the oxygen flask, recommended by REVERCHON (1965) has also been reported to result in low values for Sn (MARR 1975); the same holds true for decomposition with $H_2SO_4 + HClO_4$ (KOHAMA 1963).

GUENTHER et al. (1969) used X-ray analysis for tin, avoiding the destruction of the organic molecule.

b) Analysis of traces of phenyltin compounds

1. General.—Analysis of phenyltin compounds in extremely low concentrations serves to determine residues on plants, in animal tissue and
in food, as well as to study the metabolism of triphenyltin compounds in
biologic systems (KUMPULAINEN and KUOVISTOINEN 1977). This can be
done in different ways:
(1) Analysis of the overall tin content of the sample.
(2) Analysis of the ^{113}Sn (after application of ^{113}Sn-labeled com-
    pounds).
(3) Analysis of phenyl groups as benzene.
(4) Analysis of single phenyltin compounds after concentration and
    separation.
The first method, although rather simple, yields only the maximum
possible content of organotin compound in the sample; in addition, the
obtained Sn-value has to be corrected for the tin content naturally occurring in the sample.

Analysis of tin by measuring the radioactivity of ^{113}Sn also results in
the maximum content of organotin compound, but without the need for
corrections (see above).

Analysis of phenyl groups as benzene shows the total of the phenyltin
compounds; other forms of tin resulting from decomposition and the
naturally occurring inorganic tin content of the sample are not accounted
for. From the value obtained, the maximum possible content of triphenyltin compound in question can be calculated. The same result can be
obtained from ^{14}C- or ^{3}H-analysis after application of respectively labeled
phenyltin compounds.

The last method (analysis of separate phenyltin compounds) is most
efficient as well as elaborate and the concentration process usually leads
to considerable losses.

2. Analysis of total tin; analysis of ^{113}Sn.—Oxidative decomposition
of organic material and following analysis of total tin with known methods
has been done repeatedly. The main advantage of this procedure is that
with correct decomposition losses of tin can be avoided. The following
decomposition methods have been used: oxidation with $H_2SO_4 + HNO_3$
(BUERGER 1961 b, ALMEIDA FILHO and BRUNE 1972), $H_2SO_4 + HClO_4$
(GOLDBERG FEDERICO and VANDONI 1964), $HNO_3 + CH_3COOH + H_2SO_4$
$+ H_2O_2$ (BOENIG and HEIGENER 1972), or with $H_2SO_4 + HNO_3 + HClO_4$
(CORBIN 1970); the latter, however, suffices only to determine tricyclo-
hexyltin chloride. FILLER (1971) decomposes with $HNO_3 + H_2SO_4 +$
HNO_3 and totally evaporates the obtained solution. The residue is fused
with $K_2S_2O_7$ before determining the Sn to make sure that all tin is dissolved. OELSCHLAEGER (1960) dry ashes celery leaves in a closed muffle
at 450° to 600°C; however, this method is less recommendable because
of the formation of SnO_2 which is difficult to dissolve (BOENIG and
HEIGENER 1972). BUCCI *et al.* (1959) fuse the residue after dry ashing
with $KCN + Na_2CO_3$.

Counting radioactivity of ^{113}Sn in liquid samples (blood and urine) can be done directly. Milk, organs, and excrements are first decomposed with $H_2SO_4 + H_2O_2$ (HEROK and GOETTE 1964).

3. **Analysis of phenyl groups as benzene.**—After BUERGER (1961 a), phenyltin compounds are decomposed with strong hydrochloric acid and metallic zinc at 60°C; the resulting benzene is treated with HNO_3/H_2SO_4 to obtain *m*-dinitrobenzene. This, according to HANCOCK and LAWS (1956), reacts with methylethylketone and alkali to a purple nitron acid dye which is analyzed photometrically at 565 nm. Errors with this method amount to 10 to 20%; blank readings can be caused by other compounds in the plant.

4. **Analysis of different phenyltin compounds after concentration and separation.**—

α) *Concentration method.*—If phenyltin compounds have to be analyzed separately in biologic material, concentration at first is necessary from the bulk of plant and animal substance.

Solid samples are usually extracted with organic solvents, rarely with aqueous solutions. Organic liquids are able to dissolve tetraphenyltin, triphenyltin compounds (acetate, chloride, hydroxide, or oxide), and diphenyltin compounds from these samples; the behavior of monophenyltin compounds is not yet understood, whereas inorganic tin (IV) supposedly remains undissolved. Occasionally, the sample is stirred to a paste with some NaOH solution in order to prevent dissolving of diphenyltin compounds and inorganic tin (and presumably also monophenyltin compounds) in the subsequent organic solvent extraction (GORBACH and BOCK 1958).

If the concentration of all phenyltin compounds together with inorganic tin(IV) is required, extraction of the sample with diethylether is recommended after adding HCl and NH_4SCN (10 ml, 2N HCl + 3 g NH_4SCN, 20 ml diethyl ether) (AKAGI *et al.* 1972). For the same purpose, FREITAG and BOCK (1974 a and b) extracted with a citrate/phosphate-buffer of pH 8.5 (42.3 g $Na_2HPO_4 \cdot 12$ H_2O + 12 g citric acid monohydrate/1 L solution), adding 0.2% EDTA, if needed.

Table CLXVI shows a summary of extraction methods for solid samples.

Liquids can usually be extracted directly with organic solvents. AKAGI *et al.* (1972) extracted vinegar with diethyl ether after adding 10 ml 2N HCl and 3 g $NH_4SCN/30$ ml sample, hereby transferring the phenyltin compounds and inorganic tin to the organic phase. FREITAG and BOCK (1974 b) extracted urine concentrated at room temperature with CH_2Cl_2 or $CHCl_3$. By adjusting the pH value and adding suitable reagents the phenyltin compounds are subsequently eliminated from the aqueous solution (see below, "separations").

Emulsions occurring during extraction of solid or liquid samples can usually be broken by centrifuging. Ether/water emulsions, especially difficult to eliminate, are centrifuged first, then a paper filter is placed on

Table CLXVI. *Extraction of phenyltin compounds and inorganic tin (IV) from solid organic material and from soil samples.*

Material	Preparation	Extraction medium	Compound extracted and concentration	Yield (%)	References
Leaves (tomato, celery)	—	$CHCl_3$, cold	$(C_6H_5)_3SnOCOCH_3$ 10–28 ppm	80–120	GORBACH & BOCK (1958), NANGNIOT & MARTENS (1961)
Leaves (tomato, cherry, elderberry)	—	$CHCl_3 + 0.1$ N NaOH (50 + 3), cold	$(C_6H_5)_3SnOCOCH_3$	100–110	GORBACH & BOCK (1958), VOGEL & DESHUSSES (1964), WITT & VAN LIER (1960)
Leaves (celery)	—	CH_2Cl_2, cold	$(C_6H_5)_3SnOCOCH_3$ 0.3–40 ppm	—	HARDON et al. (1960)
Leaves (celery)	—	CH_2Cl_2, (Soxhlet) 4.5 hr	$(C_6H_5)_3SnOCOCH_3$ 0.01–3 ppm	—	KROELLER (1960)
Leaves (potatoes)	—	CH_2Cl_2, reflux 30 min	$(C_6H_5)_3SnOCOCH_3$ 7–48 ppm	93 (average)	LLOYD et al. (1962)
Leaves (sugar beets)	Shredding	Citrate/phosphate-buffer, pH 8.5; cold	Phenyltin compounds and Sn(IV)	77–99	FREITAG & BOCK (1974 b)
Plant material leaves, stems, tubers	Drying at 50°C	$CHCl_3$, reflux	$(C_6H_5)_3SnOCOCH_3$ ca. 0.5–30 ppm	—	BUERGER (1961 a)
Potato (tubers)	Shredding, extraction of H_2O with Acetone + H_2SO_4	CH_2Cl_2, cold	$(C_6H_5)_3SnOCOCH_3$ $(C_6H_5)_3SnOH$; 1–8 ppb	70–92	THOMAS & TANN (1971), VERNON (1974)
Potato (tubers)	Cutting up	CH_2Cl_2, reflux 30 min	$(C_6H_5)_3SnOSn$ $(C_6H_5)_3$ <0.3 ppm	95 (average)	LLOYD et al. (1962)
Rice	—	CH_2Cl_2	Phenyltin compounds	—	GOLDBERG & VANDONI (1964)

Table CLXVI. (*continued*)

Material	Preparation	Extraction medium	Compound extracted and concentration	Yield (%)	References
Tomato	—	$CHCl_3$	$(C_6H_5)_3SnOCOCH_3$	—	Yoneda et al. (1966)
Vegetable, tomato	Addition of HCl + NH_4SCN	Diethylether	Phenyltin compounds and Sn(IV)	up to 100	Akagi et al. (1972)
Celery (tubers)	Shredding	CH_2Cl_2, (Soxhlet) 4.5 hr	$(C_6H_5)_3SnOCOCH_3$	—	Kroeller (1960)
Apple	Shredding	CH_2Cl_2	$(C_6H_5)_3SnOCOCH_3$ 0.07–0.3 ppm	—	Hardon et al. (1962)
Cacao pods— hulls, leaves	—	CCl_4, cold	Phenyltin compounds	—	Massaux (1971)
Cacao beans	—	CCl_4, (Soxhlet)	Phenyltin compounds	—	Massaux (1971)
Potato	Drying at 90°C, 1 hr	Acetonitrile	$(C_6H_5)_3SnOCOCH_3$ 0.001–0.5 ppm	>90 at 0.4 ppm	Booth & Fleet (1970)
Animal organs	Drying at 50°C, pulverizing	$CHCl_3$, reflux	Phenyltin compounds 0.4–1.7 ppm	—	Brueggemann et al. (1964 a)
Feces	Drying at 50°C, pulverizing	$CHCl_3$, reflux	Phenyltin compounds	—	Brueggemann et al. (1964 a)
Feces	Adding of citrate phosphate-buffer, pH 8.5	$CHCl_3$, cold	$(C_6H_5)_3SnOSn(C_6H_5)_3$	—	Freitag & Bock (1974 b)
Animal tissue	—	Grinding with $CHCl_3$ + CH_3OH + conc. HCl (200/100/2 Vol.)	$(C_6H_5)_3SnOCOCH_3$	—	Heath (1963 a)
Soil samples	Air drying	CCl_4, cold	Phenyltin compounds	—	Massaux (1971)
Soil samples	Screening	Ethanol (95%; + diethylether (1:1)	$(C_6H_5)_3SnOCOCH_3$ $(C_6H_5)_3SnOH$	—	Cenci & Cremonini (1969)

top of the emulsion separating the two liquid phases. Another disc of polytetrafluoroethylene of equal diameter and 0.25 mm thickness is placed on top of the filter paper. Centrifuging is repeated at 3,000 rpm. The polytetrafluoroethylene disc with its higher density (d = 2.2) compresses the emulsion and separates the lighter organic phase almost completely (AKAGI et al. 1972).

KROELLER (1960) eliminated emulsions appearing in distribution processes between aqueous EDTA solutions and CH_2Cl_2 by adding Zephirol®.

Milk, which contains highly emulsifying compounds, is best dehydrated to powder in a spray-drum dryer (BUERGER 1961 a) or by freeze-drying and then extracting it with an organic solvent.

Traces of organotin compounds in the air are enriched by absorption on silica gel at −70°C and subsequent desorption at higher temperatures (LUSKINA and SYAVTSILLA 1969).

β) Separation by distribution between two solvents.—To analyze for triphenyltin compounds in biologic material, it is first extracted with an organic solvent such as CH_2Cl_2, $CHCl_3$, benzene, or others. Then the di- and monophenyltin compounds and the inorganic tin are extracted from the organic phase with aqueous weak alkaline tartrate or phosphate/citrate solution (pH about 8.5) (BOCK et al. 1958).

Aqueous starting solutions of phenyltin compounds are adjusted to pH 8.5 by addition of tartrate (or citrate + phosphate + EDTA) and then extracted with CH_2Cl_2, $CHCl_3$, or benzene.

Both methods yield an organic solution which contains only tetraphenyltin and triphenyltin (as hexaphenyldistannoxane), and an aqueous solution with di- and monophenyltin compounds and inorganic tin.

In the organic phase, the hexaphenyldistannoxane can usually be analyzed directly, because tetraphenyltin in most cases is of no concern (tetraphenyltin does not respond to photometric analysis with dithizone). If separation of both compounds is required, the triphenyltin compound can be eliminated by repeated extraction with aqueous $0.2N$ H_2SO_4, which leaves the tetraphenyltin in the organic phase (see Table CLXVII).

From the aqueous solution of di- and monophenyltin compounds, both substances can be extracted at pH 8.0 with $CHCl_3$ after adding EDTA and pyrrolidinedithiocarbamate (BOCK et al. 1958). Inorganic tin(IV) remains in the water layer (see also Fig. 25/o).

To separate di- and monophenyltin compounds from each other and from inorganic tin, distribution methods in the presence of different organic reagents are applied. The basis for these methods can be derived from the pH distribution curves shown in Figure 25 (FREITAG 1972, FREITAG and BOCK 1974 a); about the distribution of the pyridylazonaphthol complex of diphenyltin compounds see also PILLONI (1967).

In the graphs, the %-distribution P, calculated as % $P = 100\alpha/(1 + \alpha)$ from the distribution coefficient α, is shown as function of the pH-value of the water layer (measured before equilibrium). The values

Table CLXVII. *Distribution of bis-(triphenyltin)oxide between aqueous H_2SO_4 solutions and benzene depending on pH value* (after Bock *et al.* 1962).

pH value of aqueous starting solution	Triphenyltin compound in benzene (volume ratio 1:1) (%)
0.63	19.4
1.13	30.3
1.6	71.6
2.1	86.0
2.6	96.0
3.1	98.2
4.4	99.9

show the % of Sn compound in the organic layers after distribution of the total amount in equal volumes of both phases.

The preparations consisted of 10 ml aqueous buffer solution each and 10 ml organic phase. Citrate/phosphate solutions after McIlvaine were used as buffers. High pH values were reached by adding NaOH. $CHCl_3$ was mainly used as organic solvent, occasionally also 2-methylbutanol(4).

The phenyltin compounds were dissolved in amounts of 500 μg of chloride in 0.5 ml of the organic solvent; the volume was then adjusted to 10 ml with 9.5 ml of the same solvent, containing as a rule the organic reagent. Inorganic tin was added to the aqueous solution as sulfate in amounts corresponding to 500 μg of $SnCl_4$.

The distribution curves (Fig. 25) suggest numerous possibilities for separation. It is advisable to start with aqueous citrate/phosphate solutions (pH 8.5) and extract the diphenyltin compounds with dithizone/ $CHCl_3$ [or α-benzoinoxime/$CHCl_3$, dithiol/$CHCl_3$, acetylacetone/$CHCl_3$, phenylfluorone, 2-methylbutanol-(4), and others]. Then, the monophenyltin compounds are isolated with tropolone/$CHCl_3$ or Na-diethyldithiocarbamate/$CHCl_3$, and finally the inorganic tin(IV) is separated with Na-diethyldithiocarbamate/$CHCl_3$, after adjusting the pH to 4.

A shaking period of 15 min is sufficient for almost complete equilibrium. High concentrations of phenyltin compounds interfere with the separation process; concentrations of 100 μg/ml tri- and monophenyltin chloride each and of approximately 40 μg/ml of diphenyltin chloride should not be exceeded.

In a variation of this isolation method the water layer is acidified to pH 4 after extracting the diphenyltin compounds. Thereafter, Na-diethyldithiocarbamate is added and the monophenyltin compound together with inorganic tin(IV) are extracted with $CHCl_3$. The inorganic tin is then isolated from the organic phase by reextraction with an aqueous NH_3/NH_4Cl solution (pH 8.5).

γ) *Isolation of tetraphenyltin, tri-, di-, and monophenyltin compounds and inorganic tin (IV).*—By combining the above mentioned distribution

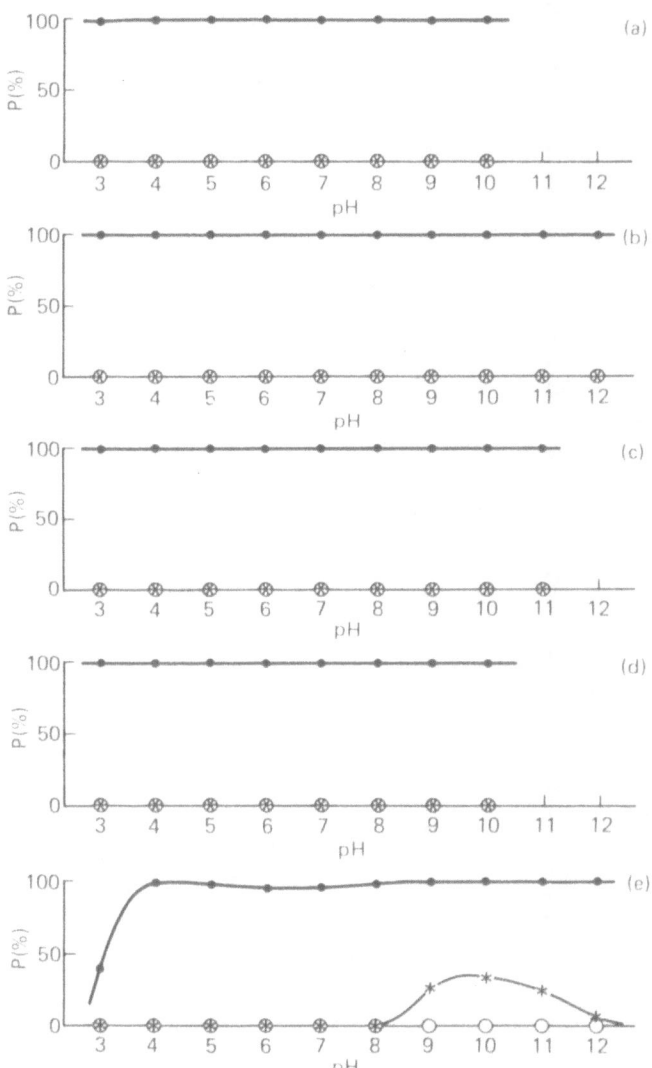

Fig. 25. Distribution of complexes of di- and monophenyltin and inorganic tin(IV) between aqueous citrate/phosphate buffer solutions and organic solvents, depending on pH value; shaking period 30 min (after Freitag 1972): dots = diphenyltin complexes, × = monophenyltin complexes, and ○ = complexes of inorganic tin(IV).

Organic reagents for Figure 25
(a) 0.1% acetylacetone in $CHCl_3$; (b) 1% toluene-3,4-dithiol in $CHCl_3$; (c) 1% 1.10-phenanthroline in $CHCl_3$, 1% dipyridyl in $CHCl_3$; (d) 0.5% 3,3',4',5,7-pentahydroxyflavone (quercetin) in 2-methylbutanol(4); 0.1%

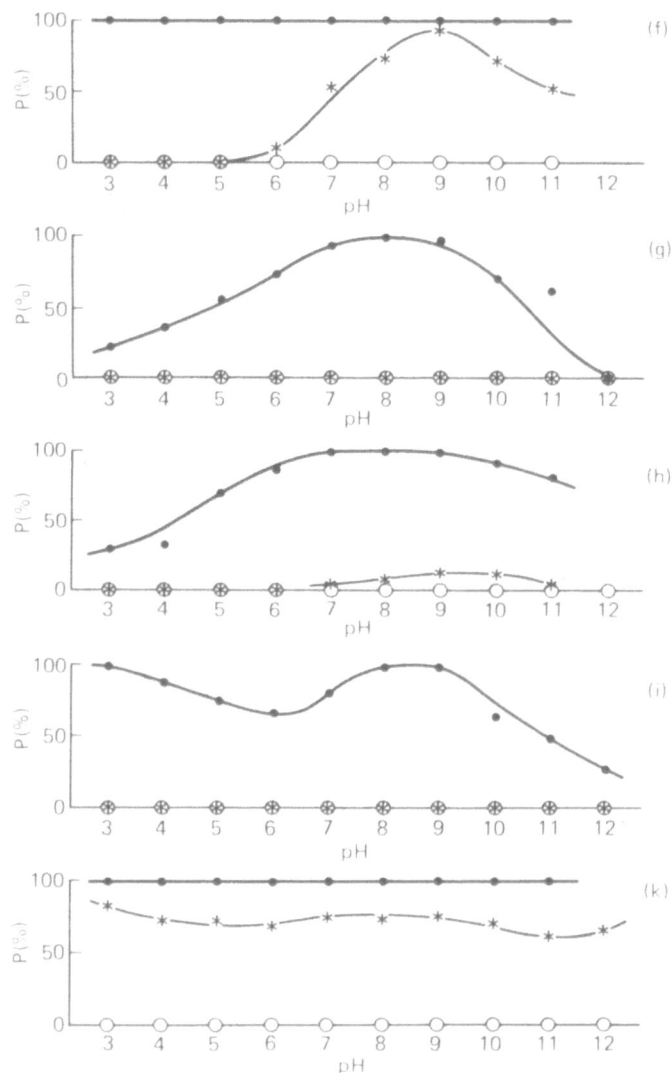

2',3,4',5,7-pentahydroxyflavone (Morin) in 2-methylbutanol(4); 0.1% 2,6,7-trihydroxy-9-phenyl-3-H-xanthenone (phenylfluorone) in 2-methylbutanol (4); (e) 0.25% α-benzoin oxime in $CHCl_3$; (f) 1% 2-nitroso-1-naphthol in $CHCl_3$; (g) 10^{-2} M 1-(2-pyridylazo)-2-naphthol (PAN) in $CHCl_3$; (h) 10^{-2} M diphenylcarbazone in $CHCl_3$; (i) 10^{-2} M 2-pyridylazoresorcine, Na-salt (PAR) in aqueous buffer solution; organic phase: 2-methyl-

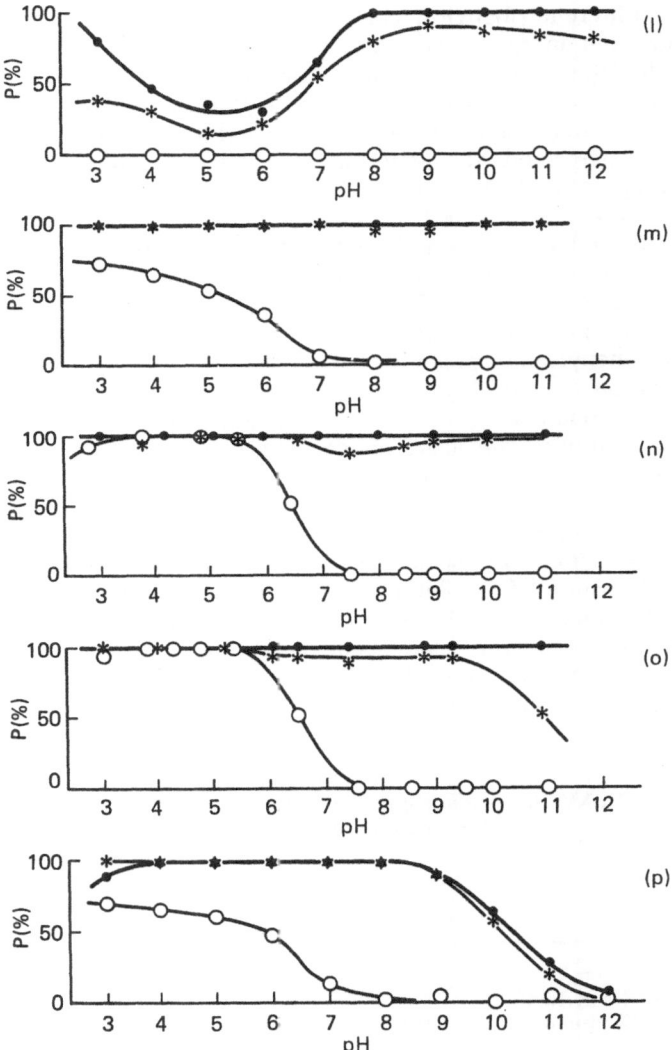

butanol(4); (k) 1% 8-hydroxyquinoline in CHCl$_3$; (l) 1% 1-nitroso-2-naphthol in CHCl$_3$; (m) 1% tropolone in CHCl$_3$; (n) 1% Na-diethyl-dithiocarbamate in aqueous buffer solution, organic phase: CHCl$_3$; (o) 1% Na-pyrrolidinedithiocarbamate in aqueous buffer solution, organic phase: CHCl$_3$; (p) 1% NH$_4$-N-nitrosophenylhydroxylamine (Cupferron) in aqueous buffer solution; organic phase: CHCl$_3$.

processes, different phenyltin compounds can be separated from each other and from inorganic tin. At first, tetraphenyltin and triphenyltin compounds (the latter as hexaphenyldistannoxane) are extracted with $CHCl_3$ from the aqueous citrate/phosphate phase (pH 8.5), without adding organic reagents. In the organic phase, both organotin compounds are analyzed as described, the water layer is treated following the above explained separation technique.

Another route is followed by SODERQUIST and CROSBY (1978). Tetraphenyltin as well as tri-, di- and monophenyltin compounds are extracted with CH_2Cl_2 from aqueous solution buffered to pH 4.7 with acetate buffer. The extract is divided into 3 parts for the determination of phenyltin compounds by gas chromatography, determination of total extracted tin, and of tetraphenyltin, respectively.

δ) *Isolation by thin-layer and paper chromatography.*—Isolation of phenyltin compounds by thin-layer chromatography (TLC) has been conducted mainly on silica gel plates, occasionally also on Al_2O_3 and in one case on polycarbonate.

Rf-values of phenyltin compounds, obtained with different eluants on plates with silica gel G (activated for 30 min at 120°C) are compiled in Table CLXVIII after BUERGER (1963) and JITSU et al. (1967). JITSU applied chlorides in $CHCl_3$ solution. However, variation of the anion of the phenyltin compound actually is unimportant, because the hydroxides or oxides are formed on the plate when basic eluants are used. The differences between the Rf-values of both authors can probably be attributed to varying adsorption properties of the silica gel in use.

Table CLXVIII. Rf-*values of phenyltin compounds on silica gel plates* (after BUERGER 1963 and JITSU et al. 1967).

Eluants	Rf-values[a]			
	$(C_6H_5)SnCl_3$	$(C_6H_5)_2SnCl_2$	$(C_6H_5)_3SnCl$	$(C_6H_5)_4Sn$
$CHCl_3$	0.0	0.0	0.0	0.9
7.5 Butanol + 3,5 pyridine (mixture H_2O saturated)	0–0.2	0.0 (0.0)	0.6 (0.55)	0.0 (1.0)
3 Butanol + 1 ethanol (mixture H_2O saturated)	0–0.5	0–0.1 (0.0)	0.6 (0.55)	0.0 (1.0)
Butanol (with 25% ammonia saturated)	0–0.2	0.0 (0.0)	0.9 (0.84)	0.0 (1.0)
8 Butanol + 2 ammonia (2.5%)	0–0.2	0.0 (0.0)	0.8 (0.60)	0.0 (1.0)
2 i-Propanol + 1 acetate-buffer (1 N CH_3COOH + $1N$ CH_3COONa, 1:1)	0–0.2	0.2 (0.29)	0.8	0.0

[a] Values in parentheses after BUERGER (1963).

Rf-values of phenyltin compounds on silica gel G with different eluants were mentioned by Braun and Heines (1968); isolation of tri-, di- and monophenyltin compounds succeeded with slightly acidified lower alcohols (Table CLXIX).

Table CLXIX. Rf-*values of phenyltin compounds on silica gel plates* (after Braun and Heines 1968).

Eluant	Rf-values			
	$(C_6H_5)_4Sn$	$(C_6H_5)_3Sn^+$	$(C_6H_5)_2Sn^{2+}$	$(C_6H_5)_2Sn^{3+}$
CH_3OH	0	0.24	0	0
$CH_3OH + 5\%$ CH_3COOH	0	0.66	0.40	0
$C_2H_5OH + C_4H_9OH +$ H_2O (1:3:2.35)	0	0.52	0	0
$C_2H_5OH + C_4H_9OH +$ H_2O (1:3:2.5) + 2% CH_3COOH	0	0.72	0.54	0

Cenci and Cremonini (1969) used silica gel H (activated for 30 min at 120°C) and n-butanol/ethanol/H_2O in 4:2:1 ratio as eluant, but did not give details about the obtained separations.

Nefedov et al. (1968) separated mixtures of tetraphenyltin and phenyltin chlorides with different eluants on plates of Al_2O_3 (activity level III after Brockmann); satisfying separations were mainly achieved with dichloroethane (Table CLXX).

Table CLXX. Rf-*values of phenyltin compounds on Al₂O₃ plates* (after Nefedov et al. 1968).

Eluant	Rf-values			
	$(C_6H_5)_4Sn$	$(C_6H_5)_3SnCl$	$(C_6H_5)_2SnCl_2$	$(C_6H_5)SnCl_3$
Ethyl acetate	0.94 ± 0.02	0.82 ± 0.05	0.04 ± 0.02	0.12 ± 0.08
Dichloroethane	0.90 ± 0.03	0.50 ± 0.05	0.08 ± 0.02	0
$CHCl_3$	0.94 ± 0.02	0.33 ± 0.05	0.05 ± 0.03	0.05 ± 0.02

Akagi et al. (1970) recommended plates with silica gel H (Merck + ammonium sulfate (18:2 g), which were activated for 1 hr at 110°C. The addition of ammonium sulfate improved the separation of monophenyltin compounds. Phenyltin chlorides, tetraphenyltin and $SnCl_4$, applied in $CHCl_3$ solution, were completely separated by eluation with benzene/acetone/CH_3COOH (20 + 2 + 1). Akagi and Sakagami (1971) listed Rf-values for different tin compounds and eluants for the above-mentioned plates (Table CLXXI).

Table CLXXI. R*f-values of phenyltin compounds on plates of silica gel with 10%
ammonium sulfate* (after AKAGI and SAKAGAMI 1971).

Eluant	Rf-values[a]				
	SnCl$_4$	(C_6H_5) SnCl$_3$	$(C_6H_5)_2$ SnCl$_2$	$(C_6H_5)_3$ SnCl	$(C_6H_5)_4$Sn
Benzene/acetone/ CH$_3$COOH (20 + 2 + 1)	0.00	0.08	0.36	0.72	ca. 0.80[a]
CCl$_4$/CH$_3$COOH (14 + 6)	0.00	0.18	0.75	—	—
i-Octane/acetone/ CH$_3$COOH (16 + 3.5 + 0.5)	0.00	0.00	0.07	0.36	—
i-Octane/acetone (17 + 3)	0.00	0.00	0.00	0.18	—

[a] After AKAGI et al. (1970).

KIMMEL *et al.* (1977) used silica gel 60 (Merck) TLC plates; with
di-i-propyl ether/CH$_3$COOH 49:1 or 99:1 and with hexane/CH$_3$COOH
9:1 as eluant, compounds R$_3$SnX and R$_2$SnX$_2$ were isolated (with R$_4$Sn in
the frontal zone and RSnX$_3$ + SnX$_4$ at the starting point). With CCl$_4$/
acetylacetone/CH$_3$COOH 20:1:1 or acetylacetone/acetone/CH$_3$COOH/
H$_2$O 15:10:2:1, R$_2$SnX$_2$ and RSnX$_3$ were isolated (with R$_4$Sn + R$_3$SnX in
front and SnX$_4$ at the starting point).
Separations on TLC-plates with either silica gel G + 20% kieselguhr G
(system A) or magnesium silicate with 40% kieselgur G (system B) were
performed by VASUNDHARA and PARISHAR (1979). Rf-values for elution
by isobutanol-water-propionic acid 7:1:4 (solvent I) or isoamyl alcohol-
acetic acid 9:1 (solvent II) are given in Table CLXXII.

Table CLXXII. R*f-values of different phenyltin compounds*
(excerpt from VASUNDHARA and PARISHAR 1979).

Tin compound	Coating	Rf-values	
		Solvent I	Solvent II
$(C_6H_5)_2$SnCl$_2$	A	0.49	0.46
	B	0.43	0.39
$(C_6H_5)_3$SnCl	A	0.55	0.43
	B	0.72	0.64
$(C_6H_5)_3$SnOCOCH$_3$	A	0.62	0.80
	B	0.52	0.70
$(C_6H_5)_4$Sn	A	0.18	0.20
	B	0.25	0.22

Qualitative data about separation of triphenyltin hydroxide and its decomposition products on polycarbonate TLC-plates (Kodak K 511 V) have been published by MASSAUX (1971). A 2:1 mixture of i-propanol and acetate buffer solution ($1N$ CH_3COOH + $1N$ CH_3COONa, 1:1) was used as eluant.

CH_2Cl_2 extracts from animal substance were examined by MAAS et al. (1972) on TLC plates of silica gel G; pretests with fortified solutions, using benzene/CH_3COOH (9:1) as eluant, yielded the following Rf-values: Sn^{4+} = 0, $(C_6H_5)Sn^{3+}$ = 0, $(C_6H_5)_2Sn^{2+}$ = 0.41 — 0.48, and $(C_6H_5)_3Sn^+$ = 0.58 — 0.68.

Triphenyl- and tributyltin compounds can be separated on silica gel G thin-layer plates with i-octane/di-i-propyl ether/CH_3COOH (80:3:8); the Rf-values are 0.77 respectively 0.57 (WOGGON and JEHLE 1973).

FREITAG (1972) and FREITAG and BOCK (1974 a) used silica gel (precoated plates by Merck Co.), washed with formic acid/H_2O (6:4) and then activated at 120°C for 1 hr. Phenyltin compounds were applied as chlorides in CH_2Cl_2 solution. With n-butanol/H_2O/HCOOH (10:40:50) as eluant, the following Rf-values were obtained: $(C_6H_5)_4Sn$ = 1.00; $(C_6H_5)_3SnCl$ = 0.98; $(C_6H_5)_2SnCl_2$ = 0.93; $(C_6H_5)SnCl_3$ = 0.43, and $SnCl_4$ = 0.10.

Better separations were obtained on plates of basic Al_2O_3 H, type E, with $CHCl_3$ containing 1% ethanol as eluant; the Rf-values were: $(C_6H_5)_4Sn$ = 1.00; $(C_6H_5)_3SnCl$ = 0.90; $(C_6H_5)_2SnCl_2$ = 0.20; (C_6H_5)-$SnCl_3$ = 0.05, and $SnCl_4$ = 0.05.

Also, phenyltin compounds were separated by thin-layer chromatography of their complexes with organic reagents. The tin compounds were extracted from their aqueous citrate/phosphate buffer solution (pH 8.5) (in the case of acetylacetone compounds from NH_3/NH_4Cl solution, pH 8.5) with $CHCl_3$, and the chloroform solutions were applied to the thin-layer plate. The resulting Rf-values are shown in Table CLXXIII.

BARNES et al. (1973) were able to separate mono-, di-, and triphenyltin compounds on silica gel GF-plates (Merck), using butanol/acetic acid (100:1); the monophenyltin compounds, together with inorganic tin, remained at the starting point.

Paper-chromatographic separations of mixtures of $SnCl_2$ and several phenyltin compounds were conducted by BARBIERI et al. (1958) on paper No. 2043 a (Schleicher and Schuell Co.). With butanol, saturated with $1N$ or $3N$ HCl, tetraphenyltin could be separated from the other compounds; Rf-values: $SnCl_2$ = 0.80 — 0.90, $(C_6H_5)_2SnCl_2$ = 0.98, $(C_6H_5)_3$-$SnCl$ = 0.98, and $(C_6H_5)_4Sn$ = 0.00.

WILLIAMS and PRICE (1960) were only able to separate di- and triphenyltin compounds on Whatman No. 1 paper; all tested eluants produced long streaks of monophenyltin trichloride (Table CLXXIV).

After HEROK and GOETTE (1963), $(C_6H_5)_3Sn^+$ and inorganic tin(IV) can be separated on Whatman No. 1 paper using 9.5 parts 80% methanol

Table CLXXIII. Rf-values of di- and monophenyltin complexes with organic reagents (after Freitag and Bock 1974 a).

Compound	TLC plate	Eluant	Rf-values $(C_6H_5)_2Sn^{2+}$	$(C_6H_5)Sn^{3+}$	Reagent
Diethyldithio-carbamate	Silica gel	CHCl$_3$/ C$_2$H$_5$OH 9:1	0.80	0.05	—
Diethyldithio-carbamate	Al$_2$O$_3$H, Type E	CHCl$_3$/ C$_2$H$_5$OH 9:1	0.85	0.20	—
Acetylacetonate	Silica gel	Acetone	0.60	0.10	—
Acetylacetonate	Al$_2$O$_3$H, Type E	Acetone	0.80	0.20	—
8-Hydroxyquino-line compound	Silica gel	Acetone	0.80	0.10	0.70
Tropolone compound	Silica gel	Acetone	0.85	0.10	0.80

Table CLXXIV. Rf-values of phenyltin compounds on Whatman No. 1 paper (after Williams and Price 1960).

Eluant	Rf-values SnCl$_4$	$(C_6H_5)_2$ SnCl$_2$	$(C_6H_5)_3$ SnCl	$(C_6H_5)_3$ SnOCOCH$_3$	$(C_6H_5)_4$Sn
60% (^{V}/v) Pyridine in H$_2$O	0.00	0.00	0.85	0.85	0.00
Butanol/pyridine/H$_2$O (7.5 + 3.5, saturated with H$_2$O)	0.00	0.00	0.95	0.95	0.00
Butanol/ethanol/H$_2$O (3 + 1, saturated with H$_2$O)	0.00	0.00	0.97	0.97	0.00
Butanol/NH$_3$/H$_2$O (1N in NH$_3$, satu-rated with H$_2$O)	0.00	0.95	0.97	0.97	0.00

and 0.5 parts n-butanol as eluant; the Rf-values are 0.9 respectively 0.7. In a radio paper chromatographic application they used the same eluant proving the purity of a radioactively labeled triphenyltin acetate sample.

Chapman and Price (1972) separated inorganic tin(IV) from phenyltin compounds on Whatman No. 2 paper with butanol/H$_2$O/CH$_3$COOH (100:50:5). The phenyltin compounds run together (Rf = 0.9), inorganic tin(IV) produces 3 spots (Rf = 0.05, 0.3, and 0.6) which are evaluated together.

Further, systems with phase-reversal are described. With dinonyl-phthalate on Whatman No. 4 paper as stationary phase and ethanol/HCl as eluant, WILLIAMS and PRICE (1964) achieved separation of phenyltin chlorides and tetraphenyltin (Table CLXXV).

Table CLXXV. Rf-values of phenyltin compounds on Whatman No. 4 paper with dinonylphthalate as stationary phase (after WILLIAMS and PRICE 1964).

Eluant	Rf-values			
	$(C_6H_5)SnCl_3$	$(C_6H_5)_2SnCl_2$	$(C_6H_5)_3SnCl$	$(C_6H_5)_4Sn$
CH$_3$OH + 1N HCl (3 + 1)	0.96	0.82	0.28	0.00
CH$_3$OH + 1N HCl (1 + 1)	0.93	0.32	0.01	0.00
CH$_3$OH + 1N HCl (1 + 3)	0.84	0.14	0.00	0.00

The same authors obtained the following Rf-values with 2-phenoxy-ethanol on Whatman 3 MM paper as stationary phase and 2,2,4-trimethyl-pentane/CH$_3$COOH (92.5 + 7.5) as eluant: Sn(IV) and $(C_6H_5)SnCl_3$ = 0.00, $(C_6H_5)_2SnCl_2$ = 0.10, and $(C_6H_5)_3SnCl$ = 0.27. Tetraphenyltin was found across the whole distance.

By combining the above-mentioned isolation on Whatman No. 2 paper with butanol/H$_2$O/CH$_3$COOH as eluant (system I) and the second method with phase-reversal and 2-phenoxyethanol as stationary phase (system II), CHAPMAN and PRICE (1972) achieved complete separation of mixtures of tri-, di-, and monophenyltin compounds and inorganic tin(IV): system I separates inorganic tin from all phenyltin compounds, then the Sn(IV) can be determined. With system II, di- and triphenyltin compounds are obtained separately, and in addition all monophenyltin compounds and inorganic tin(IV); by subtracting the value first obtained for Sn(IV), the content of monophenyltin compound is found.

Results of thin-layer and paper chromatographic separations obtained with artificial mixtures of the chlorides in an organic solvent are not necessarily transferable to extracts of biological material; the phenyltin (and more so monophenyltin) compounds as well as inorganic tin(IV) contained in these extracts cannot or only partly be converted to the chlorides.

ε) *Separations by column chromatography.*—Tetraphenyltin is separated from other phenyltin compounds and from inorganic tin on Florisil columns; only $(C_6H_5)_4Sn$ can be eluted with cyclohexane (SODERQUIST and CROSBY 1978).

ζ) *Stains.*—For staining of phenyltin compounds separated by thin-layer or paper chromatography, usually solutions of dithizone or catechol violet, sometimes also of quercetin or diphenylcarbazone are used. Tetraphenyltin does not react with these compounds, but it can be made visible after decomposition under UV light. Triphenyltin compounds undergo convenient color reactions only with dithizone, but not with catechol violet, quercetin, or diphenylcarbazone (Table CLXXVI).

Table CLXXVI. *Color reactions of phenyltin compounds and inorganic tin(IV).*

Compound	Reagents		
	Dithizone	Catechol violet	Quercetin
$(C_6H_5)_4Sn$	—	—	—
$(C_6H_5)_3Sn^+$	+	—	—
$(C_6H_5)_2Sn^{2+}$	+	+	+
$(C_6H_5)Sn^{3+}$	+	+	+
Sn^{4+}	+	+	+

Dithizone is used in solutions of 0.001% in $CHCl_3$ (FREITAG and BOCK 1974 a) or 0.1% in CCl_4 (BRAUN and HEINES 1968) or in 0.1% acetone solution with 10% water. Triphenyltin compounds produce yellow, while di-, and monophenyltin compounds and inorganic tin(IV) produce yellow/red to red spots. The following detection limits have been reported: 0.01 μg Sn(IV), 0.02 μg $(C_6H_5)SnCl_3$ and $(C_6H_5)_2SnCl_2$, and 0.2 μg $(C_6H_5)_3SnCl$ (AKAGI et al. 1970).

Catechol violet is used as spraying reagent in 0.1 to 0.2% water or ethanol solution (WILLIAMS and PRICE 1960, BUERGER 1963, JITSU et al. 1967). Triphenyltin compounds can be made visible only after decomposition with Br_2-vapor or under UV-light. A detection limit of 0.1 μg has been found (MAAS et al. 1972).

WOGGON and JEHLE (1973) decomposed triaryltin compounds on thinlayer plates with UV-radiation and detected the decomposition products with a 0.1% solution of quercetin in CH_3OH.

According to REUTOV et al. (1961), a solution of diphenylcarbazone in CH_3OH/H_2O (50:50) can be used to detect diphenyltin compounds; the detection limit is approximately 0.5 μg.

BARBIERI et al. (1958) decomposed phenyltin compounds on thinlayer plates with $HgCl_2$-solution, sprayed the plate with a solution of 0.5% 8-hydroxyquinoline in $1N$ CH_3COOH, and detected inorganic tin-(IV) fluorometrically after treatment with gaseous NH_3.

Organotin compounds of the types R_2SnX_2 and $RSnX_3$ and also inorganic tin(IV) produce a golden-yellow fluorescence with a 0.1% solution of 8-hydroxyquinoline-5-sulfonic acid in 90% ethanol (PAL and RYAN 1969, KIMMEL et al. 1977).

Di- and monophenyltin complexes of acetylacetone and tropolone can be stained with 0.01% solutions of dithizone in $CHCl_3$; diethyldithiocarbamates are detected with 1% aqueous $CuSO_4$ solution, 8-hydroxyquinoline complexes can be recognized by the typical color of their own (FREITAG and BOCK 1974 a).

Phenyltin compounds separated on TLC-plates can be stained with a 0.5% haematoxylin solution in ethanol (VASUNDHARA and PARISHAR 1979).

η) *Gas-chromatographic separations.*—Gas-chromatographic separation of phenyltin chlorides was first described by TONGE (1965). He worked with a silicone oil column (5% w/w on Celite 545) of 16 cm length and 4 mm free diameter; the carrier gas was 120 ml/min N_2, temperature of the injection port 400 to 425°C, temperature of column 180°C. Similar conditions were used by FREITAG and BOCK (1974 a): 5% silicone oil DC 550 on Chromosorb WA, length of column 0.5 m, temperature 190°C, carrier gas N_2. Such methods, however, are only of minor importance for analyzing phenyltin compounds in biological material, since these (except for triphenyltin compounds) cannot be converted quantitatively into the chlorides.

In another method, phenyltin chlorides are reacted with *n*-amylmagnesium bromide after Grignard, and the resulting mixed tin-organic compounds analyzed by gas chromatography (JITSU *et al.* 1969); conditions: 0.75 m-steel column with 3 mm free diameter, 25% silicone grease DC HV on Celite 545, temperature of column 260°C, carrier gas He. This method should also be of minor importance.

SODERQUIST and CROSBY (1978) converted tri-, di-, and monophenyltin compounds in CH_2Cl_2/hexane solution into the corresponding hydrides using lithium aluminum hydride as a reducing agent. The hydrides were separated and determined by gas chromatography (1.1 m glass column with 2 mm free diameter; 4% SE-30 on 60/80 mesh Gas-Chrom Q; 20 ml/min N_2; temperature of the column 190°C for $(C_6H_5)_3SnH$, 135°C for $(C_6H_5)_2SnH_2$, and 45°C for $(C_6H_5SnH_3)$.

θ) *Quantitative analysis of small amounts of phenyltin compounds.*—After concentration and separation, the phenyltin compounds are present in organic solvents, as aqueous extracts, or as "spots" on paper strips or thin-layer plates. It is possible now to determine the organotin compound by direct quantitative analysis or analyze for tin after decomposition of the organic substance. In radioactively labeled preparations the impulse frequency is measured.

Oxidative decomposition of the organic substance can be achieved with concentrated H_2SO_4 (INGHAM *et al.* 1960, THOMAS and TANN 1971), $H_2SO_4 + H_2O_2$ (ASMUS *et al.* 1971, MAAS *et al.* 1972), $H_2SO_4 + HNO_3$ (GILMAN and KING 1929, NANGNIOT and MARTENS 1961), $H_2SO_4 + HNO_3 + H_2O_2$ (GORBACH and BOCK 1958, VOGEL and DESHUSSES 1964), concentrated $HCl + H_2O_2$ (HEROK and GOETTE 1963), or with Na_2O_2 in the Parr bomb (INGHAM *et al.* 1960, FRITZ and SCHEER 1964).

Cleavage of the phenyl groups from the tin may be done by UV-radiation, for example in 12% HCl solution (WOGGON and JEHLE 1973) or by reaction with Br_2 (GILMAN and KING 1929).

After decomposition of the organic substance, the tin traces are present in aqueous solution and may be analyzed by traditional methods, for example photometrically as dithiol complex (BUERGER 1961 a and b), or with 3'-pyridylfluorone (ASMUS *et al.* 1971) by polarography (BOCK *et al.* 1958, GORBACH and BOCK 1958), inverse voltammetry (NANGNIOT and MARTENS 1961, WOGGON and JEHLE 1973, FANO and ZANOTTI 1973), or neutron activation analysis (MAAS *et al.* 1972; see also SCHMIDT and STARKE 1969).

Direct tin analysis in ethanolic extracts without decomposition of the organic substance can be conducted by atomic absorption (FREELAND and HOSKINSON 1970) or by emission spectroscopic analysis with carbon disc electrodes (RAUTSCHKE and HEINRICH 1972).

Analysis of individual phenyltin compounds without destroying the organic substance is usually done photo- or electrometrically; fluorometric and the already mentioned gas-chromatographic methods are of minor importance.

ι) *Photometric analysis of phenyltin compounds.*—Triphenyltin compounds form a yellow complex with dithizone of the structure $(C_6H_5)_3$-SnDz (IRVING and COX 1961). For quantitative analysis, the benzene CCl_4 or $CHCl_3$ solution of the organotin compound is mixed with dithizone in the same solvents (for example $0.01N$ dithizone in benzene) and the absorbance determined at 458 nm after eliminating excessive reagent with $0.1N$ NaOH (BOCK *et al.* 1962). Measuring the decrease in absorbance of the reagent itself at 610 to 612 nm after reaction with the tin compound is more convenient (HARDON *et al.* 1960, FREITAG and BOCK 1974 a). The triphenyltin dithizone compound does not absorb at this wavelength (see Fig. 26). Formation of the complex is accelerated by shaking the $CHCl_3$ solution with an aqueous citrate/phosphate buffer (pH 8.5).

With this method, the calibration curve in the 1 to 18 μg $(C_6H_5)_3$-SnCl/ml range is straight. The detection limit in 1 cm cuvettes is approximately 0.5 μg/ml.

Several photometric methods have been described for diphenyltin compounds (FREITAG and BOCK 1974 a). Some of the same reagents are used as described under section **VIII** *b*, also separating methods to extract the diphenyltin compounds from aqueous citrate/phosphate solution (pH 8.5); then the absorbance is measured in the organic phase (usually $CHCl_3$) (Table CLXXVII).

Prior separation of triphenyltin compounds is advisable; monophenyltin compounds interfere very little or not at all when dithizone, PAN, PAR, and quercetin are used. However, monophenyltin compounds (but not inorganic tin) also dissolve in the organic phase. With 8-hydroxyquinoline or Na-diethyldithiocarbamate colored complexes are formed which interfere with the photometric evaluation of diphenyltin compounds.

Monophenyltin compounds are analyzed photometrically in a similar way by extraction with various reagents from aqueous citrate/phosphate

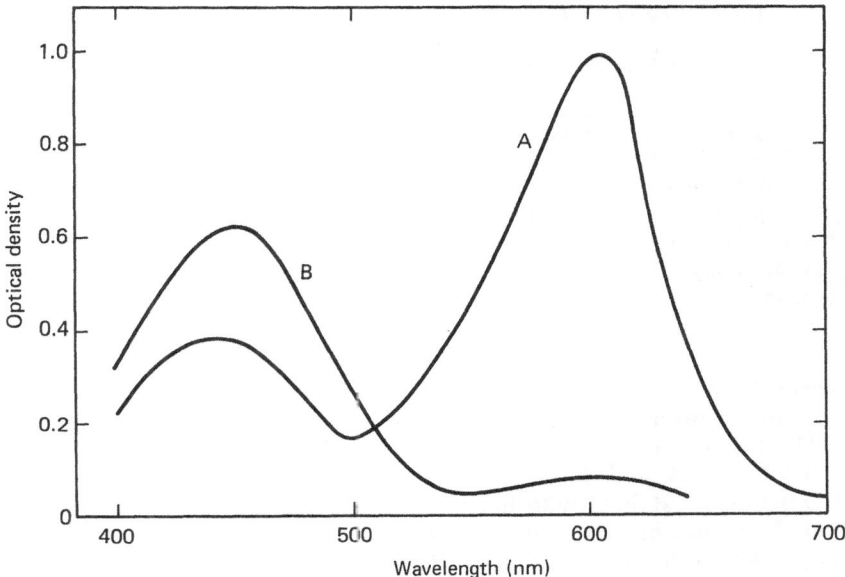

Fig. 26. Absorption spectra of (A) dithizone (67 μg in 10 ml CHCl₃) and (B) triphenyltin dithizone complex (67 μg dithizone + excess triphenyltin acetate in 10 ml CHCl₃); 1-cm cuvettes (after HARDON *et al.* 1960).

Table CLXXVII. *Photometric analysis of diphenyltin compounds* (after FREITAG and BOCK 1974 a).

Reagent	Organic solvent	Wave-length (nm)	Calibration curve straight at	Lower detection limit (1-cm cuvette)
Dithizone	CHCl₃	612[a]	0–10 μg/ml	ca. 0.2 μg/ml
Diphenylcar-bazone	CHCl₃	546	0–8 μg/ml	ca. 0.2 μg/ml
PAN	CHCl₃	556	0–30 μg/ml	ca. 0.25 μg/ml
PAR	2-Methyl-4-butanol	518	0–15 μg/ml	ca. 0.13 μg/ml
Quercetin	2-Methyl-4-butanol	442	0–20 μg/ml	ca. 0.5 μg/ml
8-Hydroxyquino-line	CHCl₃	380	0–80 μg/ml	ca. 2 μg/ml
Na-Diethyldithio-carbamate	CHCl₃	255	0–20 μg/ml	ca. 0.3 μg/ml

[a] Decrease of reagent absorbance.

Table CLXXVIII. *Photometric analysis of monophenyltin compounds* (after FREITAG and BOCK 1974 a).

Reagent	Organic solvent	Wave-length (nm)	Calibration curve straight at	Lower detection limit (1-cm cuvette)
8-Hydroxyquinoline	CHCl₃	384	0–100 µg/ml	ca. 2 µg/ml
5,7-Dibromo-8-hydroxyquinoline	CHCl₃	400	0–100 µg/ml	ca. 1.5 µg/ml
Na-Diethyldithio-carbamate	CHCl₃	259	0–15 µg/ml	ca. 0.3 µg/ml

buffer solution (Table CLXXVIII). Tri- and diphenyltin compounds must be removed prior to extraction, inorganic tin(IV) does not interfere.

The photometric determination of phenyltin compounds is described by VASUNDHARA and PARISHAR (1979). The separated compounds on TLC-plates are stained with haematoxylin and the spots are dissolved in ethanol; the absorbance is measured at 575 nm. The authors claim to be able to determine tri- and diphenyltin compounds and in addition tetraphenyltin.

Further, phenyltin compounds (tetraphenyltin only after decomposition) can be determined semiquantitatively from their spot size on paper or thin-layer plates after the spots have been developed.

κ) *Fluorometric analysis.*—For fluorometric analysis of triphenyltin compounds, the benzene solution of the sample is completely evaporated and the residue dissolved in a solution of 5 ml 0.01% hydroxyflavone and 1 ml saturated sodium acetate. Fluorescence is excited at 415 nm, the intensity is measured at 497 nm (VERNON 1974).

The diphenyltin/quercetin compound can be analyzed photometrically and fluorometrically (FREITAG and BOCK 1974 a). This method, however, does not offer significant advantages because of high interference potential; the detection limit is approximately 0.1 µg (C_6H_5)$_2SnCl_2$ in 1 cm cuvettes.

For the described photometric and fluorometric methods it should not be overlooked that extracts from biologic material, mainly from plants, usually contain other absorbing or fluorescent components. These impurities have to be eliminated by additional operations, for example through column chromatography (HARDON et al. 1960) or thin-layer chromatography (FREITAG and BOCK 1974 a).

λ) *Electrochemical analysis.*—Reduction on mercury electrodes has been proposed repeatedly for quantitative analysis of phenyltin compounds (except tetraphenyltin) (see Section III/e, "electrochemical behavior"). Data from the literature pertaining to this method are compiled in Table CLXXIX.

However, the polarographic methods developed so far have no advan-

Table CLXXIX. *Polarographic analysis of phenyltin compounds.*

Compound	Supporting electrolyte	Halfwave potential (V)	Concentration range	Reference
$(C_6H_5)_3SnCl$ $(C_6H_5)_3SnOH$	50% ($^V/_V$) ethanol/H_2O + 0.1 M NH_3 + 0.1 M CH_3COOH + 0.002% Triton X-100	−0.6; 0.9; −1.35 (against sat. calomel electrode)	$5 \cdot 10^{-6}$ to $1 \cdot 10^{-3} M$	BOOTH & FLEET (1970)
$(C_6H_5)_3SnCl$	Dimethoxyethane + tetra-butylammonium-perchlorate	−1.6; −2.9 (against 0.001 M $AgClO_4$/0.1 M Bu_4NClO_4/Ag)		DESSY et al. (1966)
$(C_6H_5)_2SnCl_2$	Dimethoxyethane + tetra-butylammonium-perchlorate	−1.6; −2.7 (against 0.001 M Ag,ClO_4/0.1 M Bu_4NClO_4/Ag)		DESSY et al. (1966)
$(C_6H_5)_3SnF$	80% ethanol + 0.1 M KCl + 0.1 M HCl + 10^{-4}% Triton X-100	−0.65; −0.92 (against sat. calomel electrode)	to ca. $10^{-3} M$	VANACHAYANGKUL & MORRIS (1968)
$(C_6H_5)_2SnCl_2$	CH_3OH/H_2O (8:2) + 0.002% Triton X-100 pH 3–7	−0.53 (against sat. calomel electrode)	10^{-3} to $10^{-5} M$	FREITAG & BOCK (1974 a), FREITAG (1972)
$(C_6H_5)_2Sn(DDTC)_2$	CH_3OH/H_2O/tetrabutyl ammonium iodide (90:8:2)	−0.80; −1.25 −1.5 (against sat. calomel electrode)	10^{-3} to $5 \cdot 10^{-6} M$	FREITAG & BOCK (1974 a)
$(C_6H_5)SnCl_3$	1 N HCl + 1% KCl + 0.002% Triton X-100	−0.28 (against sat. calomel electrode)	—	FREITAG & BOCK (1974 a)
$(C_6H_5)SnCl_3$	1% tartaric acid + 1% KCl + 0.002% Triton X-100	−0.28 (against sat. calomel electrode)	—	FREITAG & BOCK (1974 a)

tages over photometry and determination of tin after decomposition of the organic substance; they are subject to interference and limited to rather narrow concentration ranges.

Also, the inverse voltammetric analysis should be mentioned for very low amounts of triphenyltin compound in a solution of 50% (v/v) ethanol/H_2O with 0.1 M NH_3 + 0.1 M CH_3COOH + 0.002% Triton X-100 (BOOTH et al. 1970). Electrolysis takes place for 2 min at -1.0 V against the saturated calomel electrode. The radical $(C_6H_5)_3Sn^{\cdot}$ which is collected on the surface of the cathode and is sufficiently stable is then oxidized anodically. The method is applicable in the concentration range of $2 \cdot 10^{-8}$ to $1.2 \cdot 10^{-7}$ M triphenyltin acetate or hydroxide.

μ) *Gas-chromatographic analysis.*—The previously mentioned gas-chromatographic separations can also be used for analysis of phenyltin compounds. SCHWEDT and RUESSEL (1973) described an analysis of tetraphenyltin without prior separation from other phenyltin compounds; stationary phase: 10% silicone grease on "Gas Chrom P," temperature of column 220°C, FID.

ν) *Analysis by electron-spin-resonance.*—STEGEMANN et al. (1977) proposed analysis of di- and triorganotin compounds by electron-spin-resonance. The tin compounds are converted into paramagnetic complexes with substituted o-aminophenols in the presence of oxygen.

IX. Environmental behavior of triphenyltin compounds; persistence

a) Persistence on plants

Persistence of triphenyltin compounds on plants is closely related to the previously discussed question of how fast the active ingredient is decomposed (see Section VII a).

According to BAUMANN (1958), the degree of infection with *Septoria apii* in celery is 66% (controls 100%—greenhouse experiments), if the infection takes place 28 days after application of the active ingredient; the respective result for sugarbeets infested with *Cercospora beticola* is 78% after 21 days (Table CLXXX). Field tests under normal weather conditions showed similar results (HAERTEL 1962).

FOSCHI and RAPPARINI (1963) treated sugarbeets with triphenyltin acetate or hydroxide (concentration 0.25% each) and later inoculated them with *Cercospora conidia*. The degree of infection was determined 14 days later. Where application of the active ingredient had taken place 1 day before inoculation, the infestation after 14 days was 3.6% for triphenyltin acetate and 2% for the hydroxide; when the active ingredient was applied 8 days before inoculation, the infestation was 21 respectively 10.5%. Infestation on the controls was 49%.

Field trials by NEELY (1970) with triphenyltin hydroxide (concentration 1,200 ppm) showed that the persistence (rated according to germina-

Table CLXXX. *Inoculation of celery with Septoria apii and of sugarbeets with Cercospora beticola at different intervals after application of triphenyltin acetate (0.3%)* (after BAUMANN 1958).

Infection (days after spraying)	Infestation (%)[a]	
	Celery	Sugarbeets
1	6	—
4	—	38
5	16	—
7	29	—
14	31	39
21	46	78
28	66	91

[a] In reference to untreated controls.

tion inhibition of *Monilinia fructicola* conidia) depends very much on the kind of plant. For 12 different bushes and trees values of between <1 and 4 wk, and an average of 1.8 wk were found. Weather conditions (rain) had little influence.

b) Behavior in soil

Experiments about the degradation of triphenyltin acetate and hydroxide in different soils were conducted by CENCI and CREMONINI (1969). The 1,000-g samples of soil were treated with 8 mg each of active ingredient and exposed to air, but protected against rain. At intervals of 12 hr, samples were taken and extracted twice with ethanol (95%) + diethyl ether (1:1). Depending on the kind of soil and active ingredient, no triphenyltin compound could be detected after 12 to 240 hr (Table CLXXXI).

Table CLXXXI. *Degradation of triphenyltin compounds in soil* (after CENCI and CREMONINI 1969).

Soil	pH	Calcium carbonate content (%)	Time to complete degradation of triphenyltin compound (hr)	
			Acetate	Hydroxide
Litorina sand	8.0	7.0	72	36
Peat for improvement	7.3	8.8	100	72
Clay	7.1	16.7	72	72
Sand-clay mixture	7.3	11.0	192	36
Sand from pine forest	7.0	12.1	84	36
Soil from spruce forest	5.1	—	240	12

MASSAUX (1971) reported that within 1 wk triphenyltin chloride was mineralized in the ground to approximately 99%.

BARNES *et al.* (1971 and 1973) studied the degradation of triphenyltin acetate in soil with a ^{14}C-labeled preparation. Excluding light, they aerated the soil sample treated with 5 or 10 ppm active ingredient and measured the resulting $^{14}CO_2$. Until the 80th day after start of the experiment, the degradation rate was 0.44%/day; after that degradation was slower (0.21%/day). Since no $^{14}CO_2$ development has been observed for sterilized soil, degradation can be attributed to bacteria; indeed, several kinds of fungi and bacteria could be isolated which developed $^{14}CO_2$ from the organotin compounds.

The fact that the degradation times are so much longer than those mentioned by CENCI and CREMONINI (1969) may be attributed to differing experimental conditions with CO_2 being the final breakdown product.

Triphenyltin acetate could not be removed from a 25-cm layer of soil even when washing with water for weeks; the major amount had been absorbed in the top 4 cm of the soil. Experiments about the effects of active ingredient on the bacterial conversion of ammonium into nitrate showed that there is no danger of significant interference with the nitrification reaction (Fig. 27).

SUESS and EBEN (1973) studied the behavior of triphenyltin acetate in 2 specific kinds of soil for hops (Table CLXXXII), using ^{14}C-labeled triphenyltin acetate where the ^{14}C was statistically distributed over all 3 phenyl rings. Degradation of the compound was examined with active ingredient held by the soil through adsorption and being present on the surface after admixture.

Tests to establish the adsorption characteristics showed that of 478 μg ^{14}C-triphenyltin acetate, added to 10 g of soil, 31.5% could be removed from the light soil and 35.6% from the heavy soil through 3 extractions with $CHCl_3$. The remainder had been adsorbed to the soil complex which could be confirmed through incineration of the soil samples followed by ^{14}C-analysis.

Table CLXXXII. *Chemical and physical characteristics of soil samples* (after SUESS and EBEN 1973).

Experiments	Light hops soil	Heavy hops soil
pH	6.7	6.4
C(%)	2.50	2.48
N(%)	0.29	0.43
Ca (mval)	7.7	12.8
Mg (mval)	0.4	0.6
K (mval)	0.7	6.3
Na (mval)	0.1	0.9
T-value	11.25	16.82
Total silt (%)	18.8	59.4
Clay (%)	7.3	17.5

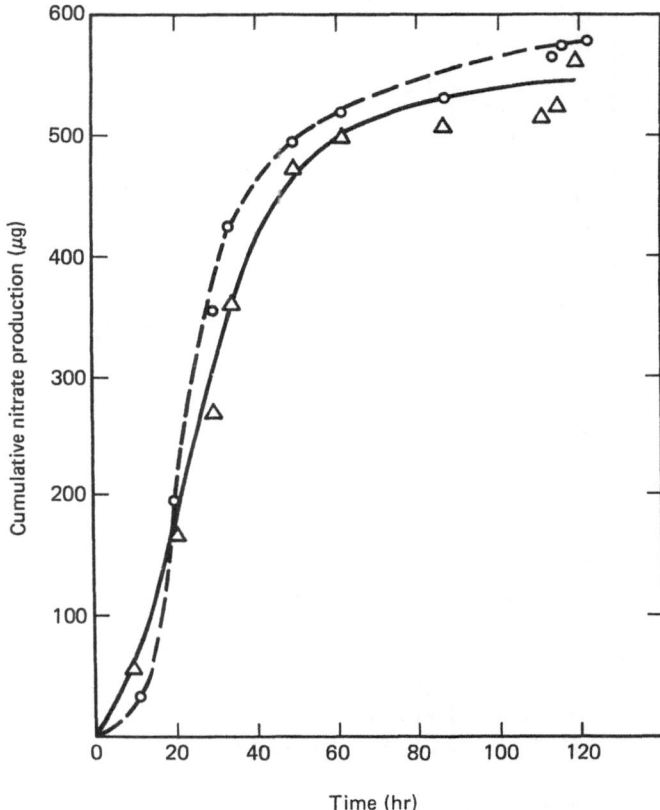

Fig. 27. Effect of triphenyltin acetate on soil nitrification; flow rate approximately 3.5 ml/hr, ammonium concentration of inflowing medium 1,000 ppm (after BARNES *et al.* 1973); △ = triphenyltin acetate added and ○ = triphenyltin acetate omitted (control).

Degradation of ^{14}C-triphenyltin acetate was studied at 30°C in closed glass containers in the incubator, from which the air was sucked twice a wk for 30 min and passed through NaOH for adsorption and subsequent analysis of CO_2. In both soil samples, degradation of triphenyltin acetate to $^{14}CO_2$ started immediately. Active ingredient adsorbed to the soil was decomposed faster than material which had been admixed. Between light and heavy soil there was no major difference in decomposition rate for the active ingredient. A peak in degradation rate was already reached after 2 to 3 wk (Fig. 28). After 9 wk, with active ingredient admixed to light soil, 27.7% of the active ingredient had decomposed; with active ingredient admixed to heavy soil 26.1%; and with soil-adsorbed active

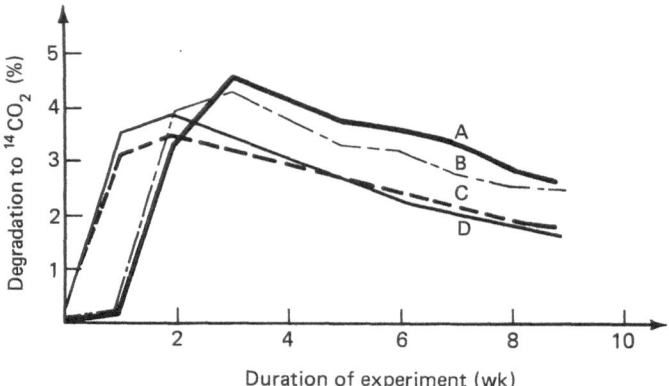

Fig. 28.   Weekly decomposition rate of ^{14}C-labeled triphenyltin acetate in hops soil
(after SUESS and EBEN 1973): A = light hops soil, triphenyltin acetate admixed; B = heavy hops soil, triphenyltin acetate admixed; C = triphenyltin
acetate, adsorbed to light hops soil; and D = triphenyltin acetate, adsorbed
to heavy hops soil.

ingredient 23.3 respectively 24.0% (Fig. 29). Degradation products could
not be detected in the soil.

The overall CO_2 production was not changed by adding triphenyltin
acetate to the soil. The initially higher CO_2 production was attributed by
the authors to acetone in which the active ingredient had been dissolved.
In heavy soil, more CO_2 was released than in light soil (Figs. 30, 31).

c) Persistence in water; purification of waste water; toxicity to fish and aquatic life

In water, triphenyltin compounds decompose only slowly; CASTEL
et al. (1963) and GRAS and RIOUX (1965) reported persistence of the
acetate (concentration 0.05 to 4 ppm) of more than 60 days against
larvae of *Culex pipiens berbericus* in city water of 21°C and pH 6.8.
DESCHIENS and FLOCH (1963) found persistence against water snails of
100 days at a concentration of 0.5 ppm of triphenyltin acetate or chloride
and of at least 8 mon at concentrations of 1.0 to 2.5 ppm.

These observations are somewhat contradicted by results obtained
from water on rice fields. CHIAPPARINI et al. (1964) noticed persistence
of triphenyltin acetate (concentration 0.67 ppm) against algae of about
9 to 10 days (see Fig. 19). HOPF et al. (1967) reported that rice seedlings
were not damaged when sowing took place 7 days after application of
active ingredient. CROSSLAND (1964) found a decrease in concentration
of the active ingredient in rice fields from 1.0 to 0.054 ppm within 15 hr.

According to HOCKING and WHITE (1967), the accelerated inactivation
of the triphenyltin compounds in rice fields is caused less by bacterial

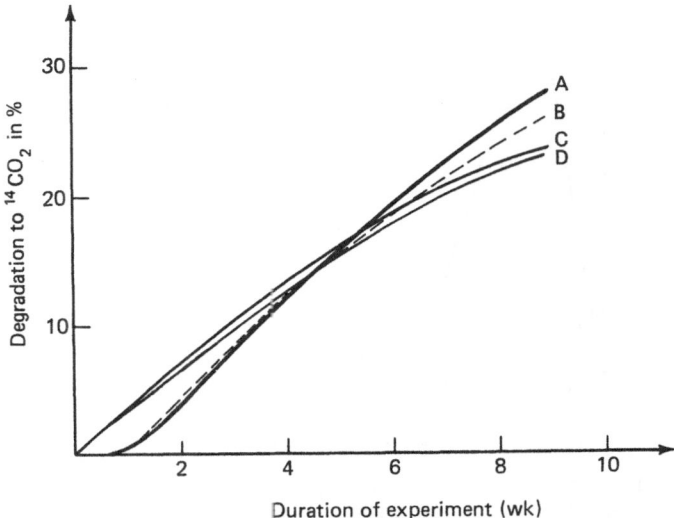

Fig. 29.   Decomposition of ^{14}C-labeled triphenyltin acetate to $^{14}CO_2$ in hops soil (after Suess and Eben 1973): A = light hops soil, triphenyltin acetate admixed; B = heavy hops soil, triphenyltin acetate admixed; C = triphenyltin acetate, adsorbed to light hops soil; and D = triphenyltin acetate, adsorbed to heavy hops soil.

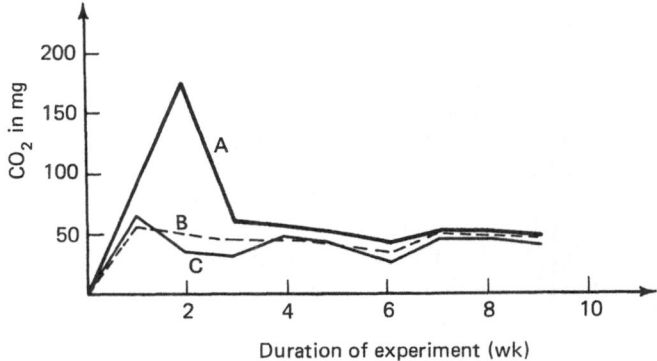

Fig. 30. CO_2 production (mg) in light hops soil (after Suess and Eben 1973): A = light hops soil, triphenyltin acetate admixed; B = triphenyltin acetate, adsorbed to light hops soil; and C = light hops soil.

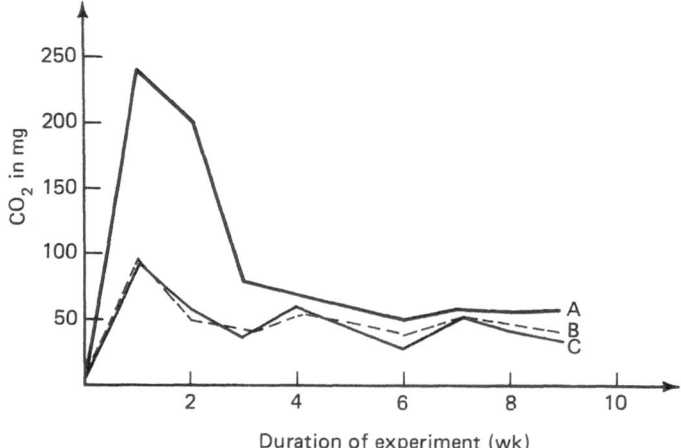

Fig. 31. CO_2 production (mg) in heavy hops soil (after SUESS and EBEN 1973):
A = heavy hops soil-triphenyltin acetate admixed; B = triphenyltin acetate,
adsorbed to heavy hops soil; and C = heavy hops soil.

and chemical decomposition, but mainly by adsorption to sediment. The
fact that organotin compounds are readily adsorbed is also indicated by
experiments of CHERNEGA *et al.* (1971) who purified waste water by
adsorption of these compounds on activated carbon. The organotin compounds were then extracted with organic solvents, and finally destroyed
by electrochemical oxidation.

Triphenyltin compounds are highly toxic for fish and other aquatic
inhabitants. Several kinds of fish (*Gambusia affinis, Carassius auratus,
Anguilla anguilla*) expire to 100% within 24 hr at a concentration of only
0.4 ppm of triphenyltin acetate (GRAS and RIOUX 1965). SEIFFER and
SCHOOF (1967) noticed a 10% mortality for sunfish in ponds with triphenyltin acetate concentration of 1 ppm, and high mortality for guppies,
tadpoles, toads, and crayfish in fresh water at concentrations of 5 ppm of
active ingredient. MOORE (1969) reported a case of unintended fish kill
in a pond when triphenyltin compounds had been sprayed from an airplane. Some data about the toxicity of triphenyltin chloride are compiled
in Table CLXXXIII.

For 6.9-cm long carp (*Cyprinus carpio*) weighing 8.5 g, UEDA *et al.*
(1961) determined a median tolerance limit (tlm) of 0.521 ppm after 24
hr and 0.32 ppm after 48 hr.

While conducting experiments to control harmful algae, MUKHERJI
and RAY (1966) also studied the toxicity of triphenyltin acetate for several kinds of fish occurring in rice fields of West Bengal. They concluded
(Table CLXXXIV) that the triphenyltin acetate concentration of 1.0
ppm, which is sufficient to control algae, is not toxic for the fish.

Table CLXXXIII. *LC₁₀₀ of triphenyltin chloride for various aquatic inhabitants* (after FLOCH and DESCHIENS 1962).

Animal	LC$_{100}$ (ppm)	
	After 24 hr	After 48 hr
Tadpoles (toads, frogs)	0.25	0.05
Carassius auratus	0.25	0.25
Stickleback (spawn)	0.25	0.25
Crab (*Daphnia longispina*)	0.25	0.05

Table CLXXXIV. *Effect of different concentrations of triphenyltin acetate on different kinds of fish* (after MUKHERJI and RAY 1966).

Brestan-60	Observation of 24 hr at concentrations of			
	1,000 ppm	100 ppm	10 ppm	1 ppm
Major carp[a] (Rui. Katla, etc.)	Expired after 15 min	Expired after 30 min	Expired after a few hr	Living under normal condition; no effect
Minor carp[a]	Expired after some hr	Expired after a few hr	Not dead (living)	Living under normal condition; no effect

[a] The following species of "carps" were tested: (a) Major carps: *Labco rohita* Ham. (Rui), *Catla-catla* Ham. (Katla), and *Cirrhina imrigala* Ham. (Mirgel) and (b) minor carps: *Ophicephalus punctatus* Bloch (Lata), *Mugilparsi* Ham. (Parse), *Mystus cavasius* Ham. (Tangra), *Trichogaster fasciatus* Bl. & Schm. (Kholsa), and *Anabas testudineus* Bl. (Koi).

KNAUF (1974) studied the toxicity of triphenyltin acetate for 3 different kinds of fish. The results from 24- and 48-hr exposures are shown in Table CLXXXV.

d) Effects on game and useful insects

Only a few reports have been published about the effects of triphenyltin compounds on hair and feather game and useful insects. According to KUTZER *et al.* (1972), any loss of game (especially hares) in areas with

Table CLXXXV. *Toxicity of triphenyltin acetate tested on 3 species of fish* (after KNAUF 1974).

Species	Length (cm)	Water temp. (°C)	Toxicity (LC$_{50}$) in ppm after	
			24 hr	48 hr
(*Idus idus melanotous*)	5–7	22	0.18	0.11
Carp (*Cyprinus carpio* L)	9–12	22	1.2	0.84
Guppy (*Lebistes reticulatus*)	1–1.5	25	0.12	0.054

sugarbeet crops is unlikely if the triphenyltin compound is applied at the recommended dosage of 240 to 360 g/ha. This can be concluded from the fact that a hare would have to eat during 1 day 15 to 20 kg or for several days 4 to 5 kg of leaves at an active ingredient concentration of 4 ppm in order to reach the lethal dose. This argument removes the suspicion that many hares died in the sugarbeet areas of Austria because of the use of triphenyltin compounds.

DRESCHER-KADEN et al. (1971) examined hares in areas of sugarbeet plantations where mainly triphenyltin compounds were applied during the period of *Cercospora* control. There was no case of substantial triphenyltin residues in the liver of the animals which could have suggested poisoning.

Studies about the oral toxicity of triphenyltin acetate were conducted by the *Wisconsin Alumni Research Foundation* in Madison (ANONYMOUS 1966 b, c, and d) with chicks of pheasants and quails and with sparrows. The young birds were fed for 6 days *ad libitum* with various dosages of triphenyltin acetate. The following LC_{50} were found: 700 ppm for pheasants, 500 ppm acetate for quail, and 515 ppm triphenyltin acetate for wild sparrows. According to these experiments, which widely coincide with test results by STONER (1966), triphenyltin compounds can be considered harmless for feathered game. Studies about the effects of triphenyltin acetate on honey bees were conducted by the *Biologische Bundesanstalt* (1956). Triphenyltin acetate, used in continuous application at concentrations of 0.1%, showed no effects of contact poisoning. Poisoning by feeding was very low with an LD_{50} >100 µg. Poisoning by inhalation was not observed. Accordingly, triphenyltin acetate was rated "not dangerous for bees."

BERAN (1958) also concluded that triphenyltin acetate represents the lowest danger index of all tested pesticides (Table CLXXXVI). In field trials, up to the third day after treatment, the loss of flying bees amounted to only 0.2%. On the fourth day, due to a severe chill, mortality of the bees increased to 2.7%.

Studies by STEINER (1960/1961) seeking spray sequences for fruit trees that avoid damage to useful insects showed triphenyltin acetate to have the least impact on the general fauna although it had been applied 6 times in concentrations of 0.2 to 0.07% (as 20% product) and at intervals commonly used for other fungicides.

BARDNER and MATHENGE (1974 a) tested the effects of triphenyltin hydroxide on parasites of leaf miners. They dipped coffee leaves into a 0.125% suspension of triphenyltin hydroxide and placed the parasites on the leaves after the active ingredient on the surface had dried. After a contact period of 24 hr, no difference could be determined between the treated parasites and the controls.

FINDLAY (1969) also reported that triphenyltin compounds have no detrimental effects on parasites, predators, and pollinators. According to BROODRYK (1967), parasitizing of larvae of the potato tuber moth (*Gnori-*

Table CLXXXVI. *Toxicity of several commonly used pesticides for bees* (excerpt from BERAN 1958).

Pesticide[a]	ED_{50} orally (μg/bee)	ED_{50} contact poisoning; application method (μg/bee)	ED_{50} contact poisoning; deposit method (μg/100 cm^2)
Methyl parathion	0.0266 ± 0.0036	0.0476 ± 0.0029	16.750 ± 0.693
Diazinon	0.0840 ± 0.0117	0.1366 ± 0.0049	9.563 ± 0.420
Lindane	0.0968 ± 0.0158	0.1834 ± 0.0184	1.947 ± 0.128
Dieldrin	0.325 ± 0.044	0.1502 ± 0.0071	13.365 ± 0.752
DDT	10.264 ± 1.532	9.192 ± 0.791	971.980 ± 90.093
Toxaphene	52.105 ± 6.879	14.530 ± 0.668	3193.0 ± 141.0
DNOC	2.039 ± 0.254	3.170 ± 0.262	520.0 ± 23.80
2-Chloro-4,6-bis-ethyleneamine-s triazine (simazine)	> 10	> 2	>25,000
Captan	91.510 ± 14.796		788.460 ± 73.261
Triphenyltin acetate	>100	>16	>25,000
Copper oxychloride	267.060 ± 39.93		>50,000

[a] The results are based on pure or technical active ingredients, even if trade names of preparations are listed.

Table CLXXXVII. *Parasitism in triphenyltin acetate-treated plots, 4 wk after beginning of spraying on potato tuber moth (Gnorimoschema operculella Zell.)* (after BROODRYK 1967).

Fungicide	Concentration (% active ingredient)	Spraying interval (days)	Parasitism locality	
			Bapsfontain (%)	Putfontain (%)
Triphenyltin acetate	0.024	7	84	93
Triphenyltin acetate	0.018	7	78	86
Triphenyltin acetate	0.024	14	78	71
Untreated	—	—	87	63

moschema operculella Zell.) is not affected by weekly treatment of the potatoes with 0.06 respectively 0.03% triphenyltin acetate (Table CLXXXVII).

X. Formulation

The formulation (particle size, addition of organic wetting and spreading agents, inerts, pH-value of the aqueous suspensions) has great influence on the biologic properties of triphenyltin compounds. Moreover, phytotoxicity can be reduced by suitable formulation (PIETERS 1962).

Fatty alcohol sulfonates or alkylaryl sulfonates can be used as wetting agents (KOOPMANS 1957 and 1958) and Spindel-oil (LUIJTEN 1960) or Triton X-100 (FORSYTH and BROADWELL 1962) can serve as sticking and spreading agent. Usually, an inert carrier is added to the mixtures, for example kaolin, chalk, clay, talcum, and others, possibly also colloidal silica (KOOPMANS 1957 and 1958, LUIJTEN 1960, KAARS SIJPESTEIJN and WIGGERINK 1961).

The influence of particle size on the toxicity is demonstrated by the following (SEIFFER and SCHOOF 1967): Triphenyltin acetate when precipitated in very small particles by adding the solution in alcohol to water is 15 times more efficient against water snails than if applied as wettable powder.

Formulations of commercial preparations are not known; however, some data for experimental preparations are listed in the literature which will be mentioned in the following (cp. EBELING 1963):

HAERTEL and BAUMANN (1956) used 59 parts $CaCO_3$, 20 parts resin, and 1 part alkylaryl sulfonate for 20 parts triphenyltin acetate.

In a patent by *Farbwerke Hoechst AG* (1960), the following formulations are listed among others:

30 parts calcium ligninsulfonate and 50 parts kaolin for 20 parts bis-(triphenyltin)-sulfide;

20 parts calcium ligninsulfonate and 60 parts kaolin for 20 parts bis-(triphenyltin)-oxide;

20 parts calcium ligninsulfonate + 59 parts kaolin + 1 part of a polyethoxylated nonylphenol for 20 parts triphenyltin acetate.

KOOPMANS (1957 and 1958) obtained a wettable powder from 50 g of active ingredient (triphenyltin-4-acetylaminobenzoate) + 7 g of sodium ligninsulfonate (as dispersing agent) + 3 g of sodiumcetylsulfonate (as wetting agent) and 40 g of kaolin. The average particle size of the mixture was given as 10 μg. For use as dust, 10 parts of this mixture are diluted with 90 parts talcum.

KAARS SIJPESTEIJN and WIGGERINK (1961) prepared a dust of bis-(triphenyltin)-disulfide from 0.5% active ingredient in a silica gel-dolomite mixture.

BRUECKNER *et al.* (1960) used mixtures of 5 parts of triphenyltin acetate and 60 parts of manganous ethylenebisdithiocarbamate with 35 parts of added kaolin, cell pitch, and wetting agent. Further, mixtures of 4.4 parts of triphenyltin acetate, 62.4 parts of manganous ethylenebisdithiocarbamate, 20 parts of cell pitch, 13 parts of silica-chalk, and 0.2 parts of wetting agent were recommended.

TAMURA (1965 b) mixed 10 parts of triphenyltin compound with 85 parts of kaolin as inert material and 5 parts of Solpol-W 150 as wetting and spreading agent.

McINTOSH (1966) improved the efficiency of triphenyltin acetate by adding 1% of a paraffin wax emulsion and nonionic surfactants.

KOULA (1971 a) prepared oily suspensions for spraying from 5% triphenyltin oleate in lubrication oil; however, the solutions are unstable and gelatinize after approximately 8 to 12 mon. Further, mixtures with 40% active ingredient + 6% urea + 12% thiourea + 15% $KClO_3$ and 27% talcum were used as aerosols (KOULA 1971 b).

XI. Discussions

Successful use of triphenyltin compounds for control of plant diseases has led to high production of these substances, especially after the problem of phytotoxicity was overcome by anion selection and suitable formulation. The most important field of application is that of root crops; control of *Cercospora beticola* in sugarbeets, *Phytophthora infestants* in potatoes, and *Septoria apii* in celery with triphenyltin compounds results in significantly increased yields mainly during years of severe infestation.

Another area of application is control of algae in rice fields. Finally, feed inhibition of various insects and their larvae is a desirable side effect. This is important mainly for the larvae of Colorado beetle and the army worm in beet cultures. Useful insects such as bees, parasites, or predators are not affected. Another advantage is the absence, so far, of fungi strains which have developed resistance to triphenyltin compounds.[4]

Toxicity of triphenyltin compounds for warm-blooded animals is rather low. The few cases where people have been harmed during application were caused by negligent handling of the active ingredients. These accidents can be easily avoided, as proven by the low accident rate compared with the widespread application.

Triphenyltin compounds have the advantageous property of decomposing almost completely within several wk after application. The end product, inorganic tin(IV), has to be considered physiologically harmless, and the intermediates (di- and monophenyltin compounds) are of low toxicity.

The residues found after correct application constitute no danger for human health as far as known to date; besides that, further decomposition takes place during food processing.

It is of further advantage that triphenyltin compounds are fixed in the soil by adsorption and that here, too, decomposition takes place at a rate depending on the soil bacteria. Gradual accumulation in the ground or dangerous pollution of ground water is not to be expected. However, only a few studies have been published about decomposition of these active ingredients in the ground.

The high toxicity of triphenyltin compounds for fish, frogs, crabs, and other aquatic inhabitants and the long persistence in water should be noted. Any application, therefore, should take into consideration that the aquatic environment could be endangered.

---

[4] Observation of possible triphenyltin-resistant *Cercospora* strains on fields in Greece by GIANNOPOLITIS (1978) still awaits decisive confirmation in laboratory experiments.

Acknowledgments

After the fungicidal qualities and the possibilities for practical application of triphenyltin compounds had been recognized, many studies were undertaken within the last 25 years defining mainly their physiologic effects in warm-blooded animals and the environmental behavior. The purpose of this book is to compile the extensive, scattered, and sometimes not easily accessible literature, reflecting the original results to the maximum extent possible. The goal was to give as complete a survey as possible over the present state of research in this field.

This project was made possible through the generous support by Hoechst AG, which provided a large amount of literature and also unpublished experimental results. Above all, I owe thanks to Mr. G. Hörlein, Mr. S. Gorbach, Mr. B. Strecker, and Mrs. U. v. Preuschen. I also want to thank the many authors and publishing houses that permitted the reprint of parts of their published material.

I am especially grateful to Mr. K. Härtel, who reviewed the manuscript and helped with supplements and improvements mainly in the toxicological and agricultural sections.

Finally, I would like to express my thanks for the translations of foreign literature to Mrs. G. Munk, La Tour de Peilz (Spanish), Mr. H. Behr, Frankfurt/M-Hoechst (Japanese), Mr. K. Beyermann, Mainz (Russian), Mr. E. Noam, St. Legier (Hebrew), Mr. V. Sarcevic, Frankfurt/M-Hoechst (Yugoslavian), Mr. J. Silbereis, La Tour de Peilz (Italian), Mr. F. Vláčil, Praha (Czechoslovakian), and Mr. J. Zarebski, Krakow (Polish).

References

ABASA, R. O.: Work in progress in coffee research. Series III, part one. Entomology. Kenya Coffee, No. 12, p. 364 (1972).
—— Laboratory studies with antifeeding compounds against larvae of *Ascotis selenaria reciprocaria* on coffee. Plant Prot. Bull. FAO 23 (2), 43 (1975).
——, and S. K. MULINGE: Effect of the fungicide "Duter" on the giant coffee looper, *Ascotis serenaria reciprocaria* Walk. Turrialba 22 (1), 99 (1972).
—— —— Effect of the fungicide "Duter" on the giant coffee looper, *Ascotis serenaria reciprocaria* Walk. Kenya Coffee, No. 8, p. 257 (1973).
ABDEL MEGEED, M. I., Z. H. ZIDAN, and R. A. KHALID: Response of the greasy cutworm, *Agrotis ypsilon* Rott. to certain antifeedants. Z. Angew. Entomol. 76, 106 (1974).
ABDEL-RAHMAN, M.: Evaluation of the interaction between cultivars of potatoes and different fungicides and programs of application on early blight control. Plant Disease Reptr. 61, 473 (1977).
ABDUL KAREEM, A., S. SADAKATHULLA, M. S. VENUGOPAL, and T. R. SUBRAMANIAM: Efficacy of two organotin compounds and neem extract against the sorghum shoot fly. Phytoparasitica 2, 127 (1974).
——, S. YAYARAJ, P. THANGAVEL, and A. V. NAVARAYAN PAUL: Studies on the effects of three antifeedants on the egg hatchability of *Corcyra cephalonica* Staint. (Galleriidae: Lepidoptera) and parasitism by *Trichogramma australicum* Gin. (Trichogrammatidae: Hymenoptera). Z. Angew. Entomol. 83, 141 (1977 a).

——, P. THANGAVEL, G. BALUSUBRAMANIAM, and M. BALUSUBRAMANIAM: Studies on the predation of the lady bird beetle, *Menochilus sexmaculatus* (F.) on bean aphid, *Aphis craccivora* Koch., treated with antifeeding compounds. Z. Angew. Entomol. 83, 406 (1977 b).

ABO ELGHAR, M. R., and H. RADWAN: Toxicological, chemosterilant and histopathological effects of triphenyltin hydroxide on *Spodoptera littoralis* Boisd. Acta Phytopathol. Hung. 6, 261 (1971).

——, E. A. ELBADRY, and H. S. RADWAN: Some toxic and antifeeding effects of Du-Ter on the cotton leaf worm *Spodoptera littoralis* Boisd. Z. Angew. Entomol. 69, 177 (1971 a).

—— —— The feeding response of the cotton leafworm *Spodoptera littoralis* Boisd. to ten different host plants treated with Du-Ter. Z. Pflanzenkr. Pflanzenschutz 78, 535 (1971 b).

ADDY, S. K., and B. K. DASH: Comparative trials of copper, dithiocarbamate and organo-tin fungicides for control of Ticca-disease of groundnut and early blight of potato in Orissa. Presented Session X—Fungicides and Nematocides, Internat. Symposium Plant Pathol., New Delhi, 12/30 (1966).

AKAGI, H., and Y. SAKAGAMI: On degradation of organotin compounds by ultra violet rays. Koshu Eiseiin Kenkyu Hokoku 20 (1), 1 (1971).

——, M. FUJITA, and Y. SAKAGAMI: Thin-layer chromatography of inorganotin and organotin compounds in foods. Shokuhin Eiseigaku Zasshi 13, 85 (1972).

——, R. TAKESHITA, and Y. SAKAGAMI: Separation and detection of inorganotin and organotin compounds by thin-layer chromatography. Koshu Eiseiin Kenkyu Hokoku 19 (3), 185 (1970).

ALDRIDGE, W. N., and J. E. CREMER: The biochemistry of organo-tin compounds. Diethyltin dichloride and triethyltin sulfate. Biochem. J. 61, 406 (1955).

ALDROVANDI, A.: L'Impiego dello stagno nella lotta contro la cercospora. Progresso Agricolo 4, 1117 (1958).

ALEXANDRESCU, S., and T. BAICU: The use of feeding inhibitors in the protection of potato crops against the attack of Colorado beetle (*Leptinotarsa decemlineata* Say.). Anal. Inst. Cercetări pentru Protectia Plantelor 9, 449 (1971; publ. 1973).

——, C. PRUNESCU, E. TRACIUC, and C. CODREANU: Mode of action of Brestan and copper oxychloride as feeding inhibitors to the Colorado beetle (*Leptinotarsa decemlineata* Say.). Anal. Inst. Cercetări pentru Protectia Plantelor 11, 231 (1973; publ. 1976).

ALL, J. N., and D. M. BENJAMIN: Potential antifeedants to control larval feeding of selected *Neodiprion* sawflies (Hymenoptera: Diprionidae). Canadian Entomol. 108, 1137 (1976).

ALLESTON, D. L., A. G. DAVIES, M. HANCOCK, and R. F. M. WHITE: Organotin chemistry. Part II. Compounds of the composition $R_4Sn_2X_2O$. J. Chem. Soc. (London), p. 5469 (1963).

ALMEIDA FILHO, J. de, and W. BRUNE: Quantitative evaluation of the active principles of tin-based fungicides. Rev. Ceres 19, 108 (1972).

AMMAL, L. S., and D. DALE: Antifeedants against the larvae of *Spodoptera mauritia* B. (Noctuidae) and *Pericallia ricini* F. (Arctiidae). Indian J. Plant Prot. 2 (1/2), 21 (1974).

—— —— Evaluation of three antifeedants against caterpillars of *Spodoptera litura* Boisd. and *Achoea janata* Linn. Agr. Res. J. Kerala 12 (1), 36 (1974; publ. 1975).

ANDES, J. O., and T. R. GILMORE: Snap beans, root rot, *Rhizoctonia solani*. Fungicide nematocide tests. Results of 1963. Amer. Phytopathol. Soc. 19, 67 (1963).

ANDRADE, A. C.: Potato foliage spot. Fungicide nematocide tests. Results of 1960. Amer. Phytopathol. Soc. 16, 56 (1960 a).

—— Early blight of tomato (*Alternaria solani*). Fungicide nematocide tests. Results of 1960. Amer. Phytopathol. Soc. 16, 63 (1960 b).

ANDRÉN, F., and B. OLOFSSON: Bekämpningsförsök mot bönfläcksjuka. Växtskydds-Notiser 23 (1), 3 (1959).

—— —— Besprutningsförsök mot potatisbladmögel 1961. Växtskydds-Notiser **26** (2), 18 (1962).

ANGELETTI, J. M., and J. C. MAIRE: Étude des spectres de résonance magnétique nucléaire de ¹⁹F et ¹H de quelques fluorophénylstannanes, germanes et silanes. II. Bull. Soc. Chim. France, p. 1858 (1969).

ANONYMOUS: Tätigkeitsbericht des Pflanzenschutzamtes der Landwirtschaftskammer Weser-Ems für das Jahr 1956, Kartoffelkrautfäule. Sonderdruck aus den Jahresberichten der Pflanzenschutzämter, Braunschweig (1957).

—— Organotins for the control of pests. Agr. Chem. **20** (8), 36 and 86 (1965).

—— Anwendung von Brestan 20 aus Flugzeugen zur Behandlung von Zuckerrüben gegen *Cercospora beticola* Sacc. Versuchsbericht 1966, Staatsgut Pik "Belje", Osijek/Jugoslavia (1966 a).

—— Assay report for toxicity testing Brestan 60 in Chinese ring-necked pheasants. Wis. Alumni Res. Foundation Lab., Madison, Wis. 26.5. (1966 b).

—— Assay report for toxicity testing Brestan 60 in Coturnix quail. Wis. Alumni Res. Foundation Lab., Madison, Wis. 27.4. (1966 c).

—— Assay report for toxicity testing Brestan 60 in wild sparrows. Wis. Alumni Res. Foundation Lab., Madison, Wis. 27.4. (1966 d).

—— Bekämpfung der Kartoffel-Knollenfäule durch Bodenbehandlung mit Brestan 60. Versuchsbericht der Stader Saatzucht vom 2.2. (1968).

—— Bekämpfung von *Helminthosporium,* sp. (turcicum ?) in Mais-Hybriden mit Brestan 60. Versuchsbericht der Jugochemia vom 26.3. (1971 a).

—— Fungicides. E. African Trop. Pest. Res. Inst. (T.P.R.I.), Arusha, Tansania, Ann. Rep., p. 52 (1971 b).

—— Control of coffee berry disease and leaf rust in 1972 (Tech. Circ. No. 19). Kenya Coffee Feb., p. 73 (1972).

D'ANS, J., and H. ZIMMER: Der Einfluss von Katalysatoren auf den Verlauf der Fries'schen Reaktion. Ber. Deut. Chem. Ges. **85**, 585 (1952).

ANTOINE, R.: Pineapple disease. Mauritius Sugar Ind. Res. Inst., Ann. Rept., p. 63 (1957).

ARONHEIM, B.: Synthese der Zinnphenylverbindungen. J. Liebigs Ann. Chem. Pharm. **194**, 145 (1878).

ASARI, P. A. R., and D. DALE: Insect antifeedants against snail *Opeas gracile* (Hutton). Curr. Sci. **43**, 803 (1974).

ASCHER, K. R. S.: Insect pest control by chemosterilants and antifeedants. Magdeburg 1966 to Milan 1969. Presented IIIᵉᵐᵉ Congress Internat. des Antiparasitaires, Milan 6.10. (1969 a).

—— Advanced approaches to insect control. Presented IIIᵉᵐᵉ Congress Internat. des Antiparasitaires. World Rev. Pest Control **8** (4), 164 (1969 b).

——, and I. ISHAAYA: Antifeeding and protease- and amylase-inhibiting activity of Fentin acetate in *Spodoptera littoralis* larvae. Pesticide Biochem. Physiol. 3, 326 (1973).

——, and J. MEISNER: The antifeedant effect of organometallics for larvae of the potato tuber moth (*Gnorimoschema operculella* Zell.): A laboratory screening on leaves of eggplant. Z. Pflanzenkr. Pflanzenschutz 76, 564 (1969).

——, and J. MOSCOWITZ: Fentins (triphenyltins) have an anti-feedant effect for house-fly larvae. Internat. Pest Control **10** (3), 10 (1968).

——, and N. E. NEMNY: A paradoxical concentration effect in the toxicity of Fentin acetate for insects. Experientia **32**, 902 (1976).

——, and S. NISSIM: Organotin compounds and their potential use in insect control. World Rev. Pest Control **3** (4), 188 (1964).

—— —— Quantitative aspects of antifeeding: Comparing "antifeedants" by assay with *P. litura.* Internat. Pest Control **7**, 21 (1965).

——, and G. RONES: Fungicide has residual effect on larval feeding. Internat. Pest Control **6** (3), 5 (1964).

——, N. AVDAT, J. MEISNER, and J. MOSCOWITZ: The effect of Fentins on the fertility

of the female housefly. Incorporating a review of their influence on insect fertility and fecundity in general. Z. Angew. Entomol. **69**, 285 (1971).
——, E. GUREVITZ, S. RENNEH, and N. E. NEMNY: The penetration of females of the fruit bark beetle *Scolytus mediterraneus* Eggers into antifeedant-treated twigs in laboratory tests. Z. Pflanzenkr. Pflanzenschutz **82**, 378 (1975).
——, and J. MEISNER, and S. NISSIM: The effect of Fentins on the fertility of the male housefly. World Rev. Pest Control 7 (2), 84 (1968).
——, J. MOSCOWITZ, and S. NISSIM: An antifeeding effect of triphenyltin compounds on housefly larvae. Tin and its Uses **73**, 8 (1967).
ASHRAFUZZAMAN, H.: Systemic control of rice blast caused by *Piricularia oryzae* Cav. Pakistan J. Sci. Ind. Res. **13**, 97 (1970).
ASMUS, E., B. KROPP, and F. M. MOCZKO: Photometrische Bestimmung von Zinn in zinnorganischen Verbindungen mit 3'-Pyridylfluoron. Z. Anal. Chem. **256**, 276 (1971).
AWODERU, V. A.: Rice diseases in Nigeria. Pest Articles and New Summaries (PANS) **20**, 416 (1974).
AWODERU, V. A., and O. F. ESURUOSO: Reduction in grain yield of two rice varieties infected by the rice blast disease in Nigeria. Nigerian Agr. J. **11**, 170 (1974).
—— —— Fungicide evaluation for the control of the blast disease of rice. Nigerian J. Plant Prot. **1**, 1 (1975).
BACHTHALER, G.: Ertragssicherung durch gezielte Krautfäulebekämpfung. Der Kartoffelbau **13**, 76 (1962).
—— Versuchserfahrungen zur Alternariabekämpfung in Kartoffeln. Der Kartoffelbau **15**, 156 (1964).
—— Mehrjährige Versuchsresultate zur Alternaria- und Phytophthorabekämpfung. Der Kartoffelbau **18**, 148 (1967).
——, and A. DAHTE: Ergebnis eines Mittelprüfungsversuches zur Bekämpfung von *Dothichiza populea*, dem Erreger des Pappelrindentodes. Pflanzenschutz **11** (10), 135 (1959).
BAEHR, G.: Gewinnung von Triphenylzinnchlorid und Triphenylzinnhydroxid aus Tetraphenylzinn. Darstellung von Triphenylzinnhydroxid aus Hexaphenylstannan. Z. anorg. allg. Chem. **256**, 107 (1940).
BAETS, A. DE: Bestrijding van *Peronospora tabacina* Adam. Internat. Symposium Fytofarm. Fytiatrie Rijslandb. Hogeschool Gent, p. 1251 (1961).
BAKER, C. J.: Coffee research notes future developments in testing fungicides for CBD and leaf rust. Kenya Coffee **7**, 191 (1971).
—— 1973 Trials with new recommended fungicides. Kenya Coffee **8**, 185 (1973).
BALLESTEROS, O. Q., J. B. CAINGCOY, and N. A. CADELINA: Snails and their control in fishponds. Philippine Fishing J. **8**, 18 (1969).
BARBIERI, R., U. BELLUCO, and G. TAGLIAVINI: Separazione cromatografica su carta di composti metallorganici di piombo e di stagno. Ann. Chimica (Roma) **48**, 940 (1958).
BARDNER, R., and W. M. MATHENGE: Organo-tin compounds against caterpillars on coffee leaves. Kenya Coffee **9**, 257 (1974 a).
—— —— First record of *Phytometra orichalcea* (F.) (Lepidoptera: Noctuidae) feeding on coffee foliage. E. African Agr. Forestry J. **40** (2), 214 (1974 b).
BARNES, G. L.: Effectiveness of some new fungicide formulations for control of pecan scab in Oklahoma. Plant Disease Reptr. **49**, 285 (1965).
—— Effectiveness of certain new fungicide formulations for control of pecan scab in Oklahoma. Plant Disease Reptr. **50**, 599 (1966).
—— Effectiveness of benomyl and thiophanate methyl for long-term control of pecan scab. Plant Disease Reptr. **58**, 687 (1974).
BARNES, J. M., and L. MAGOS: The toxicology of organometallic compounds. Organomet. Chem. Rev. **3**, 137 (1968).
——, and H. B. STONER: The toxicology of tin compounds. Pharmacol. Rev. **11**, 211 (1959).

BARNES, R. D., A. T. BULL, and R. C. POLLER: Behaviour of triphenyltin acetate in soil. Chem. Ind. (London), 204 (1971).
—— —— —— Studies on the persistence of the organotin fungicide Fentin acetate (triphenyltin acetate) in the soil and on surfaces exposed to light. Pest. Sci. 4, 305 (1973).

BATALLA, J. A.: Las Algas de los arrozales y medios para combatirlas. Levante Agricola 7 (76), 33 (1968).

BAUER, S.: Der Einsatz von Hubschraubern in Westdeutschland zur Bekämpfung von Krankheiten und Schädlingen der Forst- und Landwirtschaft. Gesunde Pflanzen 13, 65 (1961).

BAUMANN, J.: Versuche zur Bekämpfung von Pilzkrankheiten im Feldfruchtbau mit der Organo-Zinnverbindung V. P. 1940. Pflanzenschutz 9 (3), 44 (1957).
—— Untersuchungen über die fungizide Wirksamkeit einiger Organo-Zinnverbin-dungen, insbesondere von Triphenyl-Zinn-acetat, ihren Einfluss auf die Pflanze und ihre Anwendung in der Landwirtschaft. Thesis, Landwirtschaftliche Hoch-schule Hohenheim (1958).

BEATTIE, I. R.: The acceptor properties of quadropositive silicon, germanium, tin, and lead. Quart. Revs. (London) 17, 382 (1963).
——, G. P. McQUILLAN, and R. HULME: 5-Coordinate tin. Chem. & Ind., p. 1429 (1962).

BENOY, C. J., P. A. HOOPER, and R. SCHNEIDER: The toxicity of tin in canned fruit juices and solid foods. Food Cosmetic Toxicol. 9, 645 (1971).

BERAN, F.: Anwendung von Pflanzenschutzmitteln und Bienenschutz. Anzeiger f. Schädlingskunde 31 (7), 97 (1958).

BHALLA, O. P., and A. G. ROBINSON: Effects of chemosterilants and growth regulators on the pea aphid fed an artificial diet. J. Econ. Entomol. 61, 552 (1968).

BHAT, P. K., and M. J. CHACKO: Preliminary screening with antifeedants against hairy caterpillar attacking coffee. J. Coffee Res. 6 (1), 29 (1976).

BIEDERMANN, W., and E. MUELLER: Die Inaktivierung des gelösten Kupfers(II) in Fungiziden. Phytopathol. Z. 18, 307 (1952).

Biologische Bundesanstalt Braunschweig: V.P. 1940 auf Bienenunschädlichkeit, Aner-kennungsbescheid v. 21.12. (1956).

BLAKE, E. S., W. C. HAMANN, J. C. EDWARDS, T. E. REICHARD, and M. R. ORT: Thermal stability as a function of chemical structure. J. Eng. Chem. Data 6, 87 (1961).

BLUM, M. S., and J. J. PRATT: Relationships between structure and insecticidal activity of some organotin compounds. J. Econ. Entomol. 53, 445 (1960).

BOCK, R., and K.-D. FREITAG: Abbau von Triphenylzinnchlorid ("Fentinchlorid") auf der Pflanze. Naturwiss. 59, 165 (1972).
——, S. GORBACH, and H. OESER: Analyse von Triphenylzinn-Verbindungen. Angew. Chem. 70, 272 (1958).
——, H.-TH. NIEDERAUER, and K. BEHRENDS: Die Verteilung der Triphenylzinnver-bindungen von Anionen zwischen wässrigen Lösungen und Benzol. Z. Anal. Chem. 190, 33 (1962).

Boehringer Sohn, C. H.: Procédé de fabrication de composés complexes sulfoxyde-étain et leur utilisation pour combattre organismes nuisibles. Belg. Pat. 668 466, (1965/1966).

BOENIG, G., and H. HEIGENER: Papierchromatographie: Quantitative Bestimmung von Mikromengen Zinn. Landw. Forschg. 25, 378 (1972).

BOKRANZ, A., and H. PLUM: Technische Herstellung und Verwendung von Organo-zinnverbindungen. Fortschr. Chem. Forschg. 16, 365 (1971).

BOLKAN, H. A., F. P. CUPERTINO, J. C. DIANESE, and A. TAKATSU: Fungi associated with pre- and postharvest fruit rots on papaya and their control in Central Brazil. Plant Disease Reptr. 60, 605 (1976).

BONGIOVANNI, G. C.: Prova di lotta il "mal bianco" della Barbabietola. Notiz. Mal. Piante 64, 21 (1963).

—— La lotta contra la cercospora: richerche sul dosaggio del rame e dello stagno in prove di campo. Atti Convegno Nazionale per la Bieticoltura in Italia e nel M.E.C., p. 119 (1965).

—— Confronto tra anticrittogamici sperimentali per la lotta contro la cercospora della bieticola (*Cercospora bieticola Sacc.*). Annali Accademia Naz. Agricoltura 88 (4), 104 (1968 a).

—— Prime prove con un nuovo anticrittogamico contro la cercospora della bieticola (*Cercospora beticola* Sacc.). Accademia Naz. di Agricoltura, Estratto dagli Annali, III Serie, Vol. LXXVIII, Bologna, p. 1 (1968 b).

—— Lotta alla cercospora della bieticola (*Cercospora beticola* Sacc.). Undici anni di sperimentazione con rame e stagno. L'Italia Agricola 106, 694 (1969 a).

—— Risultati sperimentali ottenuti nel periodo 1958–68 con composti di stagno e nuovi anticrittogamici organici contro la cercospora della bieticola. L'Industria Saccarifera Italiana 62, 237 (1969 b).

—— Considerazioni sul numero degli interventi per combattere la cercospora della bieticola in Italia ed esame di risultati sperimentali ottenuti nel quadriennio 1965–1968. Atti Giornale Fitopatologiche, p. 371 (1969 c).

—— Accertamenti de campo sulla tolleranza di Cercospora beticola Sacc., ai derivati benzimidazolici e confronti tra prodotti sostitutivi. Conf. on Phytopathology 1975, Proceedings, Torino 12.-14.11.1975; Giornale Fitopatologiche, p. 651 (1975).

BONNEMAISON, L.: Essais de substances chimiostérilisantes II.—Action sur divers Lépidoptères. Phytiatrie-Phytopharmacie 15, 79 (1966).

BOOTH, M. D., and B. FLEET: Electrochemical behaviour of triphenyltin compounds and their determination at submicrogram levels by anodic stripping voltammetry. Anal. Chem. 42, 825 (1970).

——, M. J. D. BRAND, and B. FLEET: A fully automatic apparatus for stripping voltammetry. Application to the determination of triphenyltin compounds. Talanta 17, 1059 (1970).

BORBÉLY-KUSZMANN, A., and J. NAGY: Ueber die Herstellung von Alkyl- and Aryl-zinnhalogen-Verbindungen nach einem modifizierten Grignard-Verfahren. Period. Polytechn. Chem. Eng. 6, 127 (1962).

BOŘKOVEC, A. B.: Insect chemosterilants. Adv. Pest Control Res. VII, 26, 37–38, and 51 (1966).

BOŠCOVIĆ, M.: Suzbijanje lisne rde psenice nekim fugicidima (Control of leaf rust of wheat with some fungicides). Poseban otisak iz casopisa "Savremena poljo-privreda" broj 9, 670 (1962); through Contemporary Agr. 9, 670 (1962).

BOWEN, H. J. M.: The determination of tin in biological material by using neutron-activation analysis. Analyst 97, 1003 (1972).

BOYLE, L. W.: Peanuts, Leafspot, *Cercospora arachidicola*, *Cercospora personata*. Fungicide nematocide tests. Results of 1963. Amer. Phytopathol. Soc. 19, 78 (1963).

BRASE, O.: Gedanken zu den Kartoffelernten der letzten Jahre. Mitteilungsblatt der Landwirtschaftskammer Weser-Ems 50, 1769 (1956).

BRAUN, D., and H.-TH. HEINES: Zur Dünnschicht-Chromatographie von zinnorgani-schen Verbindungen. Z. anal. Chem. 239, 6 (1968).

BROODRYK, S.: Personal communication (May 1967).

BROWN, D. H., ALI MOHAMMED, and D. W. A. SHARP: The far infrared spectra (650–200 cm⁻¹) of some monosubstituted benzene derivatives. Spectrochim. Acta 21, 659 (1965).

BROWN, N.: The effects of two organotin compounds on C3H strain mice. Thesis, Clemson Univ. (1972).

—— The effects of two organotin compounds on C3H strain mice. Diss. Abstr. Internat. B 33 (11), 5356 (1973).

BRUECKNER, H., and K. HAERTEL (to *Farbwerke Hoechst AG*): Behandlung von Saatgut und lebenden Pflanzen. DBP 1 058 302 (1955/1959).

—— —— (to *Farbwerke Hoechst AG*): Verfahren zur Behandlung von durch Pilze oder Bakterien befallenen Substraten, insbesondere Saatgut und lebenden Pflanzen. DAS 1 061 561 (1956/1959).

—— —— (to *Farbwerke Hoechst AG*): Mittel zum Behandeln von Kulturpflanzen oder Saatgut u. dgl. zwecks Wachstumssteigerung und Bekämpfung von Pilzen und Bakterien. DAS 1 109 444 (1959/1961).

—— ——, and G. GASSNER (to *Farbwerke Hoechst AG*): Schädlichen Bewuchs verhindernde Zusätze zu Unterwasser-Anstrichmitteln für Schiffsböden. DBP 1 165 183 (1959/1967).

——, M. CZECH, and K. HAERTEL (to *Farbwerke Hoechst AG*): Fungizide Mittel. DBP 1 127 140 (1960/1962).

—— —— (to *Farbwerke Hoechst AG*): Fungizide Mittel. DBP 1 143 668 (1961/1963).

BRUEGGEMANN, J., K. BARTH, and K.-H. NIESAR: Experimentelle Studien über das Auftreten von Triphenylzinnacetat-Rückstanden in Rübenblättern, Rübenblattsilage, damit gefütterten Tieren und deren Ausscheidungsprodukten. Zentralbl. Veterinärmed. A 11, 4 (1964 a).

——, O. R. KLIMMER, and K.-H. NIESAR: Die bei Pflanze und Tier auftretenden Triphenylzinnacetat-Rückstande und ihre Bedeutung unter hygienisch-toxikologischen Aspekten. Zentralbl. Veterinärmed. A 11, 40 (1964 b).

BRUIN, T. DE: Groundbespruitingen met fungiciden ter bestrijding van de Phytopthora-Knolaantasting bij aardappelen. Jaarboek, p. 87 (1961).

BRUINSMA, J.: De Invloed van Bestrijdingsmiddelen op het Chlorophyllgehalte van Bladen en Stengels. Mededel. Landb. Hoogesch. Gent XXVI (3), 1513 (1961).

BUCCI, F., A. CESARI, and V. AMORINO: Lo stagno nelle conserve alimentari niscalotate e negli alimenti in genere. Chim. e Ind. (Milano) 41, 294 (1959).

BUCKTON, G. B.: Further remarks on the organometallic radicals and observations more particularly directed to the isolation of mercuric, plumbic and stannic ethyl. Proc. Roy. Soc. 9, 309 (1858).

—— Untersuchungen über organische Metallverbindungen. J. Liebigs Ann. Chem. Pharm. 112, 220 (1859).

BUERGER, K.: Zur Analytik von Organozinnverbindungen. Zeitschr. Lebensm.-Unters. Forschg. 114, 1 (1961 a).

—— Bestimmung von Mikrogramm-Mengen Zinn in tierischem und pflanzlichen Material nach der Dithiolmethode. Zeitschr. Lebensm.-Unters. Forschg. 114, 10 (1961 b).

—— Dünnschichtchromatographische Arbeitsweisen zum Nachweis, zur Kennzeichnung und zur Trennung von Spuren mono-, di-, tri- und nullvalenter Organozinnverbindungen. Z. Anal. Chem. 192, 280 (1963).

BULLOCK, R. C. (quoted by ASCHER and NISSIM 1964): Florida Agr. Expt. Station, Ann. Rept. (1964).

BYFORD, W. J.: Experiments on the control of sugar beet downey mildew with fungicides. Brit. Insect. Fungicide Conf., p. 169 (1965).

—— Experiments with fungicide sprays to control *Ramularia beticola* in sugar-beet seed crops. Ann. Applied Biol. 82, 291 (1976).

——, Experiments with fungicide sprays in late summer and early autumn on sugar-beet root crops. Ann. Applied Biol. 86, 47 (1977).

—— Field experiments on sugarbeet powdery mildew, *Erysiphe betae*. Ann. Applied Biol. 88, 377 (1978).

BYRDY, S., Z. EJMOCKI, and Z. ECKSTEIN: Organotin compounds as insect chemosterilants. Evaluation of the activity of some triphenyltin derivatives on the Colorado potato beetle (*Leptinotarsa decemlineata* Say) and housefly (*Musca domestica* L.). Bull. Acad. Polon. Sci., Ser. Sci. Chim. XIII (10), 683 (1965).

—— —— —— Beitrag zur Anwendungsmöglichkeit der organischen Zinnverbindungen als Pestizide mit breitem Wirkungs-Spektrum. Meded. Rijksfac. Landb. Wetensch. Gent 31, 876 (1966 a).

—— —— —— Control of Colorado potato beetle by organotin compounds. Tin and its Uses 71, 11 (1966 b).

ČAČA, Z., and J. JAŠEK: Profitableness of chemical protection of potato against late blight. Agr. Lit. Czechoslovakia 1975, 3/4, p. 561; through Rev. Plant Pathol. 56, 4650 (1977).

CADENA HINOJOSA, M. A.: Dosificación y Numero de Applicationes Optimo-Económicas del Brestan 60 para el Control de la "Quema" del Arroz, *Piricularia oryzae* Cav., en el Estado de Morelos. Agricultura Tecnica en Mexico 3, 385 (1975).

——, and A. E. RODRIGUEZ: Evaluación de Fungicides para el Control de la "Quema" de Arroz *Piricularia oryzae* Cav. en el Estado de Morelos. Agricultura Tecnica en Mexico 3, 330 (1975).

CAHEN, R., M. BOUCARD, M. LALAURIE, and C. LACOUR: Production experimentale d'oedeme cerebral par le sulfate de triethyletain. Compt. Rend. Acad. Sci. 271, Serie D, 1816 (1970).

CAHOURS, A.: Untersuchungen über die metallhaltigen organischen Radikale. J. Liebigs Ann. Chem. Pharm. 122, 48 (1962).

CALINGAERT, G., and H. A. BEATTY: The redistribution reaction. I. The random intermolecular exchange of organic radicals. J. Amer. Chem. Soc. 61, 2748 (1939).

—— ——, and H. R. NEAL: The redistribution reaction. II. The analysis of metal alkyl mixtures and the confirmation of random distribution. J. Amer. Chem. Soc. 61, 2755 (1939).

CALLBECK, L. C.: Potato, late blight, *Phytophthora infestans*. Fungicide nematocide tests. Results of 1963. Amer. Phytopathol. Soc. 19, 87 (1963).

—— Screening of potato fungicides in 1973. Canad. Plant Dis. Survey 54 (1), 22 (1974).

CAMEY, T., and E. PAULINI: Atividade Moluscicida de Alguns Compostos Organo-Estanicos. Revista Brasileira de Malariologica e Doencas Tropicais 16, 487 (1964).

CAMPACCI, C. A.: Early blight of potato (*Alternaria solani*). Fungicide nematocide tests. Amer. Phytopathol. Soc. 16, 59 (1960).

——, and C. F. O. SANTOS: Fungicidas para controle de "Queima da folha" da Batatinha. Arquivos do Instituto Biologico 26, 185 (1959).

—— —— Experiencia com fungicidas em batatinha. Arquivos do Instituto Biologico 29, 29 (1962).

CAMPION, G. G., and I. OUTRAM: Insecticidal and possible chemosterilant effects of certain organo-metal compounds against red bollworm. Internat. Pest Control 10 (2), 21 (1968).

CARLSON, L. W.: Sugar beets, Cercospora leaf spot, *Cercospora beticola*. Fungicide nematocide tests. Results of 1963. Amer. Phytopathol. Soc. 19, 80 (1963).

—— Fungicidal control of sugar beet leaf spot. J. Amer. Sugar Beet Technol. 14, 254 (1966).

CARSON, A. S., R. COOPER, and D. R. STRANKS: Measurement of very low vapour pressures using radioactive isotopes. Part 2. Latent heats of sublimation of lead tetraphenyl and tin tetraphenyl. Trans. Faraday Soc. 58, 2125 (1962).

CASIDA, J. E., E. C. KIMMEL, B. HOLM, and G. WIDMARK: Oxidative dealkylation of tetra-, tri-, and dialkyltins and tetra- and trialkylleads by liver microsomes. Acta Chem. Scand. 25, 1497 (1971).

CASTEL, P., G. GRAS, and S. BEAULATON: Recherches sur l'activité anthelmintique et la toxicité de l'oxyde d'étain diphényle. Rev. Pathol. Gen. Physiol. Clin. 715, 235 (1960).

—— ——, J.-A. RIOUX, and A. VIDAL: Activité de quelques composés organiques de l'étain sur les larves de moustiques. Trav. Soc. Pharm. Montpellier 23, 45 (1963).

Cela Landw. Chemikalien G.m.b.H., Ingelheim: CA 6830, Zinnhaltiges Fungizid (Prospectus, without year).

CENCI, P., and B. CREMONINI: Cromatografia su strato sottile di due antiparassitari stagno-organici e loro comportamento in vari tipi di terreno. Industria Saccarifera Italiana 62, 313 (1969).

CETAS, R. C.: Late blight of potato. Fungicide nematocide tests. Results for 1962. Amer. Phytopathol. Soc. **18**, 64 (1962 a).

—— Potato late blight, *Phytophthora infestans.* Fungicide nematocide tests. Results for 1962. Amer. Phytopathol. Soc. **18**, 71 (1962 b).

—— Potato late blight, *Phytophthora infestans.* Fungicide nematocide tests. Results for 1963. Amer. Phytopathol. Soc. **19**, 88 (1963).

CHALLENGER, F., and F. PRITCHARD: The action of inorganic halides on organometallic compounds. J. Chem. Soc. (London) **125**, 864 (1924).

——, and E. ROTHSTEIN: The nitration of phenyl derivatives of mercury, thallium, lead, bismuth, tin and iodine. J. Chem. Soc. (London), p. 1258 (1934).

CHALMERS, L.: The chemistry and applications of organotin compounds. Manuf. Chemist Aerosol News **38** (6), 37 (1967).

CHAMBERS, R. F., and P. C. SCHERER: Phenyltin compounds. J. Amer. Chem. Soc. **48**, 1054 (1926).

CHAPMAN, A. H., and J. W. PRICE: The degradation of triphenyltin acetate by U. V. light. Internat. Pest Control, p. 11 (1972).

CHARI, M. S., and N. G. PATEL: Antifeeding properties of triphenyltin hydroxide and triphenyltin acetate against *Heliothis armigera* Huebner. Indian J. Entomol. **35** (2), 174 (1973; publ. 1974).

CHAUHAN, S. S.: Effect of fungicides on *Alternaria carthami* Chowdhury in vitro. Labdev J. Sci. Tech. **8 B** (3), 162 (1970).

CHAWLA, S. S., J. M. PERRON, and M. CLOUTIER: Topical application effects of three chemosterilants on the potato aphid, *Macrosiphum euphorbiae* (Thomas) (Homoptera: Aphididae). Phytoprotection **55** (1), 43 (1974).

CHERNEGA, L. G., A. S. KOZYURA, A. A. KAZAKOVA, L. D. SIRAK, and L. D. KUDENKO: Purification of waste waters from the production of organotin compounds. Vodosnabzh., Kanaliz., Gidrotekh. Sooruzh. Mezhvedom. Respub. Nauck.-Tekh. Sb. **13**, 75 (1971); through Chem. Abstr. **76**, 158020u (1972).

CHIAPPARINI, L., E. BALACCI, and C. MOGILA: Sull'impiego del trifenilacetato di stagno come alghicida. Il Riso, Sept. (1964).

—— Un nuovo prodotto alghicida delle risaie. L'Informatore Agrario, No. 12 (1965).

—— Weitere Forschungen über die algizide Wirkung von Triphenylzinnacetat in Reisfeldern. Abstracts of the 6th Intern. Congr. of Plant Protection, Vienna, Aug. 30.-Sept. 6, p. 459 (1967).

CHINN, S. H. F.: Influence of fungicide sprays on sporulation of *Cochliobolus sativus* on Cypress wheat and on conidial populations in soil. Phytopathol. **67**, 133 (1977).

Codex Committee on Pesticide Residues: Fentin acetate, Fentin chloride, Fentin hydroxide. 5th Session, Sept. 28 to Oct. 6; The Hague, The Netherlands (1970).

Coffee Research Institute, Ruiru, Kenia: Control of coffee berry disease and leaf rust in 1972—Recommendations. Kenya Coffee **2**, 73 (1972).

COLE, J. R.: 1964 Spray tests to control pecan scab. Plant Disease Reptr. **49**, 703 (1965).

COLLIER, W. A.: Zur experimentellen Therapie der Tumoren. Die Wirksamkeit einiger metallorganischer Blei- und Zinnverbindungen. Zeitschr. Hyg. Infektionskrankh. **110**, 169 (1929).

COMES, I., E. ILEANA, C. GOLDEANU, and M. COSTESCU: The effectiveness of some fungicides in the control of leaf spot on sugar. Probleme de Protectia Plantelor **2**, 28 (1974).

CORBAZ, R.: Le mildiou du tabac en suisse. Revue Romande d'agriculture, de viti-culture et d'arboriculture **16**, 101 (1960).

CORBIN, H. B.: Separation and determination of trace amounts of tin present as organotin residues on fruits. J. Assoc. Official Anal. Chemists **53**, 140 (1970).

COSTA DA LIMA, A., F. A. A. COUTO, and G. M. CHAVES: Ensaios comparativos da eficiencia de fungicidas associados a espalhantes adesivos do controle a mancha purpura da cebola. Parte da tese apresentada como um dos requisitos para o grau Magister Scientian em Olericultura, a Universade Rural do Estado de Miñas Gerais (1962).

COUSSEMENT, S.: Analyse des depots de fentin-acetate sur feuilles de pommes de terre. Ann. Gembloux 78, 41 (1972).

COUTURE, L., and J. C. SUTTON: Efficacies of fungicides in controlling spot blotch of barley. Canad. J. Plant Sci. 58, 311 (1978).

COZZANI, Z., A. M. SISTO, and U. MANTAUT: Contributo ad una valutazione dell'attivita anticercosporica del'acetato di trifenilstagno. Nottizario sulle Mallatie delle Piante 67 (N.S. 46), 23 (1963).

CRAMER, C. R.: Industrial manufacture or organotin compounds. Tin and its Uses, No. 46, 7 (1959).

CROSSLAND, N. O. (1962): See HOCKING and WHITE (1967).

——, P. O. PARK, C. E. McKONE, W. M. ADAMS, and A. R. HAMSHERE: A field trial to compare the molluscicidal properties of I.C.I. 24 223 (isobutyltriphenylamine), Bayer 73 (5,2'-dichloro-4'-nitrosalicylicanilide) and triphenyltin acetate. Tropical Pest. Res. Inst., Arusha, Tansania; Misc. Rept. No. 334, Apr. (1962).

CZERWIŃSKA, E., Z. ECKSTEIN, Z. EJMOCKI, and R. KOWALIK: Correlation between chemical structure and fungicidal activity of some organotin derivatives. Bull. Acad. Polon. Sci., Ser. Sci. Chim. 15, 335 (1967).

DA COSTA, E. W. B., and L. D. OSBORNE: Laboratory evaluations of wood preservatives. VI. Effectiveness of organotin and organolead preservatives against decay and soft rot fungi. Holzforschung 25 (4), 119 (1971).

—— —— Laboratory evaluations of wood preservatives. VII. Effect of chemical structure on toxicity of organotin and organolead compounds to wood-destroying fungi. Holzforschung 26 (3), 114 (1972).

DALE, D., and S. CHANDRIKA: Antifeeding property of triphenyltin acetate against the larvae of *Spodoptera littoralis* Boisd. (Noctuidae: Lepidoptera) on different host plants. Pest. Ann., Dec., p. 92, (1972).

—— Note on the antifeeding effects of fentin acetate against the larval stages of three important pests. Indian J. Agr. Res. 7, 207 (1973).

——, and K. SARADAMMA: Laboratory evaluation of three antifeedants against the grubs of *Epilachna vigintioctopunctata*. Agr. Res. J. Kerala 11 (2), 174 (1973; publ. 1974 a).

—— —— Effect of continous feeding of fentin acetate on the biology of the Indian mealworm, *Corcyra cephalonica* S. (Pyralidae: Lepidoptera). Bull. Grain Technol. 12 (1), 66 (1974 b).

DARPOUX, H., T. STARON, A. LEBRUN, and B. DE LA TULLAYE: Action curative remarquable du Thiabendazole sur la Ercosperiose de la batterave. Phytiatrie—Pharmacie 15, 113 (1966).

DELGADOS, S., and M. CADENAH: A two-year evaluation of fungicides for the control of potato late blight under the high humid conditions of Central Mexico. Pest Articles and New Summaries (PANS) 15, 228 (1969).

DESCHIENS, R., and H. FLOCH: Les propriétés molluscicides du chlorure et de l'acétate de Triphénylétain dans le cadre de la prophylaxie des bilharzioses. Compt. Rend. Acad. Sci. 255, 1236 (1962).

—— —— Données complémentaires sur l'action molluscicide rémanente des sels de Triphényl-étain. Bull. Soc. Pathol. Exot. 56, 23 (1963).

—— —— Recherches sur l'action molluscicide du laurate de Trioctyl-étain, du chlorure du Trioctyl-étain et du Tetraphényl-étain. Bull. Soc. Pathol. Exot. 58, 1058 (1965).

DESSY, R. E., W. KITCHING, and T. CHIVERS: Organometallic electrochemistry. I. Derivatives of group IV-B elements. J. Amer. Chem. Soc. 88, 453 (1966).

DIENER, U. L., and F. E. GARRETT: Control of pecan scab in 1964 in Southwest Alabama. Proc. SE Pecan Growers Assoc. 58, 73 (1965).

—— —— Control of scab and other diseases of pecan in Alabama. Plant Disease Reptr. 51, 185 (1967).

DIERCKS, R.: Über den "Rutenbrenner" (*Physalospora miyabeana*) an Korbweiden unter besonderer Berücksichtigung zweijähriger Bekämpfungsversuche. Pflanzenschutz 9 (3), 37 (1957).

—— Zur chemischen Bekämpfung der Korbweiden-Parasiten *Fusicladium saliciperdum* (All. et Tub.) Lind. und *Glomerella miyabeana* (Fuk.) v. Arx. Pflanzenschutz 11 (9), 125 (1959).

DODGE, F. N.: Control of foliage diseases of pecans. Proc. SE Pecan Growers Assoc. 59, 72 (1966).

DOERFELT, CH., and H. GELBERT (to *Farbwerke Hoechst AG*): Verfahren zur Herstellung von Hexaorganodistannoxanen. DBP 1 084 722 (1959/1960).

DONÀ DALLE ROSE, A.: Efficacia anticercosporica di un prodotto a base di stagno. L'Informatore Agrario, 2.1. (1958).

—— Trifenilacetato di stagno nella lotta anticercosporica sui portasemi di barbabietola. Estratto da "Sementi elette" 8, No. 1-2 (1976).

——, and R. OLIMPIERI: Un nuovo principio attivo nella lotta anticercosporica. L'Informatore Agrario 22, 3 (1967).

DORMAL, S., and P. NANGNIOT: Étude de la persistence des résidus d'étain organique sur céleris frais et stérilisés. Phytiatrie-Phytopharmacie 10, 39 (1961).

DOSIOS, K., and J. PIERRI: Über Metallbestimmungen in nicht elektrolysierbaren organischen Verbindungen. Z. anal. Chem. 81, 214 (1930).

DOVAS, C., G. SKYLAKAKIS, and S. G. GEORGOPOULOS: The adaptability of the benomyl resistant population of *Cercospora beticola* in Northern Greece. Phytopathol. 66, 1452 (1976).

DRESCHER-KADEN, U., W. EISELE, and L. GRIMM: Nachweis und Bestimmung von Pflanzenschutzmitteln in den Organen von Feldhasen (*Lepus europaeus*), die in der Gemeinde Zurndorf N.O. im Zeitraum vom 20.11. bis 24.11.1970 erlegt wurden. Union Intern. des Biologistes du gibier, Actes du Xe Congr., Paris 3.-7. mai., p. 245 (1971).

DUB, M.: Organometallic compounds, Vol. II, pp. 79–235. Berlin: Springer (1961).

DURIG, J. R., C. W. SINK, and S. F. BUSH: Infrared and Raman spectra of $C_6H_5SiCl_3$, $C_6H_5GeCl_3$ and $C_6H_5SnCl_3$. J. Chem. Phys. 45, 66 (1966).

DUTT, B. L., S. RAM, P. N. MATHUR, M. C. MATHUR, and S. S. VERMA: Potato early blight (*Alternaria solani*). Fungicide nematocide tests. Amer. Phytopathol. Soc. 28, 83 (1972).

DUYFJES, W., and W. DE LANGE (to *North American Phillips Corp.*): Fungicidal composition containing triphenyltin compounds. USP 3 140 977 (1961/1964).

EBELING, W.: Analysis of the basic processes involved in the deposition, degradation, persistence and effectiveness of pesticides. Residue Reviews 3, 35 (1963).

ECKSTEIN, Z., and Z. EJMOCKI: Preparation of triphenyltin compounds. Polon. Pat. 51 771 (1964/1966).

ELBADRY, E. A., M. R. ABO ELGHAR, and H. S. RADWAN: Further studies on the antifeeding effect of Du-Ter on the cotton leafworm *Spodoptera littoralis* Boids. in the U.A.R. Zeitschr. Pflanzenkr., Pflanzenschutz 78, 530 (1971 a).

—— —— —— Chemosterilisation of larvae of the cotton leafworm *Spodoptera littoralis* (Boisd.) by Du-Ter. Zeitschr. Pflanzenkr., Pflanzenschutz 78, 700 (1971 b).

—— —— —— Laboratory cage studies on the effect of the antifeeding compound Du-Ter on the cotton leafworm *Spodoptera littoralis* Boisd. Zeitschr. Angew. Entomologie 69, 438 (1971 c).

—— —— —— Histological changes in some tissues of *Spodoptera littoralis* Boisd., following treatment with Du-Ter. Zeitschr. Angew. Entomologie 72, 34 (1972).

EL-SEBAE, A. H., and Y. M. AHMED: Factors affecting efficiency of some organotin compounds against cotton leaf and boll worms. Zeitschr. Angew. Entomologie 72, 367 (1972/73).

ESCH, J. G. VAN, and A. M. ARNOLDUSSEN: Range finding test as to the toxicity of triphenyltin hydroxide during 18 days. Unpub. Rept. Nat. Inst. Public Health, Utrecht; No. Tox 39/62 (1962); quoted by FAO/WHO Evaluations (1971).

ESTIENNE, V., and G. HENNEBERT: Growth stimulation by triphenyltin acetate. Tin and its Uses **45**, 5 (1958).

—— —— Perspectives des composés organiques de l'étain en phytopharmacie. Agricultura **7**, 483 (1959).

EVANS, C. J.: The development of organotin-based antifouling paints. Tin and its Uses **85**, 3 (1970).

EXCONDE, O. R., A. D. RAYMUNDO, and E. L. SORIANO: The economics of duter/ dithane M-45 foliar spray for the control of Philippine corn downey mildew. Philippine Agriculturist **59**, 237 (1975/76).

EYLEN, L. VAN: Bladvlekkenziekte bij Chrysanth. Tuinbouw Berichten **27**, 226 (1963).

FAHLSTROM, G. B.: Organotin compounds: Evaluation of preservative properties by soil-block-techniques. Proc. Amer. Wood Preservers Assoc. **54**, 178 (1958).

FANO, V., and L. ZANOTTI: Trace determination of tin (II) and tin (IV) in different solvents (H_2O, CH_3OH) by anodic stripping voltammetry. Microchem. J. **18**, 345 (1973).

FAO/WHO: Evaluations of some pesticide residues in food. 1971. WHO/FOOD ADD/71.42 (1971).

—— Report of the 1972 joint meeting, Annex 1 (1972).

—— Pesticide residues in food (1973).

—— Data sheets on pesticides. No. 22, Fentin compounds. (Dec. 1976).

Farbwerke Hoechst AG: Pesticidal preparations for the treatment of culture plants and seeds. Brit. Pat. 946 770 (1960/1964).

FILLER, T. D.: Fluorimetric determination of submicrogram quantities of tin. Anal. Chem. **43**, 1753 (1971).

FINDLAY, J. B. R.: The use of antifeeding compounds as protectants against insect damage to plants. M. Sc. Thesis, Fac. of Math. and Nat. Sci., Univ. Pretoria (1968).

—— A new approach to insect control on crops. Farming in S.A. **6**, 23 (1969).

—— Laboratory studies on the effects of triphenyltin acetate and triphenyltin hydroxide on the stages in the life-cycle of *Spodoptera littoralis* (Bois.). Phytophylactica **2** (2), 91 (1970).

FINKNER, R. D., D. E. FARUS, and L. CALPOUZOS: Evaluations of fungicides for the control of *Cercospora* leaf spot of sugar beets. J. Amer. Soc. Sugar Beet Technol. **14**, 232 (1966).

FISH, R. H., E. C. KIMMEL, and J. E. CASIDA: Bioorganotin chemistry: Biological oxidation of organotin compounds. In J. J. ZUCKERMAN (ed.): Organotin compounds: New chemistry and applications. Adv. Chem. Series **157**, 197 (1976).

FISHER, F. E.: Greasy spot of citrus and its chemical control. Florida Agr. Expt. Station, Ann. Rept., p. 214 (1963).

—— Chemical control of scab on citrus in Florida. Plant Disease Reptr. **53**, 19 (1969).

FLOCH, H., and R. DESCHIENS: Étude comparée de l'action molluscicide du 5, 2'-dichloro-4'-nitro-salicylanilide (Bayer 73) et des sels (acétate et chlorure) de triphényl-étain. Bull. Soc. Pathol. Exot. **55**, 816 (1962).

—— ——, and TH. FLOCH: Sur les propriétés molluscicides de l'oxyde et de l'acétate de tributyl-étain (Prophylaxie des Bilharzioses). Bull. Soc. Pathol. Exot. **57**, 454 (1964).

FOELDESI, I., and G. STRÁNER: Organotin compounds, II. Preparation of organotin oxinates and testing their fungicidal activity. Acta Chim. Acad. Sci. Hung. **45**, 313 (1965).

FORSYTH, F. R., and C. E. BROADWELL: Control of *Cercospora* leaf spot of sugar beets with protective fungicides. J. Amer. Soc. Sugar Beet Technol. **12**, 91 (1962).

FOSCHI, F., and G. RAPPARINI: Prove di lotta anticercosporica. Atti Giornale Fitopathologiche, p. 239 (1962).

—— —— —— Controllo dell'attività immediata della persistenza nel tempo e della resistenza al dilavamento di prodotti anticercosporici. Ricerca Sci., Rendiconti B, **3** (4), 341 (1963).

———, F. Laffi, B. Flamini, and F. Koch: Confronto di preparati stanorganici e rameici in trattamenti su portaseme di barbietola da zucchero. Atti Giornale Fitopathologiche Cagliari, p. 327 (1969).

Foucart, G., and L. Delcambre: Recherche d'agents chimiothérapeutiques contre le flétrissement bactérien de la pomme de terre. Parasitica 16 (4), 126 (1960).

Frankland, E.: Über eine neue Reihe organischer Körper, welche Metalle enthalten. J. Liebigs Ann. Chem. Pharm. 85, 329 (1853).

Franz, W.: Erfahrungen zur Bekämpfung der Möhrenschwärze. Gesunde Pflanzen 24 (6), 97 (1972).

Fraselle, J.: La lutte chimique contre le Mildiou (Phytophthora infestans de By.) de la pomme de terre. Rev. Agr. (Brussels) 20, 1013 (1967).

Freeland, G. N., and R. M. Hoskinson: Non-aqueous atomic-absorption spectrophotometric determination of organometallic biocides. Analyst 95, 579 (1970).

Freitag, K.-D.: Metabolismus und Rückstandsanalytik des Triphenylzinnchlorids. Thesis, Univ. Mainz (1972).

———, and R. Bock: Analyse von Gemischen aus Tri-, Di- und Monophenylzinnverbindungen und anorganischem Zinn (IV). Z. anal. Chem. 270, 337 (1974 a).

——— ——— Degradation of Triphenyltin Chloride on Sugar Beet Plants and in Rats. Pest. Sci. 5, 731 (1974 b).

Frick, L. P., and W. Q. de Jimenez: Molluscicidal qualities of three organotin compounds revealed by 6-hour and 24-hour exposures against representative stages and sizes of Australorbis glabratus. Bull. World Health Org. 31, 429 (1964).

Friebe, E., and H. Kelker: Molekularer Zustand und Analytik des Systems Triphenylzinnhydroxid/Bistriphenylzinn-oxid. Z. anal. Chem. 192, 267 (1963).

Fritz, G., and H. Scheer: Zur photometrischen Zinn-Bestimmung mit Kakothelin. Z. anorg. allg. Chem. 331, 151 (1964).

Froyd, J. D., G. Johnson, and G. F. Stallknecht: Sugar beet leaf spot (Cercospora beticola). Fungicide nematocide tests. Amer. Soc. Phytopathol. 23, 103 (1967).

Fuse, G., and K. Nishimoto: Studies on the organo-mercuric and tin compounds as the wood preservatives. VIII. On the fungicidal activity and the effect of ultraviolet irradiation on fungicidal action of the organo-tin compounds. Wood Res., Bull. Wood Res. Inst. Kyoto Univ. 32, 15 (1964) (quoted by Hof 1969).

Fuse, T: Study on the toxicity of organotin compounds against fish. Rept. Res. Lab., Sankyo Yasugawa Plant, June 17 (1971).

Fye, R. L., G. C. la Breque, and H. K. Gouck: Screening tests of chemicals for sterilisation of adult houseflies. J. Econ. Entomol. 59, 485 (1966).

Gaines, T. B., and R. D. Kimbrough: Toxicity of Fentin hydroxide to rats. Toxicol. Applied Pharmacol. 12, 397 (1968).

Galandzovskaya, L. D., and I. V. Gamin: Effectiveness of new fungicides in the control of cercosporic sugar beet. Novoe v Issled. po Sakhar, Svekle i Zern. Kul'turam. Kiev USSR (1973), p. 135; through Rev. Plant Pathol. 54, 1482 (1975).

Gálvez E. and G. E., and Z. J. Castano: Aplicatión de Productos Quimoterapéuticos al Suelo para el Control de Piricularia oryzae en Arroz. Fitopathologia 9 (1), 18 (1974).

Gardiner, B. G., and R. C. Poller: Insecticidal activity and structure of some organotin compounds. Bull. Entomol. Res. 55 (1), 17 (1964).

Georgopoulos, S. G., and C. Dovas: A serious outbreak of strains of Cercospora beticola resistant to benzimidazole fungicides in Northern Greece. Plant Disease Reptr. 57, 321 (1973).

Geyer, R., and H. J. Seidlitz: Volumetrische Zinnbestimmung in Organozinnverbindungen. Z. Chem. 4, 468 (1964).

Giannopolitis, C. N.: Occurrence of strains of Cercospora beticola resistant to triphenyltin fungicides in Greece. Plant Disease Reptr. 62, 205 (1978).

Giesen, M.: Metallorganische Verbindungen als Biocide für Anstrichstoffe. FATIPEC, VIII⁰ Congres, p. 185 (1966).

GILMAN, H., and L. A. GIST: Organotin compounds containing water-solubilizing groups. Some m-hydroxyphenyl derivatives. J. Org. Chem. **22**, 368 (1957).

——, and W. B. KING: A method for the quantitative analysis of tin in organic compounds. J. Amer. Chem. Soc. **51**, 1213 (1929).

——, and H. W. MELVIN: Some reactions of triphenylsilane and triphenyltin hydride. J. Amer. Chem. Soc. **71**, 4050 (1949).

——, and L. S. MILLER: The determination of silanols with the Karl Fischer reagent. J. Amer. Chem. Soc. **73**, 2367 (1951).

——, and S. D. ROSENBERG: Organotin compounds containing an azo linkage. J. Amer. Chem. Soc. **74**, 5580 (1952).

—— Reaction of triphenyltin hydride with methyllithium. J. Amer. Chem. Soc. **75**, 368 (1953).

GILMOUR, J. W., and A. L. VANNER: Radiata pine (*Pinus radiata*). Terminal crook disease (*Colletotrichum acutatum* f. sp. pinea). Fungicide nematocide tests. Amer. Phytopathol. Soc. **28**, 128 (1972).

GMELIN: Handbuch der Anorganischen Chemie, Ergänzungswerk zur 8. Auflage, Vol. 26, (Zinnorganische Verbindungen, Teil 1). Berlin: Springer (1975).

GOLDBERG, F. L., and M. V. VANDONI: Indagini analitiche sulla presenza di composti organici dello stagno in cariossidi, lolla, paglia di piante di riso cresciute in risaia e trattate con trifenil-acetato di stagno. Il Riso, Sept., p. 13 (1964).

GONZALEZ, J. A.: Ojo de Gallo en cafetales del Mombacho. Orientando al Agricultor (1958).

GORBACH, S.: Über Versuche in vitro, das Verhalten von Triphenylzinnhydroxid, -acetat und -chlorid im Magen-Darmtrakt von Warmblütern zu klären. Versuchsbericht des Analyt. Laboratoriums der HOECHST AG v. 30.1. (Ber. Nr. 18/68) (1968).

——, and R. BOCK: Die Bestimmung kleiner Mengen von Triphenylzinnacetat in Pflanzenmaterial. Z. anal. Chem. **163**, 429 (1958).

——, and H. HOMMEL: Analysenmethode 129/63 der FARBWERKE HOECHST AG (1963).

GOVINDARAJAN, T. S.: Uses of some newer fungicides for the control of coffee diseases. Indian Coffee **35** (8), 308 (1971).

GRABER, M., and G. GRAS: Étude de l'activité anthelmintique et de la toxicité de quelques composés organiques de l'étain. III. Oxyde d'étain diphényle. Rev. Elev. Med. Vet. Pays Trop. **17** (2), 205 (1964).

—— —— Étude de l'activité anthelmintique et de la toxicité de quelques composés organiques de l'étain. IV. Le dichlorure d'étain diphényle. Rev. Elev. Med. Vet. Pays Trop. **18** (4), 405 (1965).

—— —— Étude de l'activité anthelmintique et de la toxicité de quelques composés organiques de l'étain. VI. Comparaison entre les divers composés organiques de l'étain étudiés. Conclusions générales. Rev. Elev. Med. Vet. Pays Trop. **19** (1), 7 (1966).

GRAS, G., and J.-A. RIOUX: Relation entre la structure chimique et l'activité insecticide des composés organiques de l'étain (essai sur les larves de *Culex pipiens pipiens* L.). Arch. Inst. Pasteur Tunis **42**, 9 (1965).

GRAVES, C. H., and A. E. GOSSARD: Results of fungicide screening programs on pecans for 1964. Proc. SE Pecan Growers Assoc. **58**, 52 (1965).

GRAVES, J. B., J. R. BRADLEY, and J. L. BAGENT: Laboratory evaluation of several organotin compounds against *Heliothis* spp. J. Econ. Entomol. **58**, 583 (1965).

GRIFFITHS, V. S., and G. A. W. DERWISH: The physical chemistry of organotin compounds. Part I. The ultraviolet spectra of phenyltin compounds. J. Mol. Spectr. **3**, 165 (1959).

—— —— The physical chemistry of organotin compounds. Part II. The infrared spectra of phenyltin compounds. J. Mol. Spectr. **5**, 148 (1960).

GROOT, A. P. DE, V. J. FERON, and H. P. TIL: Short term toxicity studies on some salts and oxides of tin in rats. Food Cosmet. Toxicol. **11** (1), 19 (1973).

GROSJEAN, M., M. GIELEN, and J. NASELSKI: Chimie organométallique. Les réactions de substitution nucléophile sur les atoms du quatrième groupe. Ind. Chim. Belg. **28**, 721 (1963).

GUARDASCIONE, V., and M. MARZELLA DI BOSCO: Contributo alla conoscenza della patologia professionale da antiparassitari. Su tre casi di intossicazione acuta da fungicide a base di trifenilacetato di stagno. Lavoro Umano **19** (7), 307 (1967).

GUENTHER, F., R. GEYER, and D. STEVENZ: Röntgenspektrometrische Bestimmung des Zinns in Organozinnverbindungen. Neue Hütte **14**, 563 (1969).

GUTHRIE, J. E., and P. D. HARWOOD: Use of tin preparations for treatment of chickens experimentally infected with tapeworms. Amer. J. Vet. Res. **2**, 108 (1941).

GUTSHTEIN, I.: Effect of organic compounds on sugar beets and sugar yield in the control of the *Cercospora*. Hassadeh **43**, 8 (1963).

HAERTEL, K.: Organic tin compound as a crop fungicide. Tin and its uses, No. 43, 9 (1958 a).

—— Anwendungsmöglichkeiten organischer Zinnverbindungen in der Landwirtschaft. Angew. Chem. **70**, 135 (1958 b).

—— Triphenyltin compounds. Agr. Vet. Chem. **3**, 19 (1962).

—— Untersuchungen über den Einfluss der Kombination von Triphenylzinnacetat (Brestan®) mit DDT-Lindan-Paste auf die Pflanzenverträglichkeit. Interner Versuchsbericht der Pflanzenschutzforschung der HOECHST AG v. 26.8. (1963 a).

—— Einfluss von Netz- und Haftmittelzusatz zur Triphenylzinnacetat-Spritzbrühe auf die Pflanzenverträglichkeit. Interner Versuchsbericht der Pflanzenschutzforschung der HOECHST AG v. 22.10. (1963 b).

—— The present position regarding the use of triphenyltin compounds in agriculture. Presented Symp. Tin Research Inst., Frankfurt/M, 28.11. (1963 c).

—— The present position of triphenyltin acetate in agriculture. Tin and its Uses, No. 61, 7 (1964 a).

—— Composti di Trifenilstagno. L'Italia Agricola **101**, 379 (1964 b).

——, and J. BAUMANN: Verfahren zum Behandeln von lebenden Pflanzen. DBP 1 021 627 (1956/1965).

HAGUE, D. N., and R. H. PRINCE: The question of radical formation in hexaphenyl ditin and related compounds. J. Inorg. Nuclear Chem. **28**, 1039 (1966).

HAMILTON, J. T., and F. I. ATTIA: Effects of mixtures of *Bacillus thuringiensis* and pesticides on *Plutella xylostella* and the parasite *Thyraeella collaris*. J. Econ. Entomol. **70** (1), 146 (1977).

HANCOCK, W. and E. Q. LAWS: The simultaneous determination of traces of benzene and toluene. Analyst **81**, 37 (1956).

HANSEN, H. J., and L. R. SILLER: *Phythophthora palmivora*. Small-scale screening tests. Cacao Inform. Bull. **5**, 11 (1960).

HARDISON, J. R.: Prevention of apothecial formation in *Gloeotinia temulenta* by systemic and protectant fungicides. Phytopathol. **62**, 605 (1972).

HARDON, H. J., A. F. H. BESEMER, and H. BRUNINK: Feldversuche mit Triphenylzinn. Deutsch. Lebensm.-Rundsch. **58**, 349 (1962).

——, H. BRUNINK, and E. W. VAN DER POL: Colorimetric determination of triphenyltin residues. Analyst **85**, 847 (1960).

HARRAH, L. A., M. T. RYAN, and C. TAMBORSKI: Infrared spectra of phenyl derivatives of group IVb, Vb and VIIb elements. Spectrochim. Acta **18**, 21 (1962).

HARRIS, J. O. (to *Monsanto Chemical Co.*): Tin hydrocarbon compounds and processes for making same. USP 2 431 038 (1944/1947).

HARTMANN, E., P. KUEMMEL, and M. HARDTMANN (to *IG Farbenindustrie AG*): Mottenschutzmittel. DRP 485 646 (1925/1929).

HAYNES, J. W., N. MITLIN, T. B. DAVICH, and C. E. SLOAN: Evaluation of candidate chemosterilants for the boll weevil. Production Res. Rept. No. 120, ARS/USDA, Washington, D.C. (1971).

HAYS, S. B.: Reproduction inhibition in house flies with triphenyltin acetate and triphenyltin chloride alone and in combination with other compounds. J. Econ. Entomol. **61**, 1154 (1968).

HEATH, D. F.: Some problems in the determination of residues in plants and mammals. Symp. Use and Application Radioisotopes and Radiation in Control of Plant and Animal Insect Pests. IAEA, Athens (Greece), Apr. 22-26 (1963 a).

—— Radiation and radioisotopes applied to insects of agricultural importance. Proc. IAEA, Vienna, p. 185 (1963 b).

—— Radioisotopes in the detection of pesticide residues. Proc. Panel on Uses of Radioisotopes in Detection on Pesticide Residues. Vienna Apr. 12-15 (1966).

HEDGES, E. S.: Quoted by G. J. M. VAN DER KERK (1975).

HEIMES, H. T., and D. BRAUN: Mikroverfahren zur Bestimmung von Zinn in nieder- und hochmolekularen Organozinnverbindungen. Z. anal. Chem. **254**, 21 (1971).

HEPPERLY, P. R., and J. B. SINCLAIR: Aqueous polyethylene glycol solutions for treating soybean seeds with antibiotics. Seed Sci. Technol. **5**, 727 (1977; rcd. 1978).

HEROK, J.: Versuche zur Frage einer Aufnahme, Weiterleitung und Speicherung von Triphenylzinnacetat und Triphenylzinnchlorid bei Kartoffeln. Interner Versuchsbericht des Radiochemischen Laboratoriums der HOECHST AG v. 11.11. (1965).

——, and H. GOETTE: The radiometrical estimation of the distribution and excretion of [113Sn] triphenyl tin acetate in ruminants. Symp. Use of Radioisotopes in Animal Biology and Medical Sciences, Mexico City, Nov. 21–Dec. 1, 1961, p. 177. New York Academic Press (1961).

—— —— Radiometrische Untersuchungen über das Verhalten von Triphenylzinnacetat in Pflanze und Tier. Int. J. Applied Rad. Isotopes **14**, 461 (1963).

—— —— Radiometrische Stoffwechselbilanz von Triphenylzinnacetat beim Milchschaf. Zentralbl. f. Veterinärmed A **11**, 20 (1964).

HILLS, F. J., D. H. HALL, and D. G. KONTAXIS: Effect of powdery mildew on sugarbeet production. Plant Disease Reptr. **59**, 513 (1975).

HISLOP, E. C.: Studies on the chemical control of *Phytophthora palmivora* (Butl.) Butl. on *Theobroma cacao* L. in Nigeria. IV. Further laboratory and field trials of fungicides. Ann. Applied Biol. **52**, 465 (1963).

——, and P. O. PARK: Studies on the chemical control of *Phytophthora palmivora* Butl. on *Theobroma cacao* L. in Nigeria. I. Laboratory bio-assay of fungicides on detached pods. Ann. Applied Biol. **50**, 57 (1962).

HOCKING, D.: Determination of translaminar, curative activity against Leaf rust (*Hemileia vastatrix* Berk. et Br.). Pest Articles and News Summaries (PANS), Section B; Plant Disease Control **11** (3), 273 (1965).

—— Fungicides for Arabica coffee. II. Biological assessment of protective capacity against Leaf rust (*Hemileia vastatrix* Berk. et Br.). E. African Agr. Forestry J. **32**, 356 (1967 a).

—— Fungicides for Arabica coffee. III. Curative activity against Leaf rust (*Hemileia vastatrix* Berk. et Br.) E. African Agr. Forestry J. **32**, 359 (1967 b).

——, and G. H. FREEMAN: Fungicides for Arabica coffee. Relationships among some new fungicides, leaf rust (*Hemileia vastatrix*), leaf fall and yield. Tropical Agr., Trin. **45** (2), 141 (1968).

——, and P. J. WHITE: Phytotoxicity of triphenyltin acetate to rice seed. E. African Agr. Forestry J. **32**, 380 (1967).

Hoechst AG: Unpublished summary (1975).

Hoechst Holland: Pamphlet (1964).

HOESSLIN, R., VON: Dreijährige Feldversuche zur Bekämpfung der Blattfleckenkrankheit des Sellerie mit Brestan. Deutsch. Gartenbauwirtschaft **8**, 231 (1960).

HOF, T.: Review of literature concerning the evaluation of organotin compounds for the preservation of wood. J. Inst. Wood Sci. **4** (5), 19 (1969).

HOLMES, T. D., and I. F. STOREY: Comparative trials of copper, dithiocarbamate and organo-tin fungicides for potato blight control in Lincolnshire 1958–1961. Plant Pathol. **11**, 139 (1962).

HOPF, H. S., and R. L. MULLER: Laboratory breeding and testing of *Australorbis glabratus* for molluscicidal screening. Bull. World Health Org. **27**, 783 (1962).

——, J. DUNCAN, J. S. S. BEESLEY, D. J. WEBLEY, and R. F. STURROCK: Molluscicidal properties of organotin and organolead compounds. Bull. World Health Org. **36**, 955 (1967).

HORÁČEK, V., and K. DEMČIK: Group poisoning at applying Brestan 60 (triphenyltin acetate) in agriculture. Prac. lek. **22**, 61 (1970).

HORN, N. L.: Cucumber, anthracnose, *Colletotrichum obiculare*. Fungicide nematocide tests. Results for 1963. Amer. Phytopathol. Soc. **19**, 72 (1963).

——, F. N. LEE, and R. B. CARVER: Effects of fungicides and pathogens on yields of soybeans. Plant Disease Reptr. **59**, 724 (1975).

——, G. WHITNEY, and T. FORT: Yields and maturity of fungicide-sprayed and unsprayed disease-free soybean plants. Plant Disease Reptr. **62**, 247 (1978).

HUECK, H. J., and J. L. A. BRIJN: Mothproofing properties of insecticides. II. Organotin compounds. Material and Organisms **7** (4), 307 (1972).

——, and J. G. A. LUIJTEN: Organo-tin compounds as textile preservatives. J. Soc. Dye and Color **74**, 476 (1958).

HULL, R.: Sugar beet diseases. Bull. Min. Agr., Fisheries and Food, London, No. 142 (1960).

—— *Ramularia* leaf spot. Rept. Rothamsted Expt. Station for 1961, p. 206 (1962).

HULME, R.: The crystal and molecular structure of chloro(trimethyl)pyridinetin(IV). J. Chem. Soc. (London), p. 1524 (1963).

HUTCHINSON, R. W.: Investigations into the control of potato blight (*Phytopthora infestans*). Part I. Volume and time of spray applications. Record Agr. Res. **22**, 35 (1974).

—— Investigations into the control of potato blight (*Phytophthora infestans*). Part II. Type of fungicide. Record Agr. Res. **23**, 45 (1975).

INABA, T., and S. WATANABE (to *Tokyo Fine Chemicals Co. LTD.*): Preparation of organotin compounds. Japan. Pat. 41-10.102 (1963/1966).

INGHAM, R. K., S. D. ROSENBERG, and H. GILMAN: Organotin compounds. Chem. Reviews **60**, 459 (1960).

IRVING, H., and J. J. COX: Studies with dithizone. Part VIII. Reactions with organometallic compounds. J. Chem. Soc. (London), p. 1470 (1961).

ISÁK, H.: Cercosporabekämpfung an Rüben-Samenträgern. Die Bodenkultur **13** (3/4), 360 (1962).

ISHAAYA, I., and J. E. CASIDA: Phenyltin compounds inhibit digestive enzymes of *Tribolium confusum* larvae. Pest. Biochem. Physiol. **5**, 350 (1975).

JACKSON, L. F., R. K. WEBSTER, C. M. WICK, J. BOLSTAD, and J. A. WILKERSON: Chemical control of stem rot of rice in California (USA). Phytopathol. **67**, 1155 (1977).

JANCARIK, V.: Control of *Dothistroma pini* in forest nurseries. Research Leaflet, Forest Res. Inst. Rotorna, New Zealand Forest Service No. 24, Dec. (1969).

JANSSEN, M. J., and J. G. A. LUIJTEN: Investigations on organotin compounds. Part XVIII. The basicity of triorganotin hydroxides. Rec. Trav. Chim. Pays Bas **82**, 1008 (1962).

JARVIS, R. H., J. L. SHORT, and F. E. SHOTTON: Copper, dithiocarbamates and organotin compounds for the control of potato blight, 1962–65. Plant Pathol. **16** (2), 49 (1967).

JEHRING, H., and H. MEHNER: Wechselstrompolarographische Kapazitätseffekte von Organozinnverbindungen in Alkoholen. Z. anal. Chem. **224**, 136 (1967).

JEN-HSI CHO, HAN-SHENG HSU, HSU-YU CH'ENG, CH'ANG-LIEH FAN, YUAN-LIEH MUO, and CHI-FU YU: Organotin compounds I. Preparation and fungicidal properties of some triphenyltin compounds. Hua Hsueh Hsueh Pao **32** (2), 196 (1966); through Chem. Abstr. **65**, 13 753 (1966).

JENKINS, S. F.: Personal communication (quoted by ASCHER and NISSIM 1964).

JERCHEL, D.: Organozinnverbindungen mit fünfbindigem Zinn. Lecture Bonn 21.10. (1969).

JERMY, T., and G. MATOLCSY: Antifeeding effect of some systemic compounds on chewing phytophagous insects. Acta Phytopathol. Acad. Sci. Hung. 2 (3), 219 (1967).

JITSU, Y., N. KUDO, and T. SUGIYAMA: Determination of triphenyltin compounds and their thermal stability. Noyaku Seisan Gijutsu 17, 17 (1967).

—— ——, and T. TESHIMA: Determination of organotin compounds by gas-liquid chromatography. Japan Analyst 18, 169 (1969).

JOHNSON, E. W., and J. M. CHURCH (to Metal and Thermit Corp.): Process for making organotin halides. USP 2 599 557 (1948/1952).

JOHNSON, W. A.: Recent applications of organotin chemicals. Tin and its Uses, No. 54, 5 (1962).

JOLYET, F., and A. CAHOURS: Recherches sur l'action physiologique des stannéthyles et des stannméthyles. Compt. Rend. Acad. Sci. 68, 1276 (1869).

JONES, R. G., and H. GILMAN: Methods of preparation of organometallic compounds. Chem. Reviews 54, 835 (1954).

JOSHI, B. G., G. RAMAPRASAD, and C. L. NARAYANA: Antifeeding properties of triphenyltin acetate against the tobacco caterpillar. Indian J. Entomol. 29 (1), 18 (1967).

—— ——, and M. SUBRAHMANYAM: Effect of Fentin acetate as an antifeedant against tobacco caterpillar Prodenia litura Fabricius. Indian J. Agr. Sci. 41, 1110 (1971).

—— ——, and G. NARASIMHAYYA: Note on Fentin acetate as a reproduction inhibitor of the tobacco caterpillar, Spodoptera litura Fabricius. Indian J. Agr. Sci. 43, 324 (1973).

—— ——, and S. V. V. SATYANARAYANA: Relative efficacy of Neem Kernel, Fentin acetate and Fentin hydroxide as antifeedants against tobacco caterpillar, Spodoptera litura Fabricius, in the nursery. Indian J. Agr. Sci. 48, 19 (1978).

KAARS SIJPESTEIJN, A.: Organic tin compounds as potential agricultural fungicides. Meded. Landb. Hogeschool Ghent 24, 850 (1959).

——, and G. L. WIGGERINK (to North American Phillips Co.): Bis(triphenyltin)polysulfides and method of combating fungi. USP 3 113 069 (1961/1963).

——, F. RIJKENS, J. G. A. LUIJTEN, and L. C. WILLEMSENS: On the antifungal and antibacterial activity of some trisubstituted organogermanium, organotin and organolead compounds. Antonie van Leeuwenhoek J. Microbiol. Serol. 28, 346 (1962).

——, J. G. A. LUIJTEN, and G. J. M. VAN DER KERK: In D. C. TORGESON (ed.): Fungicides, an advanced treatise, Vol. I, p. 331. New York: Academic Press (1969).

KALCHSCHMID, W., and C. KRAUSE: Die Bekämpfung des Weissrostes an Meerrettich Albugo candida Pers. O. Ktze. Gesunde Pflanze 28 (2), 39 (1976).

KAMEL, A. A. M., S. H. MITRI, M. ABO ELGHAR, and M. B. ATTIA: Further studies on the effect of antifeedants against Spodoptera Larvae under natural field conditions (Lepidoptera: Noctuidae). Bull. Entomol. Soc. Egypt, Econ. Series 8, 227 (1974; publ. 1975).

KENAGA, E. E.: Triphenyltin compounds as insect reproduction inhibitors. Proc. XIIth Internat. Congress Entomol., London, p. 517 (1965 a).

—— Triphenyltin compounds as insect reproduction inhibitors. J. Econ. Entomol. 58, 4 (1965 b).

KERK, G. J. M. VAN DER: Bekämpfung von Mikroorganismen, wie Fungi, Bakterien, Protozoen. DBP 950 970 (1952/1956).

—— Address before the Soc. Ind. Microbiol., Bloomington, Ind. (1958); (quoted by INGHAM et al. 1960).

—— Tien jaar organotin onderzoek. Chem. Weekbl. 56, 339 (1960).

—— New developments in organic fungicides. In: Fungicides in agriculture and horticulture; Soc. Chem. Ind. Monograph No. 15, p. 67. London: Soc. Chem. Ind. (1961).

—— Present trends and perspectives in organotin chemistry. Chem. & Ind., p. 644 (1970).

—— Derzeitiger Stand der Anwendungen von Organozinnverbindungen. Chem. Ztg. 99, 26 (1975).

——, and J. G. A. Luijten: Investigations on organotin compounds. III. The biocidal properties of organotin compounds. J. Applied Chem. 4, 314 (1954).

—— —— Investigations on organotin compounds. IV. The preparation of a number of trialkyl- and triaryltin compounds. J. Applied Chem. 6, 49 (1956 a).

—— —— Investigations on organotin compounds. VI. The preparation of cyanomethyltin compounds by the decarboxylation of organotin cyanoacetates. J. Applied Chem. 6, 93 (1956 b).

—— —— Développements récents dans la recherche des composés organiques de l'étain. Ind. Chim. Belge 21, 567 (1956 c).

—— Zur Chemie und zu den Anwendungsmöglichkeiten von Organozinnverbindungen. Arzneimittelforschung 19, 932 (1969).

—— ——, and J. G. Noltes: Neue Ergebnisse der Organozinnforschung. Angew. Chem. 70, 298 (1958).

—— ——, J. C. van Egmond, and J. G. Noltes: Fortschritte auf dem Organozinngebiet. Chimia 16, 36 (1962).

—— ——, J. G. Noltes, and H. M. J. C. Creemers: Anwendungs- und Forschungsaspekte der Organozinnchemie. Chimia 23, 313 (1969).

Kerr, K. B., and A. W. Walde: The anthelminthic activity of tetravalent tin compounds. Exp. Parasitol. 5, 550 (1956).

Keschi, K. C., and N. N. Mohanti: Efficacy of different fungicides and antiobiotics on control of blast of ragi. Presented Internat. Symp. Plant Pathol. (New Delhi), Dec. 27–Jan. 1 (1967).

Khosa, R. L., and S. N. Dixit: Some organotin compounds as antifungal agents. Sci. Cult. 35, 637 (1969).

Kimmel, E. C., R. H. Fish, and J. E. Casida: Bioorganotin chemistry. Metabolism of organotin compounds in microsomal monooxygenase systems and in mammals. J. Agr. Food Chem. 25, 1 (1977).

Kiss, E., and T. Hetzer: Efficacy of control of the Cercospora beticola (Sacc.) in sugar beets. I. Importance of the spraying period using Brestan. Mg. Növénynem, és Növényterm. Kut. Int. Közl. 3 (3), 13 (1967).

Kissam, J. B., and S. B. Hays: Mortality and fertility response of Musca domestica adults to certain known mutagenic or anti-tumor agents. J. Econ. Entomol. 59, 748 (1966).

Klassen, W., J. F. Norland, and A. B. Borkovec: Potential chemosterilants for boll weevils. J. Econ. Entomol. 61, 401 (1968).

Klimmer, O. R.: Zur Pharmakologie und Toxikologie der Organozinnverbindungen. Angew. Chem. 70, 135 (1958).

—— Triphenylzinnacetat—ein Beitrag zur Toxikologie organischer Zinnverbindungen. Arzneimittelforschg. 13, 432 (1963).

—— Toxikologische Untersuchungen mit Triphenylzinnacetat. Zentralbl. f. Veterinärmed. A 11, 29 (1964).

—— Die Anwendung von Organozinn-Fungiziden in der Landwirtschaft in toxikologischer Sicht. Pflanzenschutz-Ber. 37 (4, 5, 6), 57 (1968).

—— Die Anwendung von Organozinn-Verbindungen in experimentell-toxikologischer Sicht. Arzneimittelforschg. 19, 934 (1969).

——, and W. Schulemann: Gutachten über die Toxizität der organischen Zinnverbindung "V.P. 1940". Pharmakologisches Institut der Rheinischen Friedrich-Wilhelm-Universität Bonn v. 31.1. (1957).

Klotzsche, C.: Zur Toxikologie neuerer Pflanzenschutzmittel. Verh. IV. Intern. Pflanzenschutz-Kongr., Hamburg 8.–15. Sept. 1957; II, p. 1641 (1960).

Knauf, W.: Brestan 60—Fischtoxizität; interner Versuchsbericht der Pflanzenschutz-Forschung der HOECHST AG, 24.1. (1974).

Kobetz, P., and R. C. Pinkerton (to Ethyl Corp.): Manufacture of tin alkyl compounds. USP 3 028 320 (1960/1962).

Koch, F., A. Vetter, and W. Voelker: Bericht über die Feldversuche im Jahrgang 1959. Arbeitsgemeinschaft zur Bekämpfung der Zuckerrübenkrankheiten, p. 34 (1959).
———— —— Bericht über die Feldversuchsergebnisse im Jahrgang 1960. Arbeitsgemeinschaft zur Bekämpfung der Zuckerrübenkrankheiten, p. 34 (1960).
———— —— —— Bericht über die Feldversuchsergebnisse im Jahrgang 1961. Arbeitsgemeinschaft zur Bekämpfung der Zuckerrübenkrankheiten, p. 27 (1961).
Kocheshkov, K. A.: Über die Einwirkung von metallischem Zinn auf Methylhalogenide. Ber. Deutsch. Chem. Ges. 61, 1659 (1928).
—— Untersuchungen über metallorganische Verbindungen. I. Mitteilung: Eine neue Klasse von Arylzinnverbindungen: Phenyl-trihalogen-stannane. Ber. Deutsch. Chem. Ges. 62, 996 (1929).
——, M. M. Nadj, and A. P. Alexandrov: Untersuchungen über metallorganische Verbindungen, VII. Mitteilung: Vereinfachte Methode zur Darstellung von Triarylzinnhalogeniden. Ber. Deutsch. Chem. Ges. 67, 1348 (1934).
—— —— —— Aryltin Hydroxy- and Halogen Compounds of Type Ar_3SnX. Zhurn. Obshch. Khim. 6, 1672 (1936).
Kochkin, D. A., V. I. Vashkov, and V. P. Dremova: Oxygen-containing organotin and -lead compounds. IV. Synthesis and insecticidal activity of stannanols and plumbanols, their acetates and methacrylates. Hexaalkyldistannoxanes and polydialkyl(aryl)stannoxanes. Zhurn. Obshch. Khim. 34, 325 (1964).
Koenig, K.: Orientierende Versuche zur chemischen Bekämpfung des Maisbeulenbrandes (Ustilago maydis [DC] Corda). Bayer. Landw. Jahrb. 46, 320 (1969).
Kohama, S.: A simple digestion method for the quantitative analysis of metal in some organometallic compounds. Bull. Chem. Soc. Japan 36, 830 (1963).
Kohlmann, J., and W. Lueders: Pflanzenschutzempfehlung im Hopfenbau. 1973. Hopfenrundschau 130 (1973).
Kolla, V. E., and V. S. Zalesov: Influence of chemical structure on the toxicity of organotin compounds. Uch. Zap. Permsk. Gos. Univ., No. 111 (1964).
Koopmans, M. J. (to North American Phillips Corp.): Method of producing organic tin compounds and preparations containing such compounds for combating noxious micro-organisms. USP 3 031 483 (1957/1962).
—— (to North American Phillips Corp.): Organo-tin compositions and method for treating plants. USP 3 097 999 (1958/1963).
Koula, V.: Physical, chemical and biological properties of aerosol solutions containing organic tin compounds. Sb. Uvti, Ochr. Rostl. (Praha) 7, 211 (1971 a).
—— Fungitoxicity of fume and warm aerosols containing organic tin compounds. Sb. Uvti, Ochr. Rostl. (Praha) 7, 219 (1971 b).
——, and O. Rajchartová: Antifeeding and repellent effect of some organic tin compounds. Sb. Uvti, Ochr. Rostl. (Praha) 7, 287 (1971).
Kovàcs, A.: Richerche sulla persistenza di fungicidi organici. Nota I. Resistenza al dilavamento. Relazione sulle Giornale Fitopathologiche, Bologna 30./31.3. (1962).
Kraak, M. (to N.V. de Bataafsche Petroleum Maatschappij): Werkwijze voor het bestrijden van insecten en voor het bereiden van een middel voor genoemd doel. Holl. Pat. 68 578 (1947/1951).
Kraemer, K.: Warndienst im Obstbau. Gesunde Pflanzen 22 (3), 44 (1970).
Král, J., Š. Neubauer, and K. Klimeš: The principal diseases of foxglove caused by fungal parasites. Naše Lieč. Rastl. 11 (3), 77 (1974).
Kranz, J.: Die Wirkung einiger Fungizide auf Mycosphaerella musicola Leach in Laboratoriums- und Mikro-Feldversuchen. Phytopathol. Z. 52 (1), 59 (1965).
Krasnoshchekov, I. M.: Efficacy of new preparations against Cercospora disease of sugar beet. Khimiya v Sel'skom Khozyaistve 10 (8), 36 (1972).
Krause, E.: Vereinfachte Darstellung von Triarylzinnhalogeniden. Ber. Deutsch. Chem. Ges. 51, 912 (1918).
—— Eine wirkungsvolle Beeinflussung des experimentellen Mäuse-Carcinoms durch organische Bleiverbindungen. Ber. Deutsch. Chem. Ges. 62, 135 (1929).

——, and R. Becker: Zweiwertiges Zinn als Chromophor in aromatischen Stannoverbindungen und die Gewinnung von Hexaaryl-distannanen. Ber. Deutsch. Chem. Ges. 53, 173 (1920).

Krexner, R.: Welche Entwicklung nimmt die Cercospora- Bekämpfung im österreichischen Zuckerrübenanbau? Der Pflanzenarzt 13 (6), 53 (1960).

—— Beobachtungen über das Auftreten von echtem Rübenmehltau (Erysiphe communis) im Zusammenhang mit der Anwendung von Brestan. Pflanzenschutzber. (Wien) 33, 41 (1965).

—— Beitrag zur Frage der Bekämpfung der Cercospora- Blattfleckenkrankheit der Rübe. Der Pflanzenarzt 24 (3), 22 (1971).

——, and H. Wenzl: Verbesserte Erfolgsaussichten bei der Bekämpfung von Krankheiten im Hackfruchtbau. Der Pflanzenarzt 11 (2), 25 (1958).

Kriegsmann, H., and H. Geissler: Infrarot- und ramanspektroskopische Untersuchungen an Phenylzinnverbindungen. Z. anorg. allg. Chem. 323, 170 (1963).

Kroeber, H., and D. Massfeller: Untersuchungen über die Blauschimmel-Krankheit des Tabaks in Deutschland. Nachr.-Bl. d. Deutsch. Pflanzenschutz-Dienstes 13 (4), 49 (1961).

Kroeller, E.: Triphenylzinnverbindungen im Pflanzenschutz und ihre Rückstandsbestimmung. Deutsch. Lebensm.-Rdsch. 56, 190 (1960).

—— Untersuchungen zum Nachweis von Brestan-Rückständen. Lebensm.-Chem. u. gerichtl. Chemie 17, 101 (1963).

Kubo, H.: Preparations of organotin-phosphorus compounds and their biological activities. Agr. Biol. Chem. (Tokyo) 29 (1), 43 (1965).

Kuethe, K.: Mehrjährige Erfahrungen bei der Bekämpfung der Kiefernschütte (Lophodermium pinastri) in Hessen. Gesunde Pflanzen 11, (8), 145 (1959).

Kumar, T. P.: Effects of some antifeedants on the caterpillars of Pericallia ricini F. (Arctiidae, Lepidoptera), a pest of cacao. J. Plantation Crops 2 (2), 21 (1974).

Kumpulainen, J., and P. Koivistoinen: Advances in tin compound analysis with special reference to organotin pesticide residues. Residue Reviews 66, 1 (1977).

Kushlefsky, B., and A. Ross: Karl Fischer reagent for determination of and differentiation between trialkyl(aryl) organotin hydroxides and corresponding oxides. Anal. Chem. 34, 1666 (1962).

——, I. Simmons, and A. Ross: Characterization of triphenyltin hydroxide and bis(triphenyltin) oxide. Inorg. Chem. 2, 187 (1963).

Kutzer, E., M. Weiser, and H. Burtscher: Der Einsatz von Brestan® im Zuckerrübenbau und seine Auswirkungen auf den Feldhasen. Oesterr. Weidwerk 4, 148 (1972).

Ladd, T. L.: Some effects of three triphenyltin compounds on the fertility and longevity of Japanese beetles. J. Econ. Entomol. 61, 577 (1968).

Ladenburg, A.: Über das Zinntriäthylphenyl. J. Liebigs Ann. Chem. Pharm. 159, 251 (1871).

Langer, H. B., and A. H. Blut: Preparation of several sulfoxide complexes with group IV organometallic compounds. J. Organomet. Chem. 5, 288 (1966).

Large, J. R.: Pecan scab, Fusicladium effusum Wint. Fungicide nematocide tests. Results of 1963. Amer. Phytopathol. Soc. 19, 65 (1963).

—— Results of two years spraying with Du-Ter (triphenyl tin hydroxide) compared with other fungicides for the control of pecan scab. Proc. SE Pecan Growers Assoc. 58, 55 (1965).

Leibelt: Bekämpfung der Primärinfektion der Hopfenperonospora mit Brestan-60. Hopfenrundschau 21, 144 (1970).

Lellis, W. T.: Resultados dos teste efetuados com o "Brestan". Test Report of the Instituto de Cacan da Bahia—Brasil. 5.12. (1959).

Le Quan Minh: Synthèse des composés organiques de l'étain possédant quatre substituants différents. Compt. Rend. Acad. Sci., Ser. C, 266, 832 (1968).

Lewis, F. B. (to Associated Lead Manufacturers LTD.): Improvements in or relating to organotin compounds. Brit. Pat. 786 545 (1955/1957).

LIEBL, H.: Zur Bekämpfung der Frühinfektion des Hopfens durch die Peronospora. Hopfenrundschau 19 (No. 6), (1968).

LINDEMANN, A.: Spritzversuch zu Sellerie in der Hamburgschen Gartenbau-Versuchsanstalt Fünfhausen. VERBEGA-Mitteilung des Versuchs- und Beratungsdienstes Hamburg v. 11.4. (1959).

LIVINGSTON, J. M., W. C. YEARIAN, and S. Y. YOUNG: Insecticidal activity of selected fungicides on Hippodamia convergens. J. Ga. Entomol. Soc. 13 (2), 148 (1978 a).

—— —— —— Insecticidal activity of selected fungicides: Effects on three lepidopterous pests of soybean. J. Econ. Entomol. 71 (1), 111 (1978 b).

LLOYD, G. A., C. OTACI, and F. T. LAST: Triphenyltin acetate residues on potato leaves in blight spraying trials. J. Sci. Food Agr. 13, 353 (1962).

LUIJTEN, J. G. A. (to N.V. Philips Gloeilampenfabrieken): Fungizides Mittel. DBP 1 122 321 (1960/1971).

——, and G. J. M. VAN DER KERK: Investigations in the field of organotin chemistry. Tin Res. Inst., Greenford, Middlesex, England (1955).

—— —— In A. G. MAC DIARMID: Organometallic compounds of the group IV elements, Vol. 1, Part 2, p. 91. New York: M. Dekker (1968).

LUSKINA, B. M., and S. V. SYAVTSILLA: Determination of organotin compounds in the air and in waste waters. Nov. Obl. Prom.-Sanit. Khim., p. 186 (1969); through Chem. Abstr. 71, 128 314 p (1969).

MAAS, H. L. VAN DER, L. VAN DEYL, and J. J. KROON: The acute and subacute toxicity of the fungicide triphenyltin hydroxide to the common pond snail, Lymnaea stagnalis L. Meded. Rijksfac. Landbouwwetensoh. Gent 37, 850 (1972).

MACKIEWICZ, S., and Z. SOSNA: The influence of germination spraying and defoliation of potato on late blight infection. Prace Naukowe Inst. Ochrony Roslin 16, 137 (1974; publ. 1975).

MAGEE, P. N., H. B. STONER, and J. M. BARNES: The experimental production of oedema in the central nervous system of the rat by triethyltin compounds. J. Path. Bact. 73, 107 (1957).

MAIRE, J.-C.: Étude des spectres de résonance magnétique nucléaire de ^{19}F et ^{1}H des (p-Fluorophényl)-chlorostannanes. J. Organomet. Chem. 9, 271 (1967).

——, and F. HEMMERT: Étude des spectres de résonance magnétique nucléaire des phenylchlorostannanes et des phénylchlorosilanes. Bull. Soc. Chim. France, p. 2785 (1963).

——, J. CASSAN, B. LEPRERTE, and J. MARROT: Spectres d'adsorption des phénylchlorosilanes, phénylchlorogermanes et phénylchlorostannanes dans l'infrarouge. Compt. Rend. Acad. Sci. 260, 5290 (1965).

MALMUS, N.: Zur Frage der Verhütung der Auswinterung durch Kleekrebs. Pflanzenschutz No. 8 (1959).

MANZER, F. E., and D. MERRIAM: Potato late blight, Phytophthora infestans. Fungicide nematocide tests. Results for 1963. Amer. Phytopathol. Soc. 19, 90 (1963).

MARFURT, T. A., and H. A. TOSCANI: Comportamiento de un derivado organico del estano en el control de las plagas del manzano Carpocapsa pomonella y Grapholita molesta. Delta del Parana 6 (10–11), 45 (1966/67).

MARIĆ, A., and D. ČAMPRAG: Contribution to the investigation of fungicides in the control of Cercospora beticola Sacc. on industrial sugar beet. Agrohemija 8, 26 (1959).

MARKS, M. J., C. L. WINEK, and S. P. SHANOR: Toxicity of triphenyltin hydroxide (Vancide KS). Toxicol. Applied Pharmacol. 14, 627 (1969).

MARR, I. L.: Microanalytical determination of tin in organotin compounds. Talanta 22, 387 (1975).

MARROT, J., J.-C. MAIRE, and J. CASSAN: Spectres d'absorption des phénylchlorostannanes et phénylchlorogermanes dans l'ultraviolet. Compt. Rend. Acad. Sci. 260, 3931 (1965).

MARTIN, D. F., P. C. MAYBURG, and R. D. WALTON: Organometallic compounds. V. Kinetics of phenyl-tin cleavage by a chelating agent. J. Organomet. Chem. 5, 57 (1966).

———— ——— ——— Kinetics of tin-phenyl cleavage by 8-quinolinol under ansolvous conditions. J. Organomet. Chem. 7, 362 (1967).

Massaux, F.: Etude des résidues de chlorure de triphénylétain (Brestanol ^{113}Sn) sur fèves de cacao. Café Cacao Thé 15, 221 (1971).

Mathur, Y. K., and R. C. Saxena: Note on the possible use of triphenyltin acetate as a crop protectant against some Lepidopterous pests. Indian J. Agr. Sci. 42, 427 (1972).

Matta, E. A. da: Microensaios com novos Fungicidas para o controle do *Phytophthora palmivora* Butl., causendor da potridao parda dos Frutos do cacaueiro. Boletim da secretaria da agricultura, Sao Paulo, p. 26 (1959).

McCombie, H., and B. C. Saunders: Toxic organo-lead compounds. Nature 159, 491 (1947).

McIntosh, A. H.: An experiment on control of potato tuber blight. Proc. 3rd Brit. Insecticide Fungicide Conf., Brighton, Nov. 4 (1965).

——— Fungicides. Rept. Rothamsted Expt. Station, p. 184 (1966).

——— Tests of aryltin compounds as potato blight fungicides. Ann. Applied Biol. 69, 43 (1971).

Meisner, J., and K. R. S. Ascher: Antifeedants against the potato tuber moth (*Gnorimoschema operculella* Zell.) and the striped maize borer (*Chilo agamemnon* Bles.): Laboratory experiments on leaves. Z. Pflanzenkrankh. Pflanzenpath., Pflanzenschutz 78, 458 (1965).

———, and H. Skatulla: Laboratory experiments with antifeedants against larvae of the gypsy moth, *Porthetria dispar* L., Z. angew. Entomol. 78, 317 (1975).

Metal and Thermit Corporation: Process for making tin hydrocarbons. Brit. Pat. 661 241 (1949/1951).

——— Purification of organotin compounds. Brit. Pat. 909 610 (1961/1962).

Milazzo, G.: Sullo spettro d'assorbimento dei tetrafenili di silicio, stagno e piombo. Gazz. Chim. Ital. 71, 73 (1941).

Miller, V. L., and C. J. Gould: Organic tin compounds as fungicides for control of blue mold of iris bulbs. Plant Disease Reptr. 47, 408 (1963).

Mitri, S. H., and A. A. M. Kamel: The sterilant effect of certain antifeedants on the moths of *Spodoptera littoralis* Boisd. (Lepidoptera: Noctuidae). Bull. Entomol. Soc. Egypt, Econ. Ser. 6, 79 (1972).

——— ——— Further studies on the sterilant effect of certain antifeedants on the adult stages of *Spodoptera littoralis* (Boisd.) (Lepidoptera: Noctuidae). Bull. Entomol. Soc. Egypt, Econ. Ser. 7, 143 (1973; publ. 1974 a).

——— ——— Studies on the antifeeding properties of five new compounds against *Spodoptera littoralis* (Boisd.) (Lepidoptera: Noctuidae). Bull. Entomol. Soc. Egypt, Econ. Ser. 7, 149 (1973; publ. 1974 b).

——— ——— Field and laboratory evaluation of certain antifeedants against *Spodoptera littoralis* Boisd. (Lepidoptera: Noctuidae). Bull. Entomol. Soc. Egypt, Econ. Ser. 8, 33 (1974; publ. 1975).

———, M. M. Zaki, and M. Abo Elghar: Laboratory evaluation of two antifeedants against the larval stages of *Spodoptera littoralis* (Boisd.). Bull. Entomol. Soc. Egypt, Econ. Ser. 4, 53 (1970).

Moedritzer, K.: Redistribution reactions of organometallic compounds of silica, germanium, tin and lead. Organomet. Chem. Reviews 1, 179 (1966).

Mohibulla: Control of *Cercospora* leaf spot disease of sugarbeet in northern zone. Agr. Pakistan 24, 167 (1973; publ. 1974).

Montgomerie, I. G., and D. M. Kennedy: Preliminary evaluation of some chemical treatments on red core of strawberry. Plant Pathol. 24, 162 (1975).

Moore, N.: Heavy metal pesticides. Metals and ecology. Ecol. Res. Comm. Bull. No. 5, p. 36 (1969).

Mukherji, S. K.: Chemical control of algae in West-Bengal paddy fields. World Crops (1968).

———, and B. K. Ray: Algal weeds of paddy fields of coastal West Bengal and their control by a new chemical. Z. Pflanzenkrankh. Pflanzenschutz 73 (1/2), 35 (1966).

MUKHOPADHYAY, A. N., and S. V. R. K. RAO: Sugar beet leaf spot (Cercospora beticola). Fungicide nematocide tests. Amer. Phytopathol. Soc. 27, 111 (1971).

—— —— Control of Cercospora leaf spot of sugarbeet with systemic and protective fungicides. Plant Disease Reptr. 58, 952 (1974).

——, and R. P. THAKUR: Systemic activity of triphenyltin chloride in sugarbeet seedlings. Plant Disease Reptr. 56, 776 (1972).

MULINGE, S. K.: Research on coffee diseases at Jacaranda. Kenya Coffee, Sept., p. 295 (1970).

MULLA, M. S.: Chemosterilants for control of reproduction in the eye gnat (Hippelates collusor) and the mosquito (Culex quinquefasciatus). Hilgardia (Berkeley) 39, 297 (1968).

MULLER, R. A., and S. E. NJOMOU: Contribution a la mise au point de la lutte chimique contre la nourriture brune des cabosses du cacaoyer (Phytophthora palmivora (Butl.) au Cameroun. Café Cacao Thé 14 (3), 209 (1970).

—— ——, and R. LOTODÉ: Appréciation de l'efficacité des fongicides contre la pourriture brune des cabosses du cacaoyer due au Phytophthora palmivora (Butl.) dans les conditions naturelles. Essai de mise au point d'une méthode rapide. Café Cacao Thé 13 (1), 34 (1969).

MURBACH, R.: Effet en plein champ de fongicides à base de fentin-acétate, de manèbe et d'oxychlorure de cuivre sur la densité de population du doryphore de la pomme de terre (Leptinotarsa decemlineata Say.). Rech. Agron. Suisse 6, 345 (1967).

—— Étude en laboratoire de l'effet de l'acétate de fentin et de l'oxychlorure de cuivre sur le doryphore (Leptinotarsa decemlineata Say.). Mitt. Schweiz. Entomol. Ges. 48, 113 (1975).

——, and R. CORBAZ: Influence de trois types de fongicides utilisés en Suisse contre le mildiou de la pomme de terre (Phytophthora infestans [Mont.] de Bary) sur la densité de population du doryphore (Leptinotarsa decemlineata Say.). Phytopathol. Zeitschr. 47, 182 (1963).

MUROMTSEV, G. S., and V. N. AGNISTOKOVA: Inhibition of dodder seedling growth by synthetic and biosynthetic fungicides. Dokl. vses. Akad. sel'-Khoz. Nauk 1969 (4); through Weed Abstr. 19 (3), 177 (1970).

NAD, M. M., and K. A. KOCHESHKOV: The interaction of organic tin and mercury salts with tinsodium alloy and tin as a method of synthesis of highly arylated organic tin compounds. Zhurn. Obshch. Khim. 8, 42 (1938).

NAGASAWA, S., H. SHINOHARA, and M. SHIBA: Sterilizing effect of Dowco-186 on the Azuki bean weevil, Callosobruchus chinensis L., with special reference to the hatchability of the eggs deposited by treated weevils. Botyu-Kagaku 30, 91 (1965).

—— —— Differential susceptibilities in sexes of Callosobruchus chinensis L. (Coleoptera, Bruchidae) to the sterilizing effects of triphenyltin hydroxide. J. Stored Prod. Res. 3, 177 (1967).

NANGNIOT, P., and P. H. MARTENS: Application de la chrono-ampérométrie par redissolution anodique au dosage de traces d'acétate de triphényl-étain. Anal. Chim. Acta 24, 276 (1961).

NEELY, D.: Persistence of foliar protective fungicides. Phytopathol. 60, 1583 (1970).

NEFEDOV, V. D., and N. A. VARSHAV: Isomorphous cocrystallisation of tetraphenyl derivatives of lead, tin and silicon. Zhurn. Fiz. Khim. 28, 981 (1954).

——, V. E. ZHURAVLEV, N. G. MOLCHANOVA, and N. N. KALININA: Chromatographic separation of phenyl derivatives of elements of group IV. Zhurn. Obshch. Khim. 38, 1219 (1968).

NETZER, D., and R. KATZIR: Combined control of Alternaria blight and powdery mildew of carrots. Plant Disease Reptr. 50, 594 (1966).

NEUMANN, W. P.: Die Organische Chemie des Zinn. 1. Aufl., F. Enke, Stuttgart (1967).

NEWKIRK, H. N.: Piezoelectric organometallic crystals: Growth and properties of tetraphenylsilicon, -germanium, -tin and -lead. J. Organomet. Chem. 44, 263 (1972).

NEWTON, D. W., and R. L. HAYS: Histological studies of ovaries in rats treated with hydroxyurea, triphenyltin acetate, and triphenyltin chloride. J. Econ. Entomol. 61, 1668 (1968).

NIENHAUS, F.: Über die chemische Bekämpfung von *Phytophthora cactorum*, dem Erreger der Kragenfäule an Apfelbäumen. Phytopath. Z. 34, 365 (1959).

NISHIMOTO, K., and G. FUSE: Fungicidal and wood preservative properties of organo-tin compounds. Tin and its Uses 70, 3 (1966).

NOLTES, J. G., J. G. A. LUIJTEN, and G. J. M. VAN DER KERK: Investigations on organo-tin compounds. XV. The antifungal properties of some functionally substituted organo-tin compounds. J. Applied Chem. 11, 38 (1961).

NUGENT, T. J.: Cucumber, downey mildew (*Pseudoperonospora cubensis*), powdery mildew (*Erysiphe cichoracearum*). Fungicide nematocide tests. Results of 1963. Amer. Phytopathol. Soc. 19, 73 (1963).

OAKES, V. (to *Pure Chemicals, Inc.*): Improvements relating to phenyl tin compounds. Brit. Pat. 1.070.942 (1964/1967).

OBST, A., and H. KEES: Neue Wege bei der chemischen Bekämpfung der Spelzenbräune des Weizens. Nachrichtenbl. d. Deutsch. Pflanzenschutzdienstes 24 (2), 17 (1972).

OELSCHLAEGER, W.: Spektralphotometrische Bestimmung sehr geringer Mengen Zinn in biologischen und anderen Materialien. Z. anal. Chem. 174, 241 (1960).

OKHLOBYSTIN, O. Yu., L. I. ZAKHARKIN, and B. N. STRUNIN: II-V Group element full aryl derivatives. USSR Pat. 144.171, (1961/1970).

OKIOGA, D. M., and S. K. MULINGE: 1973 Trials with new and recommended fungicides. Kenya Coffee Nov., p. 260 (1974).

OLOFSSON, B., and F. ANDRÉN: Bespratningsförsök mot potatisbladmogel. Växtskydds-Notiser 27 (1), 4 (1963).

Organisation Mondiale de la Santé: Lutte contre les mollusques et prévention de la Bilharziose. Genève, p. 165 (1967).

ORPHANIDIS, P. S.: Stérilisation en laboratoire de "*Ceratitis capitata*" Wied. et de "*Dacus oleae*" Gmel. au moyen d'aziridines, acaricides, fongicides et de sels mineraux. Congress Internat. des Antiparasitaires, Naples 15–17 Mars II/19 (1965).

——, and P. G. PATSAKOS: Nouvelles expériences sur la stérilisation de deux espèces de Trypetidae [*Dacus oleae* (Gmel.) et *Ceratitis capitata* Wied.] au moyen de substances chimiques, avec ou sans propriétés d'alkylation. Congress Internat. des Antiparasitaires, Milan 6–8 Oct., No. 27 (1969).

—— —— Chimiostérilisation des *Dacus oleae* (Gmel.) et *Ceratitis capitata* Wied. au moyen de substances chimiques avec ou sans propriétés d'alkylation. Extraits des Annales de l'institut Phytopathologique Benaki, Nouvelle Serie, 9 (2), 134 (1970).

PAL, B. K., and D. E. RYAN: Fluorescence and metallic valency states. Part III. Determination of tin. Anal. Chim. Acta 48, 227 (1969).

PAQUET, R., and P. WILKIN: Nouvelle association organo-stannique pour la lutte contre le mildiou de la pomme de terre (*Phytophthora infestans* de By.). 22. Internat. Symp. Fytofarm. en Fytiatrie, 5.5.1970; Mededel. Landb.—Wetenschap. Gent 36 (1), 348 (1971).

PARKIN, A., and R. C. POLLER: Brit. provisional pat. 12 168/76 (quoted by POLLER 1976).

PATE, B. D., and R. L. HAYS: Histological studies of testes in rats treated with certain insect chemosterilants. J. Econ. Entomol. 61 (1), 32 (1968).

PAULUS, A. O., F. SHIBUYA, J. NELSON, and A. O. HARVEY: Systemic fungicide interval for control of sugarbeet *Cercospora* leaf spot. Phytopathol. 60, 1541 (1970).

——, O. A. HARVEY, J. NELSON, F. SHIBUYA, and A. H. HOLLAND: Control of *Cercospora* leaf spot of sugarbeet under sprinkler irrigation. Plant Disease Reptr. **55**, 449 (1971).

—— ——, and V. MEEK: Fungicides and timing for control of sugarbeet powdery mildew. Plant Disease Reptr. **59**, 516 (1975).

PAUSCH, R. D.: A laboratory evaluation of baits and chemosterilants on the little house fly. J. Econ. Entomol. **62**, 25 (1969).

PECHINEY-PROGIL: Experimentation Brestan: Dosage de résidues sur feuilles de butteraves (juin–octobre 1960). Centre des Recherches Antiparasitaires et Laboratoire de Recherches Chimiques (1960).

PERIES, S. O., V. SIVAPALAN, and D. M. DANTANARAYANA: Methods used by the Rubber-Research Institute of Ceylon for testing watermiscible fungicides for the control of bark-rot of Hevea. Rubber-Research Inst. of Ceylon **38** (3/4), 57 (1962).

PETERSON, D. B., T. ARAKAWA, D. A. G. WALMSLEY, and M. BURTON: Energy-transfer processes in dilute solutions of organometallics in benzene. J. Phys. Chem. **69**, 2880 (1965).

PFEIFFER, P.: Beitrag zur Theorie der Doppelsalze. J. Liebigs Ann. Chem. Pharm. **376**, 310 (1910).

——, and K. SCHNURMANN: Beitrag zur Darstellung von Alkyl- und Aryl-Zinnverbindungen. Ber. Deutsch. Chem. Ges. **37**, 319 (1904).

——, B. FRIEDMANN, R. LEHNHARDT, H. LUFTENSTEINER, R. PRADE, and K. SCHNURMANN: Die Pyridinverbindungen der Zinnhalogenide. Z. anorg. allg. Chem. **71**, 97 (1911).

PICCO, D.: Prove preliminari di attività anticrittogamica del trifenilacetato di stagno. Notiz. Mal. Piante **42**, 33 (1957).

—— Prove con trifenilacetato di stagno. Notiz. Mal Piante **43**, 325 (1958).

—— Controllo comparato della *Cercospora beticola* Sacc. con trifenil-acetato di stagno ed alcuni sali di rame. Notiz. Mal. Piante **72–73**, 3 (1965).

PIEPER, G. R., and J. E. CASIDA: House fly adenosine triphosphatases and their inhibition by insecticidal organotin compounds. J. Econ. Entomol. **58**, 392 (1965).

PIETERS, A. J.: Triphenyl tin hydroxide, a fungicide for the control of *Phytophthora infestans* on potatoes, and some other fungus diseases. British Insecticide and Fungicide Conf., Brighton 1961 (No. 6–9) **2**, 461 (1962).

PILLONI, G.: Complexes of organolead and organotin ions with 1-(2-pyridylazo)-2-naphthol. Anal. Chim. Acta **37**, 497 (1967).

—— Richerche analitiche su composti metallorganici. Corsi e Seminari di Chimica **9**, 98 (1968).

PIVAR, G., L. VALENCIC, and R. VUKCEVIC: Noctuidae caterpillars on sugar beet and influence of Brestan protecting leaves from damage. Agrohemija (Beograd) **4**, 249 (1965).

PLAMADEALA, B.: Atac al ciupercii *Alternaria tenuis* Nees. La Cartof. An. I.C.C.S. Cartoful III, 359 (1972).

PLUM, H.: Emploi des composés organiques de l'étain dans la préservation des matériaux. Chim. Peintures **35** (4), 127 (1972).

POLIS, A.: Über Zinntetraphenyl. Ber. Deutsch. Chem. Ges. **22**, 2915 (1889).

POLKINHORNE, H., and C. G. TAPLEY (to *Albright and Wilson Ltd.*): Organic tin compounds. Brit. Pat. 736 822 (1955); through Chem. Abstr. **50**, 8725 a (1956).

—— —— (to *Albright and Wilson Ltd.*): Improvements relating to the manufacture of organo-tin compounds. Brit. Pat. 761 357 (1953/1956).

POLLER, R. C.: Infra-red spectra and structure of some phenyltin compounds. J. Inorg. Nuclear Chem. **24**, 593 (1962).

—— Coordination in organotin chemistry. J. Organomet. Chem. **3**, 321 (1965).

—— Infra-red spectra of phenyltin halides in the 667–222 cm^{-1} region. Spectrochim. Acta **22**, 935 (1966).

—— Some recent chemistry related to applications of organotin compounds. In J. J.

ZUCKERMAN: Organo-tin compounds, new chemistry and applications. Adv. Chem. Ser. No. 157. Washington, D.C.: Amer. Chem. Soc. (1976).

POLSTER, M., and K. HALAČKA: Beitrag zur hygienisch-toxikologischen Problematik einiger antimikrobiell gebrauchter Organozinnverbindungen. Ernährungsforschung 16, 527 (1971).

POWELL, C. C., and C. LEBEN: Epidemiology and control of Lophodermium needle-cast of Scotch pine in Ohio. Plant Disease Reptr. 57, 515 (1973).

PRASAD, R., and R. P. MOODY: Evaluation of fungicides for control of tree diseases. III. Screening against the poplar canker, Cytospora chrysosperma (Pers.) Fr. under laboratory conditions. Information Rept., Chemical Control Research Inst., Canada, No. CC-X-76 (1975).

PRESTON, P. N., L. H. SUTCLIFFE, and B. TAYLOR: NMR investigation of tetraphenyl-, vinyltriphenyl-, azidotriphenyl- and dihalodiphenyl-derivatives of fourth main group elements. Spectrochim. Acta 28 A, 197 (1972).

PRINCE, R. H.: Reaction mechanisms in organometallic compounds. Part 1. Fast reaction apparatus for the study of kinetics of hydrolysis of organometallic compounds. Trans. Faraday Soc. 54, 838 (1958).

—— Reaction mechanisms in organometallic compounds. Comparative solvolyses of organotin and organosilicon chlorides. J. Chem. Soc. (London), p. 1783 (1959).

Queensland Department of Primary Industries: Ann. Rept. (1972–73).

—— Ann. Rept. (1975–76).

RADWAN, H. S., and A. M. SHAABAN: Efficiency of certain organo-metal compounds against the Egyptian cotton leafworm Spodoptera littoralis Boisd. under field conditions. Z. Angew. Entomol. 74, 362 (1973).

RAEMAKERS, R., and G. PRESTON: Groundnut rust occurrence and foliar disease control in Zambia. Pest Articles and New Summaries (PANS) 23 (2), 166 (1977).

RAGOZZINO, A., and E. MAGLIUOLO: Prove di lotta contro la Cercospora della barbietola. Presented 2. Congresso Internazionale degli Antiparassitari Napoli 15.-17.3. (1965).

RAMASAMY, R., and A. S. MATHAR: Control of rust disease of sunflower in Tamil Nadu. Madras Agr. J. 60, 594 (1973).

RAMSDEN, H. E., and H. DAVIDSON (to Metal and Thermit Corp.): Process of preparing an alkyl- or aryltin compound. USP 2 675 398 (1950/1954).

—— Alkyl- or aryltin compounds. Brit. Pat. 701 714 (1953).

—— Preparation of organotin compounds from organomagnesium chloride complexes. Brit. Pat. 825 039 (1956/1959).

—— Chemical process and product. USP 3 010 979 (1957/1961).

——, A. E. BALINT, W. R. WHITFORD, J. J. WALBURN, and R. CSERR: Arylmagnesium chlorides. Preparations and characterizations. J. Org. Chem. 22, 1202 (1957).

RAPPARINI, G., and A. BENEVELLI: Recherche sur les charactéristiques comparées des nouveaux fongicides employés contre le Cercospora beticola. 7th Internat. Congress Plant Prot., Paris, Sect. B 312/23, p. 237 (1970).

RAUTSCHKE, R., and O. HEINRICH: Lösungsspektralanalyse von Organozinn-Verbindungen in organischen Lösungsmitteln. Spectrochim. Acta 27 B, 143 (1972).

RAZUVAEV, G. A.: Synthesis of organic compounds. Diphenyltin dichloride. Akad. Nauk SSSR, Inst. Org. Khim. Sintezy Org. Soedinenic Sbornik I (1950); through Chem. Abstr. 47, 8004 (1953).

REED, D. K., C. R. CRITTENDEN, and D. J. LYON: Acaricides screened against two rust mites of citrus. J. Econ. Entomol. 60, 668 (1967).

REGUPATHY, A.: Antifeeding properties of two Fentin compounds in the control of Pericallia ricini F. (Arctiidae) and Spodoptera litura Boisd. (Noctuidae) on castor. Madras Agr. J. 60, 32 (1973 a).

—— Antifeeding effects of Fentin compounds against Callosobruchus chinensis L. (Bruchidae: Coleoptera). Madras Agr. J. 60, 586 (1973 b).

—— Assay by different methods of the antifeeding property of Fentin chloride. Entomologia Experimentalis et Applicata 17, 447 (1964).

REINDL, E., and K. BOIDOL (to *Farbwerke Hoechst AG*): Verfahren zur Herstellung von Tetraarylzinn. DBP 1 027 669 (1957/1966).

——, and H. GELBERT (to *Farbwerke Hoechst AG*): Verfahren zur Herstellung von Triphenylzinnchlorid. DBP 1 100 630 (1959/1961).

REUTOV, O. A., O. A. PTITSYNA, and M. F. TURCHINSKII: Paper chromatography of diaryltin compounds and its use in the study of the reaction products of asymmetrical diaryliodonium salts with stannous chloride. Dokl. Akad. Nauk SSSR 139, 146 (1961).

REVERCHON, M.: Microdetermination of elementary tin in organic compounds after mineralisation by the Schöniger method. Tin and its Uses 65, 10 (1965).

RICCOBONI, L.: The electrolytic behaviour of some metallorganic compounds of tin. Atti ist. Veneto Sci., Lettere ed Arti, Classe Sci. Mat. Nat. 96 II, 183 (1937); through Chem. Abstr. 33, 7207 (1939).

RICHARDSON, B. A.: Organic solvent type preservatives. Wood, p. 57, June (1964 a).

—— Wood preservatives. The use of organotin compounds. Tin and its Uses 64, 5 (1964 b).

—— A new technique for the comparative evaluation of some organo-metallic wood preservatives. Internat. Pest Control 10 (1), 14 (1968).

RIJ, J. H. VAN (to *N.V. Philips Gloeilampenfabrieken*): Werkwijze ter bereiding van bis(trifeniltin)oxyden. Holl. Pat. 105 831 (1960/1963).

RITCHIE, L. S., L. A. BERRIOS-DURAN, L. P. FRICK, and I. FOX: Molluscicidal time-concentration relationships of organo-tin compounds. Bull. World Health Org. 31 (1), 147 (1964).

RIVETT, P.: Biological method for the assessment of leaching rates of antifouling compositions. J. Applied Chem. 15, 469 (1965).

RIZK, G. A. M., and H. S. A. RADWAN: Potency and residuality of two antimoulting compounds against cotton leaf worm and bollworms. Z. Angew. Entomol. 79, 136 (1975).

RUDKIEWICZ, F.: Some factors deciding the effectiveness of the control of *Phytophthora infestans* on potato. Nowe Rolnictwo 22 (16), 6 (1973).

RYAN, E. W., and T. KAVANAGH: Comparison of fungicides for control of leaf spot (*Septoria apiicola*) of celery. Ann. Applied Biol. 67, 121 (1971).

——, T. R. GORMLEY, and T. KAVANAGH: Effect of numbers of fungicidal sprays and dates of application on control of celery leaf spot (*Septoria apiicola*) and on chemical residues. Ann. Applied Biol. 72, 63 (1972).

SAHNI, M. L.: *Alternaria* leaf blight on roses and its control through fungicidal sprays. Indian J. Mycology Plant Pathol. 3, 150 (1973; publ. 1974).

SALEM, Y. S., M. I. ABDEL-MEGEED, and Z. H. ZIDAN: Du-Ter as inhibitor to the reproduction of the spiny bollworm, *Earias insulana* (Boisd.) (Lep., Noctuidae). Z. Angew. Entomol. 81 (2), 187 (1976).

SANTO, J. E. (to *Metal and Thermit Corp.*): Organoétains. Belg. Pat. 607.414 (1961).

SCHALLER, C. C., and G. KENKNIGHT: Fungicides reduce the incidence of stem-end blight of *Carya illinoensis*. Plant Disease Reptr. 56, 276 (1972).

SCHENK, N. C., and J. M. CRALL: Watermelon, downy mildew, *Pseudoperonospora cubensis*. Fungicide nematocide tests. Results of 1963. Amer. Phytopathol. Soc. 19, 81 (1963).

Schering AG: Anti-fouling compositions and materials. Brit. Pat. 846.687 (1958/1960).

SCHICKE, P., K. R. APPEL, and L. SCHROEDER: Neue Zinnkomplexverbindungen als Pflanzenschutz-Fungizide. Pflanzenschutzberichte (Wien) 38, 189 (1968).

SCHICKEDANZ, F., and A. HOENICK: Zur Bekämpfung der Blattfleckenkrankheit (*Septoria apii*) an Knollensellerie. Vebega-Mitteilungen 21 (6), 33 (1967).

SCHLOESSER, L. A., F. KOCH, and T. VON BOGEN: Ergebnisse eines Spritzversuches mit "Brestan" gegen *Cercospora beticola* im Schwerbefallsgebiet Oberitaliens. Pflanzenschutz 9 (8), 122 (1957).

SCHMIDT, D., and K. STARKE: Radiochemische Trennung zur Bestimmung von Zinn in Gesteinen durch Neutronenaktivierungsanalyse. Radiochim. Acta 12, 197 (1969).

SCHMIDT, T.: Über Versuche zur Bekämpfung der Brennfleckenkrankheit der Bohne. Der Pflanzenarzt 15 (4), 35 (1962 a).
—— Versuche zur Bekämpfung von Bohnenkrankheiten. I. Brennfleckenkrankheit der Bohne (*Colletotrichum lindemuthianum* Bri. et Cav.). Pflanzenschutz-Berichte 28 (5/6), 65 (1962 b).
—— Ergebnisse dreijähriger Versuche zur Bekämpfung des Blattbrandes der Karotte (Möhrenschwärze). Der Pflanzenarzt 18 (9), 99 (1965).
SCHMIDT, U., K. KABITZKE, K. MARKAU, and W. P. NEUMANN: Isolierung und Nachweis dreibindiger Zinnradikale. Ber. Deutsch. Chem. Ges. 98, 3827 (1965).
SCHMITT-STRECKER, S.: Unpublished Results (1976).
SCHMITZ-DUMONT, O.: Über das Triphenylzinnoxid (Bistriphenylzinnäther). Z. anorg. allg. Chem. 248, 289 (1941).
SCHNEIDER, C. L.: Sugar beet leaf spot (*Cercospora beticola*). Fungicide nematocide tests. Amer. Phytopathol. Soc. 24, 76 (1968).
——, and H. S. POTTER: Tests with soil treatments and crown sprays to control Rhizoctonia crown and root rot of sugarbeet. J. Amer. Sugar Beet Technol. 8 (1), 45 (1974).
SCHOLZ, J.: Antidot-Brestan. Laboratorium f. Gewerbe- und Arzneimitteltoxikologie der Hoechst AG, Bericht v. 25.11. (1959).
—— Akute orale Toxizität von Triphenylzinnacetat an Hund und Katze. Laboratorium f. Gewerbe- und Arzneimitteltoxikologie der Hoechst AG, Bericht v. 24.10. (1960 a).
—— Akute orale Toxizität von Diphenylzinnoxid, Tetraphenylzinn und Diphenylzinndiacetat an Ratte. Laboratorium f. Gewerbe- und Arzneimitteltoxikologie der Hoechst AG, Bericht v. 27.10. (1960 b).
—— Vergleichende Untersuchung über die akute orale Toxizität von Triphenylzinnacetat mit formuliertem Produkt Brestan conc. und Brestan 60 an Ratte. Laboratorium f. Gewerbe- und Arzneimitteltoxiztät der Hoechst AG, Bericht v. 9.11. (1962).
—— Toxikologische Untersuchung von Triphenylzinnchlorid und Hoe. 2840 (wettable powder) an Ratten. Laboratorium f. Gewerbe- und Arzneimitteltoxikologie der Hoechst AG, Bericht v. 18.12. (1963).
—— Toxikologische Prüfung von Triphenylzinnacetat an Ratte, Meerschweinchen und Kaninchen. Laboratorium f. Gewerbe- und Arzneimitteltoxikologie der Hoechst AG, Bericht v. 30.11. (1964).
—— Chronic toxicity testing. Nature 207, 870 (1965).
——, and C. BAEDER: Triphenylzinnacetat—Bericht über 4-Monate-Fütterungsversuche an Ratten. Laboratorium f. Gewerbe- u. Arzneimitteltoxikologie der Hoechst AG, Bericht v. 11.12. (1968).
—— —— Fortpflanzungsversuche bei Verfütterung von Triphenylzinnacetat an Ratten—3-Generationsversuch. Laboratorium f. Gewerbe- u. Arzneimitteltoxikologie der Hoechst AG, Bericht v. 6.8. (1970).
——, and R. BRUNK: Triphenylzinnacetat—chronische orale Toxizität an Hunden (120-Tage-Fütterungsversuch). Laboratorium f. Gewerbe- u. Arzneimitteltoxikologie der Hoechst AG, Bericht v. 29.8. (1968).
——, and H. HOLLANDER: Untersuchungen uber die akute orale Toxizität des formulierten Produktes Brestan 60 an männlichen und weiblichen Ratten. Laboratorium f. Gewerbe- u. Arzneimitteltoxikologie der Hoechst AG, Bericht v. 6.8. (1973).
——, and W. WEIGAND: Toxikologische Prüfung von Hoe. 2840 (Brestan 60 — Triphenylzinnacetat + Maneb — wettable powder) an Hunden und Kaninchen. Laboratorium f. Gewerbe- und Arzneimitteltoxikologie der Hoechst AG, Bericht v. 20.7. (1964).
—— —— Toxikologische Prüfung von Triphenylzinnhydroxid. Laboratorium für Gewerbe- und Arzneimitteltoxikologie der Farbwerke Hoechst AG, Bericht v. 8.1. (1965).
—— —— Triphenylzinnacetat- 4-Monate-Fütterungsversuche an Meerschweinchen.

Laboratorium f. Gewerbe- und Arzneimitteltoxikologie der Hoechst AG, Bericht v. 23.7. (1968).
—— —— Triphenylzinnhydroxyd, Vers. Nr. 2117 (akute orale Toxizität an Ratten). Bericht v. 7.7. (1969 a).
—— —— Bistriphenylzinnoxyd, Vers. Nr. 2118 (orale akute Toxizität an Ratten). Bericht v. 7.7. (1969 b).
SCHROEDER, L., K. THOMAS, and D. JERCHEL: 5-Coordination—A main characteristic of triorganostannanes. Proc. 9th Internat. Conf. Coord. Chem., St. Moritz (1966).
SCHROEDER, W. T.: Tomato, early blight (Alternaria solani), anthracnose (Colletotrichum, spp.). Fungicide nematocide tests. Amer. Phytopathol. Soc. 19, 85 (1963).
SCHUPP: Spritzversuche zur Bekämpfung der Sellerie-Blattfleckenkrankheit (Septoria apii). Der Badische Obst- u. Gartenbauer (No. 11), p. 189 (1957).
SCHWEDT, G., and H. A. RUESSEL: Gas-Chromatographie von Tl, Se, Te, Hg, As, Sb, Bi, Sn als Phenylverbindungen. Z. anal. Chem. 264, 301 (1973).
SEBBEL: Der Obstbaumkrebs wurde zum Staatsfeind Nr. 1. Mitt. f.d. Schleswig-Holst. Erwerbsobstbau 5, 57 (1975).
SEIFFER, E. A., and H. SCHOOF: Tests of 15 experimental molluscicides against Australorbis glabratus. U.S. Public Health Rept. 82, 833 (1967).
SELWYN, M. J.: Triorganotin compounds as ionophores and inhibitors of ion translocating ATPases. In J. J. ZUCKERMAN: Organotin compounds: New chemistry and applications; Adv. Chem. Ser. 157, 204 (1976).
——, A. P. DAWSON, M. STOCKDALE, and N. GAINS: Chloride-hydroxide exchange across mitochondrial erythrocyte and artificial lipid membranes mediated by trialkyl- and triphenyltin compounds. Eur. J. Biochem. 14, 120 (1970).
SENEVIRATNE, S. N. DE S.: Spraying trials on potato blight control with organo-tin and other fungicides. Trop. Agr. (Ceylon) 126 (1), 15 (1970).
SHAABAN, A. M., H. I. YOUSSEF, A. A. KAMEL, and M. R. ABULGHAR: Effect of certain chemosterilants on the larvae of the greasy cutworm, Agrotis ipsilon Rott. Z. Angew. Entomol. 78, 386 (1975).
SINGH, O. V., V. K. AGARWAL, and R. A. SINGH: Effect of fungicidal sprays on the quantum of seed-borne infection and germination of rice seeds. Oryza 9 (2), 103 (1972).
SINGH, R. A., and V. V. SHARMA: Brown spot disease (Helminthosporicum oryzae) on rice (Oryza sativa). Fungicide nematocide tests. Amer. Phytopathol. Soc. 28, 100 (1972).
—— —— Systemic fungicidal activity of triphenyltin chloride in rice seedlings. Indian J. Mycol. Plant Pathol. 3, 141 (1973; publ. 1974).
SITTERLY, W. R.: Cucumber, downey mildew (Pseudoperonospora cubensis), powdery mildew (Erysiphe cichoracearum), gummy stem blight (Mycosphaerella melonis). Fungicide nematocide tests. Amer. Phytopathol. Soc. 19, 74 (1963 a).
—— Cucumber, fruit rot, Pythium and Rhizoctonia spp. Fungicide nematocide tests. Amer. Phytopathol. Soc. 19, 75 (1963 b).
SMITH, A. L.: Low frequency vibrational spectra of group IV-A phenyl compounds. Spectrochim. Acta 24 A, 695 (1968).
SNOW, R. L.: Histological studies of recovery in the testes of albino rats treated with two insect chemosterilants. Thesis, Clemson Univ. (1970).
SODERQUIST, C. J., and D. G. CROSBY: Determination of triphenyltin hydroxide and its degradation products in water. Anal. Chem. 50, 1435 (1978).
SOLEL, Z.: A broad-range pesticidal effect of Brestan. Israel J. Agr. Res. 14 (1), 31 (1964).
—— The systemic fungicidal effect of benzimidazole derivatives and thiophanate against Cercospora leaf spot of sugarbeet. Phytopathol. 60, 1186 (1970 a).
—— The performance of benzimidazole fungicides in the control of Cercospora leaf spot of sugarbeet. J. Amer. Soc. Sugar Beet Technol. 16, 93 (1970 b).

—— Vapour phase action of some foliar fungicides. Pest. Sci. **2**, 126 (1971 a).

—— The systemic fungicidal activity of triphenyl-tin acetate against *Cercospora beticola* on sugarbeet. Phytopathol. **61**, 738 (1971 b).

——, and G. MINZ: A new compound for the control of *Cercospora* of sugar beet. Hassadeh **43**, 409 (1963).

SOSNA, Z., and W. RUDNA: Investigations on the effectiveness of new fungicides in the control of potato blight (*Phytophthora infestans* De Bary). Biuletyn Instytutu Ochrony Roslin **54**, 409 (1972).

SPARMANN, H. W. (to *Schering AG*): Schädlichen Bewuchs verhindernder Zusatz für Unterwasser-Anstrichmittel und -Baustoffe. DRP 1 042 795 (1957/1963).

—— (to *Schering AG*): Method of preventing marine growth, and antifouling compositions useful for said method. USP 2 970 923 (1958/1961).

SPURR, H. W.: Tobacco leaf injury induced by coincident applications of a sucker control agent and a fungicide. Plant Disease Reptr. **56**, 338 (1972).

——, and R. E. WELTY: Incidence of tobacco leaf microflora in relation to brown spot disease and fungicidal treatment. Phytopathol. **62**, 916 (1972).

SRIVASTAVA, T. N.: The infrared and ^{35}Cl NQR spectra of triphenyltin chloride. J. Organomet. Chem. **10**, 373 (1967).

——, and S. K. TANDON: Some triphenyl tin derivatives. Indian J. Applied Chem. **26**, 172 (1963).

—— —— Antimicrobial activity of some triphenyl tin compounds. Indian J. Applied Chem. **27**, 116 (1964).

STADER SAATZUCHT: see ANONYMOUS (1968).

STALLKNECHT, G. F., and L. CALPOUZOS: Fungicidal action of triphenyl tin hydroxide toward *Cercospora beticola* on sugar beet leaves. Phytopathol. **58**, 788 (1968).

STEGEMANN, H. B., W. UBER, and K. SCHEFFLER: Analytik von Di- und Triorgano-Zinn-Verbindungen durch Elektronen-Spin-Resonanz. Z. Anal. Chem. **286**, 59 (1977).

STEINER, H.: Über die Eignung verschiedener Pflanzenschutzmittel für eine schonende Spritzfolge im Obstbau. Z. Angew. Entomol. **47** (1), 79 (1960/61).

STERNLICHT, M.: Trials in the control of the citrus bud mite, *Aceria sheldoni* (Ewing) in Israel. J. Agr. Res. **16** (3), 115 (1966); through Rev. Applied Entomol. **57** (4), 225 (1969).

STOCKDALE, M., A. P. DAWSON, and M. J. SELWYN: Effects of trialykltin and triphenyltin compounds on mitochondrial respiration. Eur. J. Biochem. **15**, 342 (1970).

STOLZE, K. V.: Bestehen Zusammenhänge zwischen Krautfäulebekämpfung und dem Befall der Kartoffelknollen durch Braunfäule? Landwirtsch. Blatt Weser-Ems 11/12.3. (1957); through Z. Pflanzenkrankh. **65** (No. 7) (1958).

STONER, H. B.: Toxicity of triphenyltin. Brit. J. Ind. Med. **23**, 222 (1966).

——, and D. F. HEATH: The cumulative action of triphenyltin. Food Cosmetic Toxicol. **5**, 285 (1967).

——, J. M. BARNES, and J. I. DUFF: Studies on the toxicity of alkyl tin compounds. Brit. J. Pharmacol. **10**, 16 (1955).

STROHMEIER, W., and K. MILTENBERGER: Notiz über die Löslichkeiten von Tetraphenylmethan, Tetraphenyl-silicium, -germanium, -zinn und -blei in organischen Lösungsmitteln. Ber. Deutsch. Chem. Ges. **91**, 1357 (1958).

STRUFE, R.: Problems and results of residue studies after application of molluscicides. Residue Reviews **24**, 79 (1968).

SUESS, A., and CH. EBEN: Das Verhalten von Triphenylzinnacetat (TPZA) im Boden und die Aufnahme durch Pflanzen aus behandelten Böden. Z. Pflanzenkrankh. Pflanzenschutz **80** (5), 288 (1973).

SYROWATKA, T.: Investigation of the action of organotin compounds on the oxidative phosphorylation and permeability of membrane of rat liver mitochondria. Roczn. Zak. Hig. (Warsz.) **20**, 717 (1969).

—— Influence of fungicides on cellular energetic processes. Roczn. Zak. Hig. (Warsz.) **21**, 105 (1970).

TAGLIAVINI, G., and P. ZANELLA: Potentiometric titrations of organotin chloride with tetraphenylarsonium chloride in acetonitrile media. Anal. Chim. Acta **40**, 33 (1968).

TAMBORSKI, C., E. J. SOLOSKI, and S. M. DEE: Pentafluorophenyl organometallic compounds of group IV elements. J. Organomet. Chem. **4**, 446 (1965).

TAMURA, H.: Studies on the fungitoxicity of organo-tin compounds against *Piricularia oryzae* causing rice blast disease. Nogyo Gijutsu Kenkyusho Hokoku Byori Konchu **18**, 135 (1965 a).

—— Effect of organotin fungicides applied for the control of rice blast, and rice sheath blight. Nogyo Gijutsu Kenkyusho Hokoku Byori Konchu **19**, 47 (1965 b).

TAUBERGER, G.: Tierexperimentelle Analyse der Triphenylzinnwirkung nach intravenöser Injektion. Med. Exp. **9**, 393 (1963).

TER HORST, K.: Enige Proeven ter Bestrijding van de *Cercospora*-Bladvlekkenziekte van de Aardnoot. De Surinaamse Landbouw. **9** (4), 103 (1961).

THOMAS, A. B., and E. G. ROCHOW: The conductimetric behaviour in nonaqueous solvents of organometallic chlorides from group IV B. J. Inorg. Nuclear Chem. **4**, 205 (1957 a).

—— —— Conductance studies of organometallic chlorides of group IV B and of hydrogen chloride in N,N-dimethylformamide. Some observations about the purification of the solvent. J. Amer. Chem. Soc. **79**, 1843 (1957 b).

THOMAS, B., and H. L. TANN: Pesticide residues in foodstuffs in Great Britain. XV. Triphenyltin residues in potatoes. Pest. Sci. **2**, 45 (1971).

THOMAS, H. A.: Quantitative bioassay used to study inhibitory effects of two organotins in Pales weevil feeding. U.S. Dept. Agr., Forest Service Res. Note SE-118; SE Forest Expt. Station, Asheville, NC, Aug. (1969).

THURSTON, H. D., R. O. BARRIGA, and L. E. HEIDRICK: Late blight of potato (*Phytophthora infestans*). Fungicide nematocide tests. Results for 1960. Amer. Phytopathol. Soc. **16**, 61 (1960).

TICHY, V.: Organotin compounds. Industrial application and preparation. Chem. Listy **55**, 154 (1961).

TIGNER, J. R., and J. F. BESSER: A quantitative method for evaluating chemicals as rodent repellents on packaging materials. J. Agr. Food Chem. **10**, 484 (1962).

TIL, H. P., and V. J. FERON: Triphenyltin-hydroxide range-finding test with dogs. (Unpublished TNO-Report 1965, submitted by PHILIPS DUPHAR, No. R-1994). Quoted by FAO/WHO-Evaluations (1971).

—— —— Chronic (two-year) toxicity study with triphenyltin-hydroxide (TPTH) in beagle dogs. (Unpublished TNO-Report 1968, submitted by PHILIPS DUPHAR, No. 2717). Quoted by FAO/WHO-Evaluations (1971).

—— ——, and A. P. DE GROOT: Reproduction study with triphenyltin-hydroxide in three generations of rats. (Unpublished TNO-Report 1967, submitted by PHILIPS DUPHAR, No. 2476). Quoted by FAO/WHO-Evaluations (1971).

—— —— —— Observations on a possible effect of TPTH on testicular development in rats. (Unpublished TNO-Report 1968, submitted by PHILIPS DUPHAR, No. R-2620). Quoted by RAO/WHO Evaluations (1971).

—— —— —— Chronic toxicity study with triphenyltin-hydroxide in rats for two years. (Unpublished TNO-Report 1970, submitted by PHILIPS DUPHAR, No. R-3138). Quoted by FAO/WHO-Evaluations (1971).

TISDALE, W. H. (to *E. I. Du Pont de Nemours and Co.*): Antifouling Marine paints and enamels. Brit. Pat. 578 312 (1943/1946).

TODERI, G., and G. C. VIROLI: Influenza di trattamenti anticercosporici con prodotti sperimentali, eseguiti in numero diverso e in epoche diverse, sulla produzione della bietola da zucchero. Rivista di Agronomia **4** (1-2), 35 (1970).

TONGE, B. L.: The gas chromatographic analysis of butyl-, octyl-, and phenyl-tin halides. J. Chromatogr. **19**, 182 (1965).

TOSCANI, H. A., and T. A. MARFURT: Quoted by ASCHER (1969 b).

T. P. R. I. (*Tropical Pesticides Research Institute*): see ANONYMOUS (1971 b).

TRINCI, A. P. J., and K. GULL: Effect of actidione and triphenyltin acetate on the kinetics of fungal growth. J. Gen. Microbiol. **60** (Part 3), 287 (1970).

UEDA, K., and K. IIJIMA: Acute toxicity of Brestan, technical (triphenyltin-acetate), for mice. Unpublished rept. Laboratory of Hygienics, Tokyo Dental College (1961).

—— ——, and S. YOSHIDA: Acute oral toxicity of organotin compounds for mice. Tokyo Dental College (1961).

ULERY, H. E.: Cathodic synthesis of tetraalkyltin compounds. J. Electrochem. Soc. 119, 1474 (1972).

VANACHAYANGKUL, A., and M. D. MORRIS: Polarographic reduction of triphenyltin fluoride. Anal. Letters 1, 885 (1968).

VASUNDHARA, T. S., and D. B. PARISHAR: Microdetermination of organotin compounds by TLC and spectrophotometry using haematoxylin as reagent. Z. anal. Chem. 294, 408 (1979).

VENTAKATARAMIAH, G. H.: Assay of fungicides for the control of brown eye spot disease of arabica coffee seedlings. Pesticides 5 (10), 16 (1971).

VENTURA, E., and J. J. HERVÉ: L'efficacité de triphénylétain sur le mildiou et ses effets sur la végétation et le rendement de la pomme de terre. Phytiatrie-Phytopharmacie 11, 27 (1962).

VERDONCK, L., and G. P. VAN DER KELEN: The proton magnetic resonance spectra of phenyltin chlorides. Bull. Soc. Chim. Belg. 74, 361 (1965).

VERMA, J. P., and P. C. JAIN: Efficacy of triphenyl tinacetate as a crop protectant for some lepidopterus pests. Indian J. Agr. Sci. 42, 529 (1972).

——, A. R. UPADHYAY, and K. C. AGRAWAL: Studies on chemical control of blast disease of rice with application of fungicides and antibiotics. Riso 25 (2), 141 (1976).

VERNON, F.: The fluorimetric determination of triphenyltin compounds. Anal. Chim. Acta 71, 192 (1974).

VERSCHUUREN, H. G., R. KROES, H. H. VINCK, and G. J. VAN ESCH: Short-term toxicity studies with triphenyltin compounds in rats and guinea-pigs. Food Cosmetic Toxicol. 4, 35 (1966).

——, E. J. RUITENBERG, F. PEETOM, P. W. HELLEMAN, and G. J. VAN ESCH: Influence of triphenyltin acetate on lymphatic tissue and immune responses in guinea pigs. Toxicol. Applied Pharmacol. 16, 400 (1970).

VETTER, A., and W. VOELKER: Bericht über die Feldversuchsergebnisse im Jahrgang 1965. Arbeitsgemeinschaft zur Bekämpfung der Zuckerrübenkrankheiten, Regensburg, p. 46 (1965).

VILLERS, J. P. DE, and J. G. MACKENZIE: Organo-tin and Organo-lead molluscicides. Organisation Mondiale de la Santé, Mol/Inf/13, p. 63 (1963).

VIND, H. P., and H. HOCHMAN: An evaluation of organotin compounds as preservatives for marine timbers. Proc. Amer. Wood Preservers Assoc. 58, 170 (1962).

—— —— Organotin compounds as preservatives for marine timbers. Tin and its Uses 57, 10 (1963).

VINE, B. H., and P. A. VINE: Some practical implications from fungicides trials 1967–1970. Kenya Coffee 7, 181 (1971).

VOGEL, J., and J. DESHUSSES: Dosage polarographique des résidus d'acétate de triphénylétain (Brestan) sur les légumes. Helv. Chim. Acta 47, 181 (1964).

V'RBANOV, V.: Chemical control of Cercosporosis of sugar beet. Rastitelna Zashchita 22 (12), 28 (1974).

WADSWORTH, D. F., H. C. YOUNG, and R. E. McCOY: Progress-report: Peanut disease-research—1966—Processed series P., p. 559, April (1967).

WALTERS, P. J.: Susceptibility of three Stethorus spp. (Coleoptera: Coccinellidae) to select chemicals used in N.S.W. apple orchards. J. Australian Entomol. Soc. 15 (1), 49 (1976).

WARDELL, J. L.: Equilibria between organotin trichlorides and nitrogen donors in ether and the comparison of their acceptor strengths with stannic chloride, J. Organomet. Chem. 9, 89 (1967).

WATLING-PAYNE, A. S., and M. J. SELWYN: Inhibition and uncoupling of photophosphorylation in isolated chloroplasts by organotin, organomercury and diphenyliodonium compounds. Biochem. J. **142**, 65 (1974).

WEIGAND, W.: Akute orale Toxizität von Triphenylzinnhydroxid und Triphenylzinnacetat an Meerschweinchen. Laboratorium f. Gewerbe- und Arzneimitteltoxikologie der Hoechst AG, Bericht v. 30.8. (1962).

—— Special investigations of the toxicity and metabolic balance of triphenyltin acetate. Presented 8th Internat. Plant Prot. Congress, Moscow (1975).

WEIGAND, W., and H. KIEF: Triphenylzinnacetat—2-Jahres-Fütterungsversuch an Meerschweinchen. Laboratorium f. Gewerbe- und Arzneimitteltoxikologie der Hoechst AG, Bericht v. 25.5. (1965).

WELLNER, H., and A. VETTER: Ist eine gezielte Bekämpfung der Blattfleckenkrankheit schon möglich? Zucker **15**, (13) (1962).

WELTZIEN, H. C.: Über die Wirkungen von Triphenylzinnacetat auf den echten Mehltau der Zuckerrüben, *Erysiphe betae* (Van.) Weltzien. Zucker **21** (9), 241 (1968).

WHITE, T. P.: Über die Wirkungen des Zinns auf den thierischen Organismus. Arch. Expt. Pathol. Pharmakol. **13**, 53 (1881).

—— The action of tin on the animal organism. Pharm. J. **17**, 166 (1886).

WHITESIDES, G. M., J. G. SELGESTAD, ST.P. THOMAS, D. W. ANDREWS, B. A. MORRISON, E. J. PANEK, and J. SAN FILIPPO: (p → d) π Bonding in phenylchloro-tin, -germane, and -silane. J. Organomet. Chem. **22**, 365 (1970).

WILLIAMS, D. J., and J. W. PRICE: Paper chromatography of some organotin compounds. Analyst **85**, 579 (1960).

—— —— Paper chromatography of some organotin compounds. Part II. Reversed-phase systems. Analyst **89**, 220 (1964).

WILLPATTE-STEINERT, L., and J. NASIELSKI: The photochemistry of aromatic compounds. IV. Photochemical behaviour of hexaphenylditin. J. Organomet. Chem. **24**, 113 (1970).

WIT, S. L., and K. L. VAN LIER: Residu's van trifeniltinacetaat op bladselderij. Mededelingen betreffende de Volksgezondheid, No. 1 (1960).

WITTENBERG, D. (to *Badische Anilin- und Sodafabrik AG*): Verfahren zur Herstellung von Arylzinnverbindungen, die gegebenenfalls Halogen enthalten. DBP 1 124 947 (1960/1965).

—— Metallorganische Synthesen in der Arylreihe. J. Liebigs Ann. Chem. Pharm. **654**, 23 (1962).

WOGGON, H., and D. JEHLE: Zur inversvoltammetrischen Bestimmung von bioziden Organozinn-Verbindungen. Die Nahrung **17**, 739 (1973).

WOIDICH, H., and W. PFANNHAUSER: Vergleich dreier Methoden zur Bestimmung von Zinn in Konserven. Z. Lebensm. -Unters. Forschg. **151**, 114 (1973).

WOLFENBARGER, D. A., E. E. GUERRA, and W. L. LOWRY: Effects of organometallic compounds on lepidoptera. J. Econ. Entomol. **61**, 78 (1968).

WULF, R. G., and K. H. BYINGTON: On the structure-activity relationships and mechanism of organotin induced nonenergy dependent swelling of liver mitochondria. Arch. Biochem. Biophys. **167**, 176 (1975).

WYSONG, D. S., M. L. SCHUSTER, R. E. FINKNER, and E. D. KERR: Chemical control of *Cercospora* leaf spot of sugar beets in Nebraska, 1965. J. Amer. Sugar Beet Technol. **15** (3), 221 (1968).

YONEDA, H., L. O. C. REZENDE, and J. C. PIEDADE: Resíduos de fungicidas a base de estanho, em tomates. Biologico **32**, 275 (1966).

ZAHIR, S.: Krautfäule-Bekämpfung. Der Kartoffelbau **6**, 170 (1968).

ZAKHARKIN, L. I., O. YU. OKHLOBYSTIN, and B. N. STRUNIN: Application of organomagnesium compounds to the synthesis of organic derivatives of elements of groups II–V in nonetheral medium. Izvest. Akad. Nauk SSSR, Otd. Khim. Nauk, p. 2002 (1962).

ZANELLA, P., and G. TAGLIAVINI: The behaviour of organotin chlorides as chloride-ion acceptors in non-aqueous solvents. J. Organomet. Chem. **12**, 355 (1968).

ZATTLER, F., and H. LIEBL: Abschliessende Versuche im Jahre 1969 zur Bekämpfung der Frühinfektion des Hopfens durch die *Peronospora*. Hopfenrundschau **21**, 131 (1970).

ZECK, W.: Untersuchungen über den Einfluss des Kupfers auf die Kartoffelpflanze und über seine fungizide Wirksamkeit im Pflanzeninneren. Phytopathol. Z. **27**, 353 (1956).

—— Untersuchungen über Infektionsverhältnisse bei der Phytophthora-Knollenfäule und über deren Beeinflussbarkeit durch Kupferdüngung. Phytopathol. Z. **29**, 233 (1957).

ZEDLER, R. J.: Organotins as industrial biochemicals. Tin and its Uses **53**, 7 (1961).

ZIMMER, H., and H. W. SPARMANN: Organozinn-Verbindungen des Fluorens und Indens. Ber. Deutsch. Chem. Ges. **87**, 645 (1954).

Manuscript received April 31, 1980; accepted May 19, 1980.

Subject Index

———— compounds, phytotoxicity and weather 165

———— compounds, phytotoxicity to celery 167

———— compounds, phytotoxicity to potatoes 166 ff.

———— compounds, phytotoxicity to rice 168

———— compounds, phytotoxicity to sugarbeets 167

———— compounds, polarography 17

———— compounds, purification 11

———— compounds, purification of waste water 220

———— compounds, purity tests 191

———— compounds, repellent effects 147 ff.

———— compounds, residues in animal food 190

———— compounds, residues in beans 175, 176

———— compounds, residues in cacao 176, 186

———— compounds, residues in carrots 184 ff., 187

———— compounds, residues in celery 174, 175, 182 ff., 188

———— compounds, residues in coffee 186, 188

———— compounds, residues in human food 190, 191

———— compounds, residues in milk 190

———— compounds, residues in peanuts 185, 186, 188

———— compounds, residues in pecans 185, 187

———— compounds, residues in potatoes 174, 175, 180 ff., 187

———— compounds, residues in rice 176, 185, 188

———— compounds, residues in sugarbeets 179 ff., 187

———— compounds, residues in tomatoes 186

———— compounds, residues on plants 179 ff.

———— compounds, secretion 51 ff.

———— compounds, skin tests 36 ff.

———— compounds, sterilization effects 143

———— compounds, stimulation of plant growth 176

———— compounds, storage stability 24

———— compounds, synergism 163 ff.

———— compounds, synthesis 4 ff., see also specific compounds

———— compounds, systemic properties 169 ff.

———— compounds, thermal stability 22

———— compounds, titrimetric analysis 191

———— compounds, tolerances 186 ff.

———— compounds, tolerances, various countries 192, 193

———— compounds, tolerated dose 33

———— compounds, toxicity to fish and other aquatic life 115, 220 ff.

———— compounds, use on apples 125 ff.

———— compounds, use on beans 130 ff.

———— compounds, use on cacao 116 ff.

———— compounds, use on carrots 106 ff.

———— compounds, use on celery 103 ff.

———— compounds, use on coffee 118

———— compounds, use on hops 120 ff.

———— compounds, use on marine organisms 138

———— compounds, use on onions 106 ff.

———— compounds, use on peanuts 123

———— compounds, use on pecans 107 ff.

———— compounds, use on potatoes 73 ff.

———— compounds, use on rice 110 ff.

———— compounds, use on snails 132 ff.

———— compounds, use on sugarbeets 89 ff.

———— compounds, uses in agriculture 73 ff.

———— compounds, waiting periods 186 ff.

Triphenyltin fluoride as bactericide 72

Triphenyltin hydroxide 1 ff., see also Triphenyltin compounds

———— hydroxide and hexaphenyldistannoxane mixtures, analysis 194

———— hydroxide as bactericide 72

———— hydroxide, effect on reproduction 44

———— hydroxide, effects on mitochondria, chloroplasts, erythrocytes, ferments 45

———— hydroxide, efficiency spectrum 70 ff.

———— hydroxide, long-term toxicity 42, 43

———— hydroxide, photolysis 31

———— hydroxide, short-term toxicity 40, 41, 43

Triphenyltin oxalate 83

Triphenyltin oxide 1 ff., see also Triphenyltin compounds

Triphenyltin-phosphorus compounds, fungicidal activity 69

Triphenyltin poisoning, treatment 51

———— residues, cleanup 204

INFORMATION FOR AUTHORS

RESIDUE REVIEWS

(A BOOK SERIES CONCERNED WITH RESIDUES OF PESTICIDES AND OTHER CONTAMINANTS IN THE TOTAL ENVIRONMENT)

Edited by

Francis A. Gunther

Published by

Springer-Verlag New York · Heidelberg · Berlin

The original (ribbon) copy and one good xerox or other copy of the manuscript, complete with figures and tables, are required. Manuscripts will normally be published in the order in which they are received, reviewed, and accepted. They should be sent to the editor:

Professor Francis A. Gunther
Department of Entomology
University of California
Riverside, California 92502
Telephone: (714) 787-5804/5810 (office)
(714) 688-6666 (home)

1. Manuscript

The manuscript, in English, should be typewritten, double-spaced throughout, on one side of 8½ x 11 inch blank white paper, with at least one-inch margins. The first page of the manuscript should start with the title of the manuscript, name(s) of author(s), with author affiliation(s) as first-page starred footnotes, and "Contents" section. Pages should be numbered consecutively in arabic numerals, including those bearing figures and tables only. In titles, in-text outline headings and subheadings, figure legends, and table headings only the initial word, proper names, and universally capitalized words should be capitalized.

Footnotes should be inserted in text and numbered consecutively in the text using arabic numerals.

Tables should be typed on separate sheets and numbered consecutively within the text in roman numerals; they should bear a descriptive heading, in lower case, which is underscored with one line and which starts after the word "Table" and the appropriate roman numeral; *footnotes in tables* should be designated consecutively within a table by the lower-case alphabet. *Figures* (including photographs, graphs, and line drawings) should be numbered consecutively within the text in arabic numerals; each figure should be affixed to a separate page bearing a legend (below the figure) in lower case starting with the term "Fig." and a number.

2. Summary

A concise but informative summary (double-spaced) must conclude the text of each manuscript; it should summarize the significant content and major conclusions presented. It must not be longer than two 8½ × 11 inch pages of double-spaced typing. As a summary, it should be more informative than the usual abstract.

3. References

All papers, books, and other work cited in the text must be included in a "References" section (also double-spaced) at the end of the manuscript: If comprehensive papers on the same subject have been published, they should be cited but only for exceptional reasons should the bibliographic citations extend farther back than to these papers.

The references used *in the text* should consist of the COMPLETELY CAPITALIZED author's or authors' last name(s) where one or two authors are concerned; should there be more than two authors, only the first is named and *"et al."* is added. The publication year in parentheses should follow the name. If more than one paper by one author published in the same year is cited, the letters a, b, c, etc., should follow the year, e.g., "MEIER (1958 a) found . . .", or "This method is nonspecific (MEIER 1958 a)."

In the References section, the papers cited should appear in alphabetical order according to the last name of the first author; if more than one paper by an author or authors published in the same year is cited, the papers should be listed according to the year of publication followed by a, b, c, etc., as necessary. Papers published in periodicals should be cited with COMPLETELY CAPITALIZED names and initials of all authors, together with the *full title of the paper and preferably in its original language,* title of the periodical (abbreviated in accordance with *Chemical Abstracts'* "List of Periodicals Abstracted"), number of the volume (wavy underlined), initial page, and the year in parentheses. References to unpublished papers that have been submitted for publication should be cited in the same manner as other papers except the abbreviated journal name is followed by the words "In press" or "Accepted for publication" and the year in parentheses; personal communications are to be cited similarly.

In text and in the References section, citation of *governmental agencies, educational and research institutions and foundations, professional associations, and industrial companies* should consist of the full name as used by the organization completely underscored with one line and with initial capital letters only, followed by the appropriate reference information as specified above.

Examples:

EDWARDS, C. A., and E. B. DENNIS: Some effects of aldrin and DDT on the soil fauna of arable land. Nature 188, 767(1960).

GUNTHER, F. A., J. H. BARKLEY, and W. E. WESTLAKE: Worker environment research. II. Sampling and processing techniques for determining dislodgable pesticide residues on leaf surfaces. Bull. Environ. Contam. Toxicol. Accepted for publication (1974).

HESSLER, W.: Eine einfache Nachweismethode für Paraffin in Wachsgemischen. II. Mitt. Fette, Seifen, Anstrichmittel 58, 602(1956).

MELZER, H.: The qualitative and quantitative colorimetric determination of captan. Nachrbl. deut. Pflanzenschutzdienst 14, 193(1960).

Shell Chemical Co.: Letter to EPA's "Hazardous Materials Advisory Committee," Oct. 28(1971).

U.S. Environmental Protection Agency: Proposed toxicology guidelines. Fed. Register 37(183), 19383(1972).

Books should be cited with COMPLETELY CAPITALIZED name(s) and initials of the author(s), full title, edition or volume, page number(s), place of publication, publisher, and year of publication in parentheses.

Examples:

BEVENUE, A.: Gas chromatography. In G. Zweig (ed.): Analytical methods for pesticides, plant growth regulators, and food additives. Vol. I, p. 189. New York: Academic Press (1963).

DORMAL, S., and G. THOMAS: Répertoire toxicologique des pesticides, p. 48. Gembloux: J. Duculot (1960).

HARTE, C.: Physiologie der Organbildung, Genetik der Samenpflanzen. In:

Fortschritte der Botanik. Vol. 22, p. 315. Berlin-Göttingen-Heidelberg: Springer (1960).

METCALF, R. L.: Organic insecticides, their chemistry and mode of action. 2 ed., p. 51. New York-London: Interscience (1961).

4. Illustrations

Illustrations of any kind may be included only when indispensable for the comprehension of text; they should not be used in place of concise, clear explanations in text. Schematic line drawings must be drawn carefully and clearly. For other illustrations, clearly defined black-and-white glossy photographic prints are required. Should precisely placed indication darts (arrows) or letters be required on a photograph or other type of illustration, they should be marked neatly with a soft pencil on a duplicate copy or on an overlay, with the end of each dart (arrow) indicated by a fine pinprick; darts and lettering will be transferred to the illustrations by the publisher.

Photographs should be not less than five × seven inches in size. Unimportant and indistinct strips or areas on the edges of photographs should be marked on the back of the glossy print (pattern) with pencilled down-strokes, in order that the reproduction surface will not be unnecessarily large; alterations of photographs in galley-proof stage are not permitted. *Each photograph or other illustration should be marked on the back, distinctly but lightly, with soft pencil with first author's name, figure number, manuscript page number, and the side which is the top.*

If illustrations from published books or periodicals are used, the exact source of each should be included in the figure legend; if these "borrowed" illustrations are copyrighted by others, permission of the copyright holder to reproduce the illustration must be secured by the author.

5. Nomenclature

All pesticides and other subject-matter chemicals should be identified according to *Chemical Abstracts,* with the full chemical name in text in parentheses or brackets the first time a common or trade name is used. If many such names are used, a table of the names and their precise chemical designations should be included as the last table in the manuscript, with a numbered footnote reference to this fact on the first text page of the manuscript.

6. Miscellaneous

Abbreviations. Common units of measurement and other commonly abbreviated terms and designations should be abbreviated as listed below; if any others are used often in a manuscript, they should be written out the first time used, followed by the normal and acceptable abbreviation in parentheses [e.g., Acceptable Daily Intake (ADI), Angstrom (Å), picogram (pg), parts per trillion (ppt)]. Except for inch (in.) and number (no., when followed by a numeral), abbreviations are used without periods. Temperatures should be reported as "°C" or "°F" (e.g., mp 41° to 43°C).

Abbreviations

A	acre	kg	kilogram(s)
bp	boiling point	L	liter(s)
cal	calorie	mp	melting point
cm	centimeter(s)	m	meter(s)
cu	cubic (as in "cu m")	μg	microgram(s)
ft	foot (feet)	μl	microliter(s)
gal	gallon(s)	μm	micrometer(s)
g	gram(s)	mg	milligram(s)
ha	hectare	ml	milliliter(s)
hr	hour(s)	mm	millimeter(s)
in.	inch(es)	m$\underline{M}$	millimolar
id	inside diameter	min	minute(s)

Abbreviations

<u>M</u>	molar	lb	pound(s)
mon	month(s)	psi	pounds per square inch
ng	nanogram(s)	rpm	revolutions per minute
nm	nanometer(s) (millimicron)	sec	second(s)
<u>N</u>	normal	sp gr	specific gravity
no.	number(s)	sq	square (as in "sq m")
od	outside diameter	vs.	<u>versus</u>
oz	ounce(s)	wk	week(s)
ppb	parts per billion	wt	weight
ppm	parts per million	yr	year(s)
/	per		

Numbers. All numbers and fractions or decimals are arabic or roman (table numbers only) numerals. Numerals should be used for a series (e.g., "0.5, 1, 5, 10, and 20 days"), for pH values, and for temperatures. When a sentence begins with a number, write it out.

Symbols. Special symbols (e.g., Greek letters) must be identified in the margin, e.g.,

$$A = \beta/2\lambda$$

Percent should be % in text, figures, and tables.

Style and format. The following examples illustrate the style and format to be followed (except for abandonment of periods with abbreviations):

> KAEMMERER, K., and S. BUNTENKÖTTER: The problem of residues in meat of edible domestic animals after application or intake of organophosphate esters. Residue Reviews 46, 1 (1973).
> The Chemagro Division Research Staff: Guthion (azinphosmethyl): Organophosphorus insecticide. Residue Reviews 51, 123 (1974).

7. Page proof (Galley proof is no longer sent)

Corrected proof must be returned, within two weeks of receipt, to the editor. Author corrections should be *clearly* indicated on proof with soft pencil or with ink and in conformity with the standard "Proofreader's Marks" accompanying each set of proofs. In correcting proof, new or changed words or phrases should be carefully and legibly handprinted (*not* handwritten) in the margins.

8. Reprints

Senior authors receive 30 complimentary reprints of a published article. Additional reprints may be ordered from the publisher at the time the principal author receives the proof.

9. Page charges

There are no page charges, regardless of length of manuscript. However, the cost of alterations (other than corrections of typesetting errors) attributable to authors' changes in the page proof, in excess of ten % of the original composition cost, will be charged to the authors.

If there are questions that are not answered in this leaflet, see any volume of *Residue Reviews* or telephone the Editor (see p. 1 for telephone numbers). Volume 3 (Ebeling) is especially helpful.